# BIOLOGY TAKES FORM

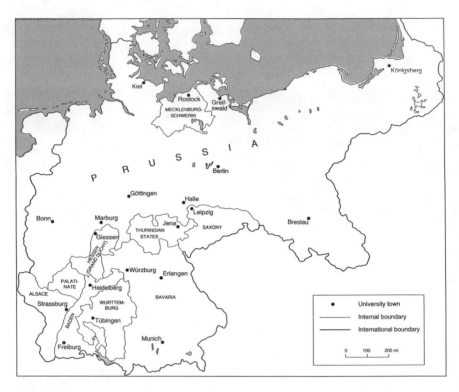

German universities, 1872–1900, showing state boundaries (prepared by University of Wisconsin—Madison Cartographic Laboratory).

*Science and its Conceptual Foundations*
A series edited by DAVID L. HULL

# LYNN K. NYHART

# Biology
# Takes Form

*Animal Morphology
and the German Universities,
1800–1900*

*The University of Chicago Press
Chicago and London*

Lynn K. Nyhart is associate professor in the Department of
the History of Science at the University of Wisconsin–
Madison.

The University of Chicago Press, Chicago 60637
The University of Chicago Press, Ltd., London
© 1995 by The University of Chicago
All rights reserved. Published 1995
Printed in the United States of America

04 03 02 01 00 99 98 97 96 95   5 4 3 2 1

ISBN (cloth): 0-226-61086-1
ISBN (paper): 0-226-61088-8

Library of Congress Cataloging-in-Publication Data

Nyhart, Lynn K.
    Biology takes form : animal morphology and the Ger-
man universities, 1800–1900 / Lynn K. Nyhart.
        p.      cm. — (Science and its conceptual foundations)
    Includes bibliographical references (p.        ) and
index.
    ISBN 0-226-61086-1 (cloth). — ISBN 0-226-61088-8
(paper)
    1. Morphology (Animals)—Study and teaching
(Higher)—Germany—History—19th century.  2. Morphology (Animals)—Germany—History—19th century.
I. Title.  II. Series.
QL799.5.N94   1995
591.4'0943'09034—dc20                              95-3227
                                                        CIP

*To my parents,*
Dan and Nina Nyhart

# CONTENTS

# LIST OF ILLUSTRATIONS

## ACKNOWLEDGMENTS

In a project that has taken as long as this one, one builds up a remarkable number of debts. The seeds of this book were originally planted in 1978, during my junior year at Princeton, where as a history of science major I had the privilege of attending Gerry Geison's graduate seminar on Darwin. I learned more about studying evolution and doing history in that class than in any other single experience of my undergraduate and graduate education. A few years later, as a graduate student at the University of Pennsylvania, I returned to Princeton to attend a semester-long seminar on science in the nineteenth-century German universities run by R. Steven Turner, who presented his deep knowledge of the subject along with large doses of humor and kindness that helped make the then nearly impenetrable German go down more easily. Since that time, I have continued to be thankful for Steve's careful reading of various parts of this work, his thoughtful criticisms, and his unflagging support and encouragement. While conducting dissertation research in Heidelberg on a fellowship from the Deutsche Akademische Austauschdienst, I benefitted from weekly conversations with Professor Hans Querner, whose knowledge of the German zoological community of the nineteenth and twentieth centuries is vast and detailed. My advisor at Penn, Mark Adams, never failed to offer interesting new ideas to explore. He taught me more about writing than I sometimes wanted to know, but he was right.

The dissertation related just a piece of morphology's complex history. For publication it seemed important to broaden the scope of the project to include the first half of the nineteenth century and zoology as well as anatomy. During a year spent at Yale while I struggled to understand the nature of morphology and physiology in the first part of the century, I profited from many conversations with Larry Holmes, who gently helped me discover some of my blinders and encouraged me to remove them. At Michigan State University and more recently at the University of Wisconsin—Madison, I have been fortunate to have supportive, friendly colleagues willing to share their wisdom on book writing, publishing, and balancing the demands of teaching and research.

For their thoughtful comments on various parts of the manuscript at various stages, I am grateful to Jane Camerini, Diane Edwards, Thomas Junker, Robert Kohler, Richard Kremer, Jane Maienschein, Marsha Richmond, Louise Robbins, the members of the Yale Section for the History of Medicine during 1989–90, and the graduate students and faculty of the Princeton program in history of science during spring 1993. I owe a special debt to John Carson, who devoted far more hours of reading and conversation to my project than even a good friend should be expected to do. David Hull, Fred Churchill, and Jonathan Harwood read the penultimate version of the entire manuscript with great care and offered knowledgeable suggestions that helped me greatly in clarifying and fine-tuning my argument.

An abundance of technical support helped get this book done. For providing funds for travel, salary, and time off from teaching, I thank the American Council of Learned Societies, the National Endowment for the Humanities, the University of Wisconsin Graduate School and the National Science Foundation (Award no. 8910873). I made use of the services of countless librarians and archivists; most of these must remain nameless, but I would especially like to thank Herr Dr. H. Rohlfing of the Handschriftenabteilung of the Universitätsbibliothek Göttingen and Frau G. Schwendler at Leipzig's Universitätsarchiv, who made a foreign scholar feel especially welcome and at home. I am grateful for the careful research assistance provided by Matthew van der Veen, Robert Goodrich, Judy Houck, and Doug Keen, and for the expert clerical support provided by Lori Grant and Eileen Ward in preparing various versions of the manuscript.

Tom Broman has lived through every part of this project with me, sometimes in closer proximity to it than I'm sure he would have liked. I thank him for his patients as I tried every idea out on him (usually in more than one incarnation), for his critical eye, his deft editorial pencil, and his sound judgment. I could not have wished for a better partner and colleague. Tim Broman, who was born while this book was still in process, is a constant reminder of the richness of life; his presence keeps scholarship in perspective.

# ABBREVIATIONS

Full citations to works listed below and to works cited by short titles in the footnotes will be found in the bibliography.

### Reference Works

| | |
|---|---|
| ADB | Allgemeine Deutsche Biographie |
| Biogr. Lex. I | Biographisches Lexikon der hervorragenden Ärzte aller Zeiten un Völker |
| Biogr. Lex. II | Biographisches Lexikon der hervorragenden Ärzte der letzten fünfzig Jahre |
| DSB | Dictionary of Scientific Biography |

### Archival Sources

| | |
|---|---|
| BGLA | Badische Generallandesarchiv, Karlsruhe |
| BHStA München | Bayerische Hauptstaatsarchiv, Munich |
| GStA Merseburg | Geheimer Staatsarchiv Merseburg |
| Medhist. Inst. Zürich | Medizinhistorisches Institut der Universität Zürich |
| SB | Senckenbergisches Bibliothek |
| SBPK | Staatsbibliothek Preussischer Kulturbesitz, Berlin |
| StaBi München | Staatsbibliothek München |
| UA | Universitätsarchiv |
| UB | Universitätsbibliothek |

## CHAPTER ONE

## Situating Morphology

In 1800, in a pair of footnotes in an obscure medical textbook, the aspiring physiologist Karl Friedrich Burdach introduced two neologisms that would become increasingly central to the life sciences over the course of the nineteenth century: "Morphologie" and "Biologie." Morphology, the study of form, was to be a principal division in a larger systematic view of life "that can be encompassed by the name of biology." Just over a century later, in 1908, the philosopher-biologist Hans Driesch was more insistent: "It is *form* particularly which can be said to occupy the very centre of biological interest; at least it furnishes the foundation of all biology."[1] Between Burdach's first tentative classificatory pronouncement and Driesch's confident assertion lies a complex story of the changing meaning and place of animal morphology within German biology; this book tells that tale.

On the face of it, the study of form may not appear the most auspicious area of biology for historical examination. For many scientists and historians it connotes a backward-facing enterprise, enmeshed with a long-discredited idealistic view of nature and its associated bankrupt methods. Morphologists might be expected to sit among dusty museum cabinets, puzzling over mange-ridden speci-

---

1. K. F. Burdach, *Propädeutik zum Studium der gesammten Heikunst* (1800), quoted in G. Schmid, "Über die Herkunft der Ausdrücke Morphologie und Biologie" (1935), p. 605; H. Driesch, *Science and Philosophy of the Organism* (1908), p. 17.

mens in an effort to make them conform to an Ideal Type or perching hypothetical ancestors on evolutionary trees, while more progressive scientists uncovered the real secrets of nature in their laboratories. While the latter were engaged in discovering the scientific laws and causes of animal life, morphologists could at best provide descriptions of anatomical structures, while failing to produce satisfactory explanations. Thus for some, a history of morphology must ultimately be a story of a failed way of doing science that has quite rightly been surpassed.

To imagine the story this way, however, would mean writing off a century's worth of success. Morphology was one of the main streams of biological investigation in the nineteenth century, and with good reason. More than just the study of anatomical structures, it engaged some of the central philosophical mysteries of biology. In what ways did organization capture the essence of an animal's life? What was the relationship between the animal as a unified whole and its parts? Between the body's structures and its mode of life? Was there one or a small number of basic plans to which all animals conformed, and were these distinct from one another or did they merge? What was the relationship between an organism's adult form and its earlier stages? In what sense can one say the earlier stages "caused" the adult to come about? How did an egg "know" to turn into a chick or a salamander, anyway?

To answer these questions morphologists throughout the nineteenth century studied animals across various stages of development, examining them whole and alive, or preserved, stained, and sectioned. They dissected animals of all kinds and looked at their tissues, internal organs, and bones with the naked eye and under microscopes. They traced the course of an individual's life cycle, in the case of vertebrates, from the fertilized egg through the embryonic and juvenile stages to the adult; in the case of invertebrates, from the egg through various different stages more complex than those possessed by the more familiar vertebrates. Sometimes this could be done by actually following a single individual through its various phases; more often a group of specimens from the same species would be halted at different stages of development to build a composite picture of the typical path. Morphologists compared anatomical structures, their physiological purposes, and their developmental sequences among animals within the same family or across families and even classes, to discern the degrees of similarity among animal forms. They also studied the relationship of the organism to its surroundings to determine the effects of the "conditions of existence" upon an animal's structures and functions.

Given this range of theoretical questions and analytical approaches, it is not surprising that methodological questions also abounded in the study of form. Which anatomical structures or relationships should provide the keys to classification? Were the comparison of adult structures and of developing ones equally reliable guides to understanding the relationships among different groups of organisms? What was the appropriate *practical* level at which one should seek the causes of form —in the cells, the tissues, the organs, the organ systems, the organism as a whole? Although there were certainly prominent areas of investigation into the organic world—such as much of physiology—for which such questions were beside the point, it would not be going too far to claim that these problems engaged the largest number of life scientists through the course of the nineteenth century. Their efforts not only produced a remarkable quantity of reliable empirical information, but also showed them to be wrestling intelligently with philosophical and methodological questions that still attract biologists today.

The significance of nineteenth-century morphology is not confined to intellectual issues, however. If we understand "biology" in its social sense as the scholarly enterprise of studying the living world, the history of animal morphology as a research program reflects the changes in that enterprise as the nineteenth century churned onward. This is especially true for the German-speaking academic community, where the study of form constituted a prominent aspect of university-based life science throughout the century. From the struggles of the early part of the century over the properly scientific approach to studying life and its incorporation into the universities, through the institutionalization of a variety of biological research programs at mid-century and the incorporation of a Darwinian intellectual framework in the 1860s and 1870s, to the noisy but institutionally fraught advent of new manipulative approaches touted as, alternatively, "experimental morphology," or a "new biology," the history of animal morphology traces the larger institutional and intellectual changes in German biology during its first century.

This side of the story is all the more important for historians and sociologists of science because morphology never took on most of the trappings we usually associate with academic success. There were no professors of morphology within the German university system, nor was morphology a subject of instruction: German universities had no standard courses with "morphology" in their titles. A single effort to found a morphological society in the late 1880s never went beyond a suggestion.[2]

One might characterize morphology's story, then, following Robert Kohler's description of biochemistry in Germany, as a "paradox of intellectual success and institutional failure."[3] But to do this would be to limit unnecessarily what it might mean to succeed institutionally. The very persistence of the morphological research enterprise over the whole of the nineteenth century suggests a level of institutional accommodation that we should certainly interpret as success, despite the absence of the expected institutional labels. It was success of a different kind, however: one of colonizing and maintaining a niche (at times a large one) within existing disciplines, rather than creating a new one. This type of success story is rarely written about but may, in fact, be far more common than the more dramatic one of creating a new discipline.

In telling this story, I have appropriated the term "orientation" to describe morphology and other approaches to animal biology. This is a rough translation of the word *Richtung,* a term Germans use to encompass an area of study—primarily related to research but not excluding issues of teaching—the group of people engaged in that area, and the philosophical attitudes accompanying the cluster of problems they are working on. It is similar to "school" used loosely; the problem with the word "school" is that it is also used narrowly, to mean a teacher and his students. When I use the word "school," I mean it in its narrow sense. An orientation can be generated by a school but also includes people who take up a given set of problems and attitudes whether or not they had the same teacher. Similarly, the term "research program" suggests something more codified than what I intend; it also directs our attention away from teaching, whereas "orientation" can include teaching concerns. Where I do talk about "programs" I mean self-consciously articulated plans for research, as in Haeckel's program of evolutionary morphology, or

2. This situation contrasts with that of Britain and the United States where, by the late nineteenth century, there were indeed professorships of morphology. At the University of Cambridge, for example, a professorship in animal morphology was created for F. M. Balfour in 1882 (G. L. Geison, *Michael Foster* [1978], p. 128). In the United States, William Keith Brooks was appointed associate professor of morphology at Johns Hopkins in 1883; Charles Otis Whitman was named professor of morphology at the University of Chicago; on Brooks's title, see J. Maienschein, *Transforming Traditions* (1991), p. 29; on Whitman's appointment, see H. H. Newman, "History of the Development of Zoology in the University of Chicago," *BIOS* 19 (1948):215. My thanks to Jane Maienschein for directing me to this source.

3. R. E. Kohler, *From Medical Chemistry to Biochemistry* (1982), p. 9.

Roux's program for *Entwicklungsmechanik*. A particular research program could lie at the heart of an orientation, but again, the latter term is intended to be both broader and looser. "Orientation" is also more suitable than "subdiscipline" or "subspecialty" because these seem too institutionally solidified. In my view disciplines are defined institutionally, by chairs, institutes, and teaching appointments, and subdisciplines would refer to areas within a discipline defined in this way. "Orientation" is looser, because it does not necessarily imply institutional boundaries, nor does it suggest the level of bureaucratic organization that "subdiscipline" does. Finally, although the idea of orientation bears some similarities to F. L. Holmes's attractive notion of "investigative streams" in its fluidity and its emphasis on specific areas of investigation, it differs in considering the people involved as a cohesive (if temporary) group, usually with an identifiable philosophical approach to their investigations.[4]

To understand morphology's peculiar sort of success, it is essential to examine it in relation to both its intellectual and institutional environments. Only when viewed in relation to these surroundings are the meaning and significance of morphology thrown into relief. This is not a matter of providing broader background to enrich the picture; my argument is that morphology's history becomes clear only if we set it next to the other biological research orientations available at the time; as those alternatives changed, so, too, did morphology. Thus I see writing the history of German morphology as analogous to constructing a series of intellectual and institutional maps of the life sciences and showing how both the internal configurations and borderlines of this biological territory changed over time. This approach, which employs a combination of spatial and relational metaphors most often associated with ecology and geography, has guided my study. The text is therefore salted with such terms as "territory," "location," and "shifting boundaries."

As attractive as such metaphors are, however, they do carry the risk of reifying morphology and other terms such as "zoology" and "anatomy"—of presenting them as solid things rather than as the always changing creations of groups of people. To counter this risk, I have sought to identify the people who were especially engaged in studying animal form, and to keep in the foreground their language and their ideas concerning the aims and content of morphology. I

---

4. On the changing historical significance accorded to "schools," see J. W. Servos, "Research Schools" (1993). F. L. Holmes makes use of the idea of investigative streams in *Between Biology and Medicine* (1992).

have thus viewed my primary tasks as locating animal morphologists; examining how they were situated intellectually, institutionally, and demographically in their surrounding community; and following them as they responded to new ideas, new institutional forms, and the ever-changing membership of their disciplines.

In pursuing these aims, I have sought to keep in mind the fact that "location" is always defined with respect to some set of landmarks. I have therefore attempted to situate morphologists from three different vantage points. The first looks at broad intellectual markers: what did morphology's proponents think made their enterprise a *Wissenschaft*, the highest level of systematic and synthetic knowledge? What metaphysical, epistemological, and methodological assumptions did they make, how did these change over time, and in what relation did they stand to the views of other biological researchers? Second, I locate morphologists with regard to the place they held in the institutionally sanctioned order of knowledge. That, too, changed over time, as the universities became the dominant research institutions and developed an ever stronger disciplinary structure. The third perspective attends to what we might call the morphologists' generational location. Since the scientific community comprised people of different intellectual generations, the position of a morphologist with respect to members of his own generation and others also forms an important aspect of his location in the map of life scientists.

Although these three ways of locating or situating morphologists by no means exhaust the possibilities, I would argue that they provide a necessary starting point for understanding morphology's history. When combined, they lend a certain dynamic to the history of morphology, for in their interaction we gain a preliminary understanding of the process of change, the route by which the study of form gained and eventually (despite Driesch's claim) lost preeminence in German biology. Each of these types of location, therefore, deserves some elaboration.

## MORPHOLOGY AS A THEORETICAL ENTERPRISE

The intellectual aspects of morphology's history have long held a prominent place in the history of biology, dating at least to 1916, when E. S. Russell's magisterial book, *Form and Function,* set the framework upon which much later work has been built. Russell's deepest underlying theme was the existence of a fundamental dichotomy between those who see the body's structures as the primary, independent factor controlling animal life and those who view such structures as resulting from the functional needs of the organism. In describing morphology's historical development, however, he took a different tack, outlining three phases of research on animal morphology in the nineteenth century. A transcendental, "idealistic" brand of morphology inspired by Johann Wolfgang von Goethe and Etienne Geoffroy St. Hilaire dominated the life sciences in the first half of the nineteenth century. This was overtaken by an "evolutionary" approach to morphology after 1859, an orientation which in its turn was thrust into the shadows near the end of the century by a new approach that Russell called "causal morphology," which later writers have tended to identify with experimental embryology. Following a mode of analysis classic in the history of ideas, Russell accorded the greatest significance in his discussion to intellectual styles that transcend national or institutional borders. Nevertheless, German scientists play a prominent role in each of his periods: Goethe and Karl Ernst von Baer in the first, Ernst Haeckel and Carl Gegenbaur in the second, and Wilhelm Roux (among a host of other Germans) in the third.

In recent years each of Russell's three phases has been subjected to revisionist analyses. Interest in morphology's "transcendental" phase has mushroomed in recent years, with new studies focusing on the different national settings of England, France, and Germany. These studies have reinterpreted Russell's own idealist differentiation between the pure formalists (exemplified by Geoffroy) and those morphologists who viewed form as a result of function (exemplified by Aristotle and Georges Cuvier). Toby Appel's authoritative examination of the Cuvier-Geoffroy debate places it into a French political, philosophical, and institutional context, showing how the protagonists' leanings toward formalism or functionalism were bound up with other philosophical and political commitments that developed over the course of their lives through particular historical circumstances and accidents of personality. For England, Adrian Desmond has located the hot debates over morphology at the center of strife over the reform of medical education in London in the 1830s, arguing that Geoffroy's ideas were associated with political radicalism while Cuvierian comparative anatomy was linked to more moderate or conservative beliefs. A somewhat different interpretation is offered by Phillip R. Sloan in providing background for the 1837 Hunterian Lectures by Richard Owen, who, by the late 1840s, would become Britain's leading morphologist; Sloan has attended carefully to the routes of transmission

of morphological ideas from the Continent to Britain, elegantly showing the twists, turns, and recouplings that Continental ideas took
along the way.[5]

German idealistic morphology, too, has come in for an important
revision. Timothy Lenoir has recast the split between formalists and
functionalists as one between *Naturphilosophen,* who cleaved to
Schelling's philosophy and speculative method, and an ultimately
more successful group committed to an approach Lenoir has dubbed
"teleomechanism," which, drawing upon Kantian epistemology and
Cuvierian method, developed a mechanistic program of empirical research within a more broadly teleological philosophical framework.
This interpretation has the advantage over an older approach that
lumped all the early German writers on form under the heading of
"romantic science" and that smoothed over significant disagreements;
nevertheless, I believe that Lenoir's effort to separate scholars into
clearly defined philosophical camps draws their differences too
sharply. Most early nineteenth-century writers on form confound cat-
egorization schemes based on rigid philosophical distinctions; they
appropriated the language of Kant, of Schelling, and of Cuvier in
different places. Few were as forthright as Johannes Spix, who credited as his two greatest inspirations Schelling and Cuvier, but many
had little apparent trouble drawing selectively from among philosophically opposed systems.[6]

Despite their differences in approach and style, these recent works
collectively have done much to elucidate various social and intellectual
contexts for early nineteenth-century morphology. The present study,
which focuses predominantly on the second half of the century, cannot approach these works in their richness of detail and social, political, and philosophical contextualization. However, in chapters 2 and
3 I hope to contribute to this larger picture by connecting the intellectual developments in early nineteenth-century German morphology
to their shifting disciplinary context in the universities.

Morphology's second, evolutionary phase has, surprisingly, received less attention than one might expect, given the quantity of
literature on Darwinism and evolution. It is generally assumed that
the program promulgated by Ernst Haeckel and Carl Gegenbaur dom-

5. T. Appel, *Cuvier-Geoffroy Debate* (1987); T. Lenoir, *Strategy of Life* (1982);
A. Desmond, *Politics of Evolution* (1989); P. R. Sloan, Introduction to *Hunterian
Lectures*, by R. Owen (1992).

6. J. Spix, *Geschichte und Beurtheilung* (1811), p. x. The Kant-Blumenbach tradition was identified earlier by J. Larson in "Vital Forces" (1979), although he did not
use the term "teleomechanism."

inated biology (or at least animal biology) during the period somewhat vaguely referred to as "the late nineteenth century." As it has
come to be characterized, this program emphasized the creation of
evolutionary trees using the methods of comparative anatomy and
comparative embryology. It relied especially on Haeckel's biogenetic
law, which stated that "ontogeny (individual development) recapitulates phylogeny (ancestral development)." Although it does appear
that recapitulationism was widely assumed and, as Stephen Jay Gould
has shown, was applied in all sorts of ways ranging from race theory
to psychological development,[7] little work has been done to examine
just how widely it was accepted and used as a tool by animal biologists
themselves. One object of this book is to rectify that, at least for the
German morphological community. Chapters 5, 6, and 7 are devoted
to analyzing the extent of success of the Haeckel-Gegenbaur program
among academic morphologists.

More revision has taken place in the relationship of morphology's
transcendental phase to its evolutionary phase. In 1976, William Coleman pointed out that neither Carl Gegenbaur's method nor his taxonomy changed dramatically from his pre-Darwinian to his Darwinian
formulation of comparative anatomy. More recently, Peter Bowler
has broadened the case for continuity, arguing that Darwin's theory
produced no revolution in biological thought, and that a fundamentally non-Darwinian, developmental model of evolution persisted until the real revolution introduced by genetics at the turn of the century.
Robert J. Richards has gone one step further, seeking to show that
Darwin himself, whom Bowler exempts from this developmental approach, was in fact a recapitulationist, a point that in Richards's view
puts Darwin into his proper setting as a developmental thinker.[8] Although they do blur Russell's boundaries, these studies have stayed
largely within the historical framework he set, taking as the central
problem the intellectual continuity or discontinuity between morphology's pre-Darwinian and Darwinian phases.

The transition from evolutionary morphology to its third, experimental phase has drawn considerably more interest in the last fifteen
years or so, beginning with the notable controversy between Garland

7. S. J. Gould, *Ontogeny and Phylogeny* (1977), esp. pp. 115–66.
8. W. Coleman, "Morphology" (1976); P. Bowler, *Non-Darwinian Revolution*
(1988), *Mendelian Revolution* (1989); R. J. Richards, *Meaning of Evolution* (1992);
for continuity of concerns between Darwin's own thought and other contemporary
issues usually viewed as "European," see also D. Ospovat, *Development of Darwin's
Theory* (1981); P. R. Sloan, "Darwin's Invertebrate Program" (1985); and M. J. S.
Hodge, "Darwin is a Lifelong Generation Theorist" (1985).

Allen and several younger historians of biology, led by Jane Maienschein, over whether or not there was a "revolt" from morphology to
experimental biology in the decades around the turn of the century.
Reflecting the changing fashions in historiography of science, most of
this recent literature, like that connecting the first and second phases
of morphology's history, has stressed continuous evolution of ideas
rather than Kuhnian-style revolutions.[9] However this change took
place, there is little disagreement that the evolutionary approach to
the study of form, using the methods popularized in the second phase,
moved out of center stage among biologists sometime between 1880
and about 1915.

In this discussion the main arena has been American biology, with
British and German biology appearing as secondary venues. The relationship between the experimental approaches of the turn of the century and the evolutionary morphological orientation has recently
come into the sights of historians of German biology, however. Most
of the resulting scholarship has concentrated on individuals. The pioneering experimental biologists August Weismann, Wilhelm Roux,
and Hans Driesch have received the most attention, with especially
admirable studies by Frederick Churchill and Reinhard Mocek leading
the way. Thanks to a number of new intellectual-biographical analyses
we are beginning to gain a better picture of the research paths tra-
versed by several other important German animal biologists in this
period as well.[10]

The only substantial historical group portrait analyzing the beginnings of the experimental phase of biology in early twentieth-century
Germany focuses not on experimental morphology per se, but on
genetics. Jonathan Harwood's masterly work, *Styles of Scientific
Thought,* provides a comprehensive picture of the period after my
study, concentrating especially on the 1920s and 1930s. A number of

9. J. Maienschein, R. Rainger, and K. R. Benson, eds., "Special Section on American
Morphology," (1981); J. Maienschein, *Transforming Traditions* (1991). For other
works comparing naturalist or morphologist and experimentalist traditions, see J. Hagen, "Ecologists and Taxonomists" (1986); and O. Amsterdamska, "Stabilizing Instability" (1991).

10. F. Churchill, "Wilhelm Roux" (1966), "August Weismann" (1968), "Weismann's Continuity" (1985), and "Weismann, Hydromedusae" (1986); R. Mocek, *Wilhelm Roux–Hans Driesch* (1974); see also P. Weindling, *Darwinism and Social Darwinism* (1991), on Oscar Hertwig; N. Jacobs, "From Unit to Unity" (1989), on Otto
Bütschli; J. M. Oppenheimer, "Curt Herbst's Contributions" (1991); and M. Richmond, "Richard Goldschmidt" (1986). Jane Oppenheimer's earlier foundational contributions to the history of experimental embryology, collected in *Essays in the History
of Embryology,* continue to merit study as well.

the themes raised in Harwood's book also appear here, suggesting some important continuities between the late nineteenth century and the 1920s, and between the fortunes of experimental morphology and those of genetics. Like geneticists, experimental morphologists had a tough time in the traditional universities in the early twentieth century; in both cases part of the problem was the absence of jobs in a period devoid of university expansion. In addition, Harwood lays great stress on the value of intellectual breadth among German biologists—a value which placed both Mendelian transmission genetics and experimental embryology at a disadvantage in the early twentieth century, since they were often perceived as "narrow" research agendas. This value was also a powerful one in the earlier period; indeed, I would contend this value extends back, at least among morphologists, to the early nineteenth century. Like Harwood, I, too, believe that this value goes a good way toward explaining the success of certain orientations relative to others in German university biology, especially within the discipline of zoology. In the last three chapters I seek to develop these themes to explicate the places of evolutionary and experimental morphology within the disciplines of anatomy and zoology between about 1880 and 1910.

As the preceding discussion suggests, Russell's three-phase framework has proven durable. With the important exception of Peter Bowler, who appears to be arguing for a complete blending of the first two phases into one, few historians have seen reason to alter the basic framework. Indeed, those who operate within Russell's own tradition of the history of ideas have adopted it almost without comment.[11] Insofar as this book provides a history of ideas about morphology, I, too, have adopted Russell's tripartite structure as a description of some fundamental changes in morphologists' ways of thinking. However, I hope to show that at every stage scholars came up with intellectual alternatives not developed in Russell's account of the dialectical struggle between "pure" and "functional" morphologists, and that the choices made were not so clear-cut.

Situating morphology intellectually requires understanding which of these choices came to dominate at any one time, and why. Russell's framework helps us little in this aim, because in focusing on the leading scientists he neglects to ask how they came to be viewed as the leaders. Thus although Russell describes the different periods brilliantly, he does not explain change very satisfactorily. The new studies that attend to politics, institutions, and social groups have reintro-

11. See esp. Gould, *Ontogeny and Phylogeny.*

duced a sense of dynamism, developing arguments about how and why these different phases came into being: what extrascientific beliefs and values were infused into morphology in the different periods by people with different agendas, where new ideas met resistance, where the power lay and how it shifted, how people made use of institutional resources or created new ones to gain or preserve intellectual dominance.

The history of German morphology has as yet remained curiously undisturbed by these currents. To the degree that scholars have paid attention to changes in morphology, they have confined themselves to philosophical and methodological battles, largely without reference to the setting in which these battles were carried out.[12] This is all the more remarkable given the quantity of writing about the development of the German university system and the disciplines within it. Whereas the histories of other sciences in Germany, such as physics, chemistry, and physiology, have been firmly embedded in their institutional settings, historians of German morphology appear to have taken their cue from their idealist subjects, as if their thought transcended the constraints of a particular setting. By situating morphology institutionally, the present study seeks to provide new insights into morphology's intellectual trajectory.

## MORPHOLOGY IN THE UNIVERSITIES

The institutional location of morphologists can be mapped according to three different scales. Most broadly, I have confined my study to scientists working in the universities. In contrast to France's Paris Academy and museums or London's private medical schools, in the German-speaking lands the state-run universities provided the niches in which morphology developed. This setting frames the next level of resolution, which focuses on morphology's place among the life-science disciplines. At a yet finer level, morphologists represented one among a number of different, often competing, subdisciplinary orientations toward teaching and research.

As Emmanuel Radl wrote in 1909, "He who says 'German science' means the universities."[13] In Germany, Radl noted, a science acquired

---

12. See Lenoir, *Strategy of Life;* F. Duchesneau, *Genèse de la Théorie Cellulaire* (1987); H. Querner, "Beobachtung oder Experiment?" (1975), "Entwicklungs-mechanik Wilhelm Roux" (1977). For an account that does attend to the larger institutional, political, and cultural setting, see Weindling, *Darwinism and Social Darwinism.*

13. E. Radl, *Geschichte der biologischen Theorien*, vol. 2 (1909), pp. 564–65.

recognition and legitimacy only if it found representation in the university setting. This is a crucial point, for the history of German morphology in the nineteenth century is intimately tied to the development of the universities.[14] The ways in which morphology did and did not find a home within the universities therefore comprise a central focus of this book.

German universities existed before Germany itself was united into one country in 1871. Although there were four German-speaking universities in Austria (Graz, Innsbruck, Prague, and Vienna) and three in Switzerland (Basel, Bern, and Zürich), I concentrate on the twenty that were united after 1871 under the German Empire: the six "traditional" Prussian universities of Berlin, Bonn, Breslau, Greifswald, Halle, and Königsberg; the Bavarian universities in Erlangen, Munich (after 1826), and Würzburg; Baden's two universities at Heidelberg and Freiburg; the single universities of various German-speaking states and principalities, including Giessen, Göttingen, Jena, Leipzig, Marburg, Rostock, Tübingen, and Kiel (1867); and the Imperial university at Strassburg (1872).[15]

An idealized story of the "Rise of the German University" has been built up over the last century or so, through countless histories of

14. As much as Radl's statement was a truism when it was written in 1909, however, we must recognize that it was a truth historically bounded by the nineteenth century: at the beginning of that century, many important German scientific contributors were not located in the universities, and neither the significance of a science nor that of an investigator was measured solely by their standing in the universities. This was true as well of scholars interested in organic form. Johann Wolfgang von Goethe, who can take credit as one of the disseminators of the word "Morphologie," could hardly have deigned to be tied down to a university teaching position. Gottfried Reinhold Treviranus, who put forth his own philosophy of nature under the term "biology," was a practicing physician. The *Naturphilosoph* Carl Gustav Carus was a professor, to be sure, but not at a university, and not specializing in morphology; he taught gynecology at Saxony's medical-surgical academy in Dresden and was personal physician to the Duke of Saxony. By the early twentieth century, significant scientific research was again being conducted outside the universities; in biology these new locations included the Kaiser-Wilhelm-Institut für Biologie, founded in 1913, and various government-sponsored and private marine and lake research stations, among others. But during morphology's nineteenth-century heyday Radl's statement would have been accepted by most scientists, including morphologists, without cavil.

15. Between 1866 and 1871, Marburg, Kiel, and Göttingen were taken over by the Prussian administration. The list of universities leaves out the *Technische Hochschulen*, which officially achieved doctoral degree-granting status only in 1900; such academies as the medical-surgical academy in Dresden and Münster's academy, which lacked a medical faculty until after 1900; the few autonomous state-run schools of forestry, agriculture, and veterinary medicine; and the new universities at Frankfurt am Main and Hamburg founded in 1914 and 1919, respectively.

individual universities and of the German higher education system as a whole. Beginning with the founding of the new University of Berlin in 1810, the story goes, the German states (especially Prussia) sought to reinvigorate the moribund universities by institutionalizing two grand, intertwined ideas: the addition of research to the university faculty's traditional mission of teaching, with the goal of developing *Wissenschaft,* or pure knowledge, and the inculcation of *Bildung,* or development of the self to its highest potential, among its students. These two ideas were realized institutionally by introducing seminars, and later, laboratories, as the vehicles of *Bildung;* simultaneously the status of the university's philosophical faculty rose to the same level as the three traditional higher faculties of law, medicine, and theology. The result for what English-language speakers call "science" was dramatic: Germans became world leaders in such fields as chemistry, physics, mathematics, physiology—and, though the story has not been told before, morphology.[16]

Although the basic outlines of this story were hammered out first in the late nineteenth century to celebrate Germany's (especially Prussia's) rise to international preeminence in the sciences, its overall contours have been refined by twentieth-century historians and sociologists. Modern scholars' desires to develop models of the university "system," discipline formation, and the internal social structure of the university, often with an eye toward contemporary policy, have brought a new set of considerations into the foreground. Most prominent in these endeavors has been the work of Joseph Ben-David and his colleagues. Ben-David spread the notion that the German-speaking universities formed a market system (though paid for by state funds), and that free-market competition among them for the best and brightest professors was what spurred Germany to its scientific greatness. His student Awraham Zloczower argued a related case for late nineteenth-century physiology, contending that the rate of significant intellectual innovation was dependent on the system's continued expansion. Once the system stopped funding professorships in new subfields, creating a backlog of aspirants with no place to go, German physiology ceased its rapid advance.[17]

16. The classic nineteenth-century history is F. Paulsen's *Die Deutschen Universitäten* (1902). The most important modern reexamination of the topic is R. S. Turner's dissertation, "The Prussian Universities" (1973).

17. J. Ben-David, "Scientific Productivity" (1960), *Scientist's Role in Society* (1971); A. Zloczower, *Career Opportunities* (1981); Ben-David and Zloczower, "Universities and Academic Systems" (1962); a similar case is made for psychology in Ben-

Embedded in Ben-David's and Zloczower's work lies a model of discipline development and its relationship to the university's internal social structure. In this aspect of the story, German universities developed a unique system in which each sanctioned body of knowledge was the province of a single full professor at each university. Proponents of new areas of knowledge gained institutional support usually first by acquiring positions as unsalaried *Privatdozenten*, or private lecturers, who had permission to teach their specialty at the universities but received no money for it from state coffers, relying entirely on student fees. Those who were successful might eventually gain appointments as *ausserordentliche* professors, also known as *Extraordinarien*, who were usually salaried (though not always) but lacked voting rights in the faculty corporation. Once again, success in attaining legitimacy for a new area of knowledge was marked by the elevation of these *ausserordentliche* professors to full professorships or *Ordinariate* at a large number of universities. In the German system, it was the *Ordinariat* that marked the arrival of a new discipline, and conversely, new disciplines are seen almost necessarily to evolve by filiation from existing ones.

Numerous parts of this story have come under criticism in the most recent wave of scholarship on German universities and science. Writers have noted that the nineteenth-century picture is largely the product of Prussian writers who presumed Prussian leadership in the sciences and, without further investigation, treated the other states as followers. The idea that it all began with the University of Berlin has been punctured several times, although it seems to reinflate itself just as often. That *Wissenschaft* and *Bildung* were the ideological fundaments of the new universities has been challenged by historians who have noted the continued desire of the state to produce citizens educated to serve its interests. Consonant with this has been greater attention to the role of activist governments in advancing their own policy agendas, rather than the idea that knowledge evolved purely through the efforts of professors to increase it. The free-market notion has also come under fire with greater knowledge of detail that reveals an absence of competition between universities, even efforts to avoid it. The very image of the German university as a research institution dedicated to the pursuit of *Wissenschaft*, which has been central to the standard story, has been counterbalanced by eloquent historians'

---

David and Collins, "Social Factors" (1968); for a modification of Zloczower's argument, see R. S. Turner, E. Kerwin, and D. Woolwine, "Careers and Creativity" (1984).

testimony to the continued importance of teaching in shaping the agendas of the universities. Similarly, the "one-professor, one-discipline" model has begun to be historicized by more careful attention to the development of particular disciplines.[18]

The story told in the following pages incorporates many of these recent directions, while not pursuing them all to an equal degree. My primary focus on the ideas and experiences of university professors who conducted morphological research has dictated which avenues I have followed. Although the study of form became important in the late eighteenth century, I have begun my story at the standard early nineteenth-century starting point for the history of university reform, in part because that is when one can begin to identify morphology by its name, in part because that is an obvious moment when dramatic institutional and intellectual changes intersected. States other than Prussia emerge as especially important to the development of morphology by midcentury, when a movement that came to be known as "scientific zoology" found a more secure home in non-Prussian universities, especially in Baden and Bavaria. Morphology's strongholds remained outside of Prussia until the 1880s, when a new round of professorial appointments opened the door there to some morphologists.

In the story of morphology, the idea of *Bildung* and the relationship between teaching and research hold important places, but as matters of contention rather than as a set of firmly or universally held principles. Among biological scientists, morphologists were perhaps the most persistent carriers of the torch of *Bildung,* arguing over and over throughout the century for the value of pure science in awakening the imagination of the student, rather than seeking to justify their pursuits in terms of developing practical skills. Similarly, their justifications for the research itself only rarely touched on socially useful results, remaining tied to the value of pure *Wissenschaft.* As will become

18. On reforms before Berlin, see F. Gregory, "Kant, Schelling, and the Administration of Science" (1989); and T. Broman, "University Reform" (1989); on practical interests, see P. Borscheid, *Naturwissenschaft, Staat, und Industrie* (1976); A. Tuchman, "From the Lecture to the Laboratory" (1988), *Science, Medicine, and the State* (1985); and Lenoir, "Science for the Clinic"; on lack of free-market system, see R. Kremer, "Building Institutes for Physiology" (1992); on the continued importance of teaching, see K. Olesko, "Commentary" (1988); and G. Schubring, "Rise and Decline" (1989). On more than one professor per discipline, see, e.g., H.-H. Eulner, *Entwicklung der medizinischen Spezialfächer* (1970), on anatomy; however, even though he acknowledges numerous occasions where there was more than one professor of anatomy, he always views these as anomalous because they do not fit into the standard model. My own view, developed in this book, is somewhat different.

evident, however, this value was less universally shared than one might expect. In addition, the story told here devotes more attention to research interests than to teaching, but it is clear that considerations of professors' teaching shaped the appointment decisions that in turn promoted certain research orientations at the expense of others.

Although these general issues in German university history, which might be applied to any science, frame my discussion of morphology's development, my primary focus is directed at the level of disciplines and orientations. One of the most important aspects of morphology's history in Germany is that it emerged at a time when the organization of university studies was in a fundamental transition. In the early nineteenth century the modern notion of discipline was just beginning to be developed and disciplinary boundaries were low; the broad program of studying form was not fixed to any given disciplinary slot. As I discuss in chapter 2, this was typical of the academic organization of biological knowledge at the time. The distribution of teaching duties was correspondingly loose. For about the first third of the century, a full professor enjoyed considerable latitude in what he could actually teach, and two professors at one university might well overlap in their teaching subjects. By the end of the second third of the century, each professor had a designated intellectual territory within his university, over which he had autonomous control. The overall territory might be parceled out slightly differently at different universities, but everywhere, much less overlap was tolerated than previously. The morphological program continued even as disciplinary consciousness grew, and the study of form, not fitting directly into any of the emerging primary disciplinary structures, came to be divided between two different disciplines, anatomy and zoology. How this one came about is the subject of chapter 3.

One consequence of this division was that morphologists did not have the luxury of wearing a single professional identity. The same people who concentrated their research on morphology were always employed in the universities under other designations. Although this mattered relatively little in the first third of the century, by the end of the second third their disciplinary identity has become a primary one. Thus in important parts of their lives morphologists also identified themselves as anatomists or zoologists, in accordance with their institutional and disciplinary home. To pursue what morphology meant after midcentury, therefore, is partly to examine how morphologists sought to define and fulfill the goals of their disciplines. This requires examining the place of morphologists within the two distinct disciplines of zoology and anatomy. Because these two settings were quite different, it is necessary to treat them separately, as I do in the second and third parts of the book.

In gauging the place of morphology within the disciplines of anatomy and zoology, I have depended mainly on the assessments of the participants and their faculty colleagues. This has meant examining various kinds of programmatic writings, such as published inaugural addresses, essays promoting or opposing changes in examinations, and occasional pieces marking the opening of a new institute or the celebration of an important anniversary in a professor's or an institute's life. Such programmatic pieces need to be read with caution, of course, with an eye toward their celebratory or self-promotional interests, but with this caveat, they are enormously useful expressions of professors' views of their field at a given moment in time.

For discerning how the place of morphology within these disciplines changed over time, I have found another sort of document especially useful. It is my view that the institutional decisions that had the most profound effect on morphology's development were those concerning professional appointments. A professorship provided the institutional base from which a scholar gained time and sanction to conduct his research; at least as important for the development of a research program, it made it possible for the professor to cultivate students in his own mold, who might then colonize positions at other universities and broaden the stream of researchers pursuing related questions.[19]

Given the importance of appointments in shaping the directions in which research grew, I have paid particular attention to the documents surrounding the appointment process. Traditionally, this process began with the university faculties. The faculty usually submitted a ranked short list of three names to the head of the faculty, the *Rektor,* or the university's chief administrator, the *Curator;* he then forwarded the faculty recommendations to the state administrator in charge of education, possibly with a cover letter adding further arguments in favor of one candidate or another. The state administra-

19. Although the differential funding of institutes was also important in promoting certain directions of research over others, such financial support depended first on the appointment of a professor who could articulate a program for his discipline and the need for increasing his funding. Even in those rare cases where funding was made available to an *ausserordentliche* professor to pursue a new research orientation, as when Wilhelm Roux was given funds for an institute of *Entwicklungsmechanik* at Breslau (see pp. 301–2), the recipient needed the support of the full professor in the parent discipline.

tor then chose one name from the list with whom to undertake negotiations. At some point in the process, usually after negotiations were complete, the appointment was officially rubber-stamped by the Prince, King, or other head of state. This process could be manipulated or short-circuited in a number of ways, more often revealed in private letters than in official documents: a powerful professor might lobby the *Curator* or the state administrator directly; the state administrator might request advice from outside the university in question, might disrupt the rankings for economic reasons, or might salt the list initially with his own suggestions for possible candidates. It was even possible for a state official to engineer the appointment of someone not on the faculty list at all, although this was such a violation of the professors' traditional privileges that officials were rarely willing to risk precipitating the inevitable ruckus.

The documents surrounding appointment decisions are also useful in revealing what faculties wanted their state administrations to think about their sciences. Like the published programmatic pieces, these arguments cannot be taken strictly at face value as the unvarnished faculty consensus concerning the needs and opportunities presented by a given discipline, of course; such statements were surely crafted with an ear toward what the administrators might want to hear. But just because they are written with this particular audience in mind does not mean the university faculties actually held a completely different hidden agenda. I believe that these documents often *do* reveal what were the important criteria by which a prospective faculty member and his work were judged by his peers and colleagues.

Although I have used this sort of hiring document fairly extensively, I have not pursued higher echelons of administrative policy-making within a *Kultusministerium* or between ministries that may have shaped the hiring decisions. In part, this was a pragmatic decision: while it was relatively easy to locate documents concerning individual university appointments, finding upper-level discussions about science and medical education that might have a reasonably direct bearing on morphologists' careers seemed like a true needle-in-the-haystack hunt. And while historians who have looked for conscious policy-making concerning science and medicine in individual German states have sometimes found it, I have wanted to study morphology in a large number of states over nearly a century; this has precluded the sort of admirable, more finely brushed study achieved by Arleen Tuchman for scientific medicine in Baden or Richard Kremer for physiology in Prussia. Nevertheless, I believe that if broad and explicit policy considerations had played a powerful role in the appointments, I would have caught some whiff of it.[20] More often than not, when it came to appointments in the disciplines colonized by morphologists—anatomy and zoology—there is every indication that state officials allowed themselves to be guided by the individual situations rather than some overarching vision of the direction toward which biological science should be moving.

## THE GENERATIONAL LOCATION OF MORPHOLOGISTS

Efforts to unite the intellectual and institutional histories of science have become quite common in the last two decades; from the beginning, I, too, have sought to understand the relationship between scientists' ideas and their institutional context. Increasingly as this project matured, however, my original conception of interacting intellectual and institutional "factors" seemed not to capture adequately the flesh-and-blood dynamics of change. As I kept returning to the question of what was at stake for the participants in the debates involving morphology, it became increasingly clear that differences and commonalities in the stakes often corresponded to the age of the participants. In this way a different dynamic began to emerge, one in which both intellectual and institutional stakes were evident and which did not require them to interact as abstract forces or entities. This dynamic, which for lack of a better term we might call a generational one, results from the inevitably mixed age composition of the community of biological researchers over time. Thus the third important way I have sought to situate morphologists is to locate them in the age structure of their scientific community.[21]

When a new idea, technique, or institutional possibility came into

20. Tuchman, "From the Lecture to the Laboratory," and *Science, Medicine, and the State;* Kremer, "Building Institutes for Physiology (1992); see also Lenoir, "Science for the Clinic," on Saxony. An important model for these arguments, which treats the role of state policy in promoting chemistry, is Borscheid, *Naturwissenschaft, Staat, und Industrie;* for one broader policy issue raised with respect to zoology, see the concerns about Darwinism discussed in chap. 10.

21. This discussion owes much to Larry Holmes and John Warner, who, in quite different ways, set me thinking about generational differences and the construction of history through participants' memories. My own approach to scientific generations is somewhat narrower than much of the existing literature on the subject, in which scientific generations tend to be defined in relation to much broader cultural generations. See, e.g., T. Lenoir, "Generational Factors in the Origin of *romantische Naturphilosophie*" (1978); and L. Feuer, *Einstein and the Generations of Science* (1984).

being, it held different meanings for people of different ages.[22] These meanings were determined in part by the stage of their careers at which the actors found themselves, in part by the earlier historical events through which they had lived. (Since most morphologists followed similar career paths, advancing up the ranks at approximately the same ages, age can serve as a convenient marker for both career stage and broader experience.) Thus for Karl Ernst von Baer, who had come of age professionally during the battles over *Naturphilosophie* and had weathered the attacks on idealistic morphology in the 1840s and 1850s, the appearance of Darwinism near the end of his life seemed to reprise these older struggles. For a young man such as Ernst Haeckel just beginning his research career, by contrast, Darwinism did not carry that historical freight, but seemed to offer a fresh way of thinking about biological problems.[23]

The converse situation is also relevant: as we follow an individual through time, his own ideas, though perhaps changing relatively little, could take on new colorations as his milieu changed.[24] Most often an idea of technique considered pathbreaking when it first appeared would almost inevitably, if successful, become the stuff of student exercises a couple of decades later. But not all ideas aged the same way. In the 1850s Rudolf Leuckart's writings on the adaptive relationship between an animal's form and its environment were denigrated as old-fashioned and teleological. Forty years later he was making remarkably similar arguments, but the intervening intellectual changes wrought by Darwinism, as well as a broad shift in audience to a new generation who did not associate adaptation with teleology, made these arguments appear very "modern."

There is more to this approach than just noticing the position of particular individuals within the group and its changes over time; what makes it a truly generational perspective in the presence of intellectually distinct age cohorts within the community. Indeed, it is possible to identify six critical cohorts of morphologists, centered at roughly fifteen- to twenty-year intervals, whose ideas and interactions shaped the main lines of nineteenth-century German morphology (see

---

22. For a recent sociologists' examination of the ways that the same events hold different meanings to people of different generations, see H. Schuman and J. Scott, "Generations and Collective Memories" (1989).

23. For further discussion of von Baer's and Haeckel's responses to Darwin's theory, among others, see chap. 4.

24. For an extended study of this sort of changing memory, see J. H. Warner, "Remembering Paris" (1991).

table 1.1). The story begins with the generation that included the leading *Naturphilosophen* and their opponents, all of whom were born within a span of a half-dozen years in the late 1770s and early 1780s. They were followed by a group of prominent morphologists born around 1800, the most eminent of whom was Berlin's professor of anatomy and physiology, Johannes Müller. The third cohort, born in the late 1810s and early 1820s, comprise students of the second cohort; although historians of biology know best the "1847 Group" of physicalist physiologists (Hermann von Helmholtz, Carl Ludwig, Ernst Brücke, and Emil Du Bois-Reymond) who opposed morphology, the same cohort produced the self-declared "scientific zoologists," the best-known of whom are Albert von Kölliker and Rudolf Leuckart, who reconstructed morphology as a new approach to zoology in the 1850s and 1860s. The fourth cohort, born between about 1829 and 1835, encompasses the first group of zoologists to grapple with Darwin's theory (including Ernst Haeckel and August Weismann), as well as a cluster of anatomists studying form from a "mechanical" perspective at both the cellular and macroscopic level (such as Wilhelm His). The fifth cohort, born around 1850, comprises a second generation of Darwinian morphologists that included the brothers Oscar and Richard Hertwig and Wilhelm Roux, for whom not only Darwin's work but also Haeckel's teachings were formative. And finally, the sixth cohort, born in the late 1860s, includes those such as Hans Driesch and Hans Spemann, who took the study of form into new, self-consciously experimental realms in the 1890s and 1900s.

The generational groupings are not perfect, of course; there are anomalies. Carl Theodor von Siebold, for example, was a contemporary of Müller but joined forces with the next cohort to found and further scientific zoology. Nor does Gegenbaur—surely one of the most important morphologists—fit into this pattern, having been born in 1826, just between two of my cohorts. Despite these caveats, however, the generational idea is useful for historical analysis because it provides a tool for situating individuals at once synchronically and diachronically, and because it helps us understand how particular events or ideas could mean different things to different groups of people.

The theory of generations has attracted recurring interest among intellectual and social historians, political scientists, and sociologists since at least the late nineteenth century. The generation theorist who has most influenced German- and English-language writers is the sociologist of knowledge Karl Mannheim, whose seminal 1927 essay

TABLE 1.1 Generations of German Morphologists and Their Contemporaries

| COHORT | LEADING FIGURES (YEAR OF BIRTH) | FOCUS* | PERIOD OF | PREDOMINATING ISSUES |
|---|---|---|---|---|
| 1. 1775–82 | Burdach (1776)<br>Oken (1779)<br>Tiedemann (1781)<br>Meckel (1781)<br>[Schelling (1775)] | A, P<br>Z, P<br>A, P<br>A, P | 1810–25 | Naturphilosophie; introducing a Wissenschaft of life into the universities. |
| 2. 1795–1805 | von Baer (1792)<br>C. A. S. Schultze (1795)<br>Bronn (1800)<br>J. Müller (1801)<br>F. Arnold (1803)<br>C. T. E. von Siebold (1804)<br>R. Wagner (1805) | A, Z<br>A, Z<br>Z<br>A, P<br>A, P<br>P, Z<br>A, P, Z | 1830s–1850s | Building and expanding morphological/physiological research and related facilities within the universities (esp. medical faculties). |
| 3. 1815–23 | Kölliker (1817)<br>F. Leydig (1821)<br>Leuckart (1822)<br>J. V. Carus (1823)<br>[C. Ludwig (1816)]<br>[DuBois-Reymond (1818)]<br>[Brücke (1819)]<br>[R. Virchow (1821)] | A, P<br>A, Z<br>Z<br>Z (ao)<br>P<br>P<br>P<br>pA | 1840s–1870 | Splintering of life-science research programs into chemical and physicalist physiology, mechanical anatomy, cell and tissue studies, systematics and scientific zoology; institutionalization of split into separate disciplines of anatomy, physiology, and zoology. |
| 4. 1829–35 | His (1831)<br>Braune (1831)<br>Semper (1832)<br>Weismann (1834)<br>Haeckel (1834)<br>Claus (1835)<br>Ehlers (1835) | A<br>A<br>Z<br>Z<br>Z<br>Z<br>Z | 1860s–1890s | Incorporating Darwinism into ongoing morphological research programs; development of new Darwinian morphologies; germ-layer studies; expanding job market early in careers. |
| 5. 1846–54 | Fürbringer (1846)<br>Wiedersheim (1848)<br>Bütschli (1848)<br>O. Hertwig (1849)<br>Roux (1850)<br>R. Hertwig (1850)<br>H. Ludwig (1852)<br>C. Chun (1852)<br>Spengel (1852)<br>Rabl (1853) | A<br>A<br>Z<br>A<br>A<br>Z<br>Z<br>Z<br>Z<br>A | 1870s–1910s | Expansion of Darwinian morphologies in new directions: experimental programs; pressure in job market intensifying early in careers. |
| 6. 1864–70 | Haecker (1864)<br>Herbst (1866)<br>Driesch (1867)<br>Braus (1868)<br>Spemann (1868) | Z<br>Z<br><br>A<br>Z | 1890s–1910s | Experimental embryology; "causal" vs. "historical" morphology; no jobs. |

Note: A = anatomy; P = physiology; Z = zoology; ao = ausserordentliche professor; pA = pathological anatomy; bracketed names refer to nonmorphologists significant for the history of morphology.
*The term "period of focus" indicates the period I am focusing on for each cohort in this book, which does not always correspond exactly to the period in which the members of the cohort did their most important work.

"The Problem of Generations" raised most of the important issues still grappled with by more recent authors on the topic.[25] Mannheim emphasized the historically contingent nature of social and intellectual generations: what Mannheim calls an "actual generation" is made up of individuals of the same age who are in a position to participate in the same set of historical events.[26] For Mannheim, the experiences that define a generation generally occur during youth and young adulthood; "all later experiences then tend to receive their meaning from this original set, whether they appear as that set's verification and fulfillment or as its negation and antithesis." A corollary to this is that people of different generations necessarily view the world differently. "While the older people may still be combating something in themselves or in the external world in such a fashion that all their feelings and efforts and even their concepts and categories of thought are determined by that adversary, for the younger people this adversary may be simply non-existent: their primary orientation is an entirely different one."[27] It is useful to contemplate where this claim might lead, especially when extrapolating from Mannheim's generations to generations in science. In my view, it does not mean that scientists have or are open to new ideas only during their youth. Rather, it means that the categories of thought that make up the broad frame of reference within which new ideas are placed are usually established early on. One need not accept Mannheim's implication that this frame of reference is wholly unalterable after its initial establishment, but it does seem reasonable to acknowledge that the enormous investment of time, energy, and thought that a scholar makes in the early part of his or her career will tend to shape the way he or she experiences later events and ideas.[28]

The generational cohorts in German morphology share many of the features of Mannheim's "actual generations." Most important, it does seem to be the case that the members of a given age cohort, because they usually shared numerous common experiences as they entered into their scientific lives, tended to view the same issues as

25. K. Mannheim, "Problem of Generations" (1927; reprint, 1952). The other leading generation theorist is José Ortega y Gasset, whose greatest influence has been among Spanish-language writers; for an extraordinarily helpful overview of the enormous literature on generations, see A. Esler, "Generation Gap in Society and History" (1984).

26. Mannheim, "Problem of Generations," pp. 303–4.

27. Ibid., 298–99.

28. My thanks to David Hull for his comments on the relationship of the age of scientists to their receptivity to new ideas, which forced me to clarify this part of my argument.

significant and interpret them along similar lines. Even when they disagreed, their disagreements tended to share a common axis: when members of one generation fought, they were fighting over the same thing. When members of two different age cohorts viewed the same scientific issue, by contrast, they would be more likely to interpret its significance in fundamentally different ways. In terms of understanding their intellectual orientations, my generations are very much like Mannheim's.

In two important ways, however, my case departs from Mannheim's and from most historical studies concerned with generational movements. In contrast to the mass political or cultural movements of concern to social historians, my groups tend to comprise five or six men, or at most a dozen. This is a consequence of my focus on those morphologists and their peers who became *Ordinarien* in anatomy or zoology. The German professoriate was not a large population, and the number of full professors of anatomy and zoology was far smaller. Among the universities under study here there were fewer than 150 *Ordinarii* for anatomy and zoology over the entire century, and at any given time the total German population of full professors of a single subject was not much over twenty. While the small size of any cohort makes statistical analysis meaningless, the correspondingly small size of the professoriate meant that a single cohort could make a big difference. Even just a half-dozen professors from one cohort, in place for thirty or forty years, could profoundly shape their field, both by promoting their own orientations and by closing off other avenues of research.

My focus on the professoriate also creates a second difference from the majority of literature on social generations. Mannheim and most social and political historians who embark on generational analyses have a problem in defining historical generations out of the surrounding population, which, of course, forms a continuous span from old to young.[29] A glance at appendix 1, which lists university professors of anatomy and zoology by birth date, shows that I have something of the same problem. In my case, however, the situation is a bit different, for the very existence of a professorial cohort is to some degree a result of the institutional structure of the German universities. At times of institutional expansion or reorganization, as in the 1850s for anatomy or the 1860s for zoology, there might be a large number of professional openings within a relatively short time span. At such moments, the new entrants into the field tended to be all in

29. See, for e.g., A. B. Spitzer, "Historical Problem of Generations" (1973).

the same age range; thus a professorial cohort was born. These groups typically held onto their professorships for over thirty years, and if they retired or died around the same time, as was the case for a group of Prussian zoologists in the early 1880s, the generational effect was repeated. At certain moments, then, the professorial community gained a fairly strong generational push, but this was not some sort of natural biological phenomenon: the professorial generations themselves were marked out from the continuous flow of potential scholars into the universities by the opening and closing of gates of institutional opportunity. Appendices 2 and 3, which chart the professors of zoology and anatomy at each German university from 1850 to 1918, show the timing of the turnovers in both disciplines. (Before 1850, the assignment of professors to these two subjects is too irregular and changeable to be charted this simply.)

While these gates of opportunity clearly shaped the demography of the professoriate, it is nevertheless important not to overdraw the institutionally constructed character of my generations. To begin with, as the appendices show, the professorial openings did not always appear in clusters, and there were always advanced students interested in moving up the academic ladder, so that some anatomy and zoology professors were born between my cohorts. Second, although a cohort could suddenly come into evidence through rapid institutional expansion or turnover, deaths and retirements tended not to occur so uniformly, so the next set of appointments might well be scattered over time, thus attenuating institutionally driven generational effects for the next set of appointments. Third and perhaps more important, the cohorts are meaningful as groups who shared intellectual and disciplinary experiences, who engaged with similar issues. The coincidence that at particular historical moments a number of individuals who shared those experiences were able *as a group* to obtain professorships is what turns them into a professorial generation. In fact, the individuals in these professorial cohorts were selected into the professoriate from a much broader group, who would have shared many of the same formative experiences, and in the part of my story concerning generations 5 and 6, where the absence of jobs becomes a significant issue, the broader circle and the process of selection itself become more important. As I hope this discussion makes clear, my professorial generations are ultimately defined through a combination of intellectual issues, disciplinary demographics, and structural features of the universities, with each generational cluster bearing a unique mix of associations.

The implications of this generational stratification for the history of morphology are significant. The concerns that predominated among a group lucky enough to be entering the profession in a period of institutional growth or turnover would then shape the direction of the discipline. Both the intellectual style and the problem areas developed by a given cohort could last for decades. Thus the interest in germ-layer development predominant among both anatomists and zoologists born in the early 1830s, who gained professorships between the late 1860s and mid-1870s, continued up to the turn of the century. In the case of the zoologists, this interest was usually combined with questions of evolutionary development as well, and that coupling, too, persisted for decades. These questions had become the hot ones in the early and mid-1860s, when these men were developing their individual research agendas, and although the heat abated somewhat, the same questions continued to provide fuel for dissertations up to the turn of the century.

This is not to say that a single generation ever had complete control over the discipline, or that members of a cohort would necessarily be averse to problem areas outside their own specialty. Intellectual change did take place even in the absence of professorial turnovers. The point is that the acquisition of professorships encouraged a budding research orientation to develop, while the denial of professorships to that orientation would make it more difficult to do so. As a generation of would-be professors born in the late 1860s became increasingly interested in experimental approaches to the study of form near the turn of the twentieth century, almost no openings were available at the level of the full professorship. Even at the time, this absence of university posts was widely perceived to have inhibited these experimental avenues of research. And when positions finally did open up, the next prospective generation—one that would have been in a position to take up Mendelian genetics as a fresh new program—were closed out by their elders who had been patiently waiting. Thus even into the 1920s there was a real dearth of professors of anatomy or zoology born in the 1870s.[30]

30. J. Harwood (*Styles of Scientific Thought* [1993]) identifies a half dozen scientists who took up Mendelian projects before World War I. The four educated as zoologists were Theodor Boveri (1862–1915), Valentin Haecker (1864–1927), Ludwig Plate (1862–1937), and Richard Goldschmidt (1878–1958). Boveri had become *Ordinarius* at Würzburg in 1893; Plate succeeded Haeckel at Jena in 1909; and Haecker moved from his professorship at the *Technische Hochschule* in Stuttgart to one at Halle in 1909. Goldschmidt is the only zoologically educated geneticist on Harwood's list born in the 1870s, and he never attained a full professorship in Germany. Most of the geneticists Harwood discusses were born from the mid-1880s on (and most of them were employed outside the traditional universities).

As the preceding discussion suggests, despite various caveats, situating morphologists according to their ages offers a number of advantages. It helps us to look beyond the most famous figures to see the ways they represented the concerns of their peers or stood out from them. It calls attention to the fact that the perception of institutional and intellectual events differed according to the layers of experience through which those events were perceived. It also reminds us that the passage of time itself, carrying with it the deaths and retirements of older scientists and the emergence of younger men into the profession, brings change to the intellectual makeup of a community. Finally, a focus on generations points to a critical nexus of intellectual and institutional change—the professorial age cohorts were derived from the institutional structure of the university, but they could bring about dramatic intellectual change because their members tended to view the scientific world along similar lines.

## THE ARGUMENT

When we pull together the three perspectives provided by intellectual, institutional, and generational analyses, the history of morphology takes on a quite different shape from that presented by Russell at the beginning of this century. A three-part chronological division remains, as is reflected in the three parts of this study, and idealistic, evolutionary, and causal-experimental approaches continue to hold critical places in the story. But I hope to demonstrate in the course of my analysis, as outlined below, that each of the three phases of morphology in Germany had more dimensions than Russell's story indicates.

Part 1 covers the period from about 1810 through the 1850s. There can be little doubt that interest in studying animal form expanded dramatically in the first quarter of the nineteenth century and continued to propel a broad stream of research for decades thereafter. However, as I argue in chapter 2, the first two generations in my study, which dominated the universities up through the 1840s, rarely viewed morphology as an independent intellectual enterprise, a distinct orientation within biology. Rather, they virtually always understood it to be an integral part of existing areas of study and teaching. Concern with animal structure and development appeared in works on anatomy, physiology, comparative anatomy, zootomy (or animal anatomy), and natural history. Before the late 1840s, these sciences were closely interconnected, both intellectually and institutionally, with physiology at their intellectual center. Thus in the earlier part of the century, the study of form was more often seen as a means to an end than as an end in itself.

As I show in chapter 3, in the 1840s and 1850s a younger generation of physiologists (in generation 3) forced a redefinition of physiology and the relationships among the cluster of sciences surrounding it. The physicalist physiologists contended that their science had been dominated by a morphological outlook that was speculative, vitalistic, and incapable of solving the fundamental problems of physiology. As is well known, these physiologists rejected the methods and goals of their predecessors in favor of a narrower but, in their view, more fruitful definition of physiology as an experimental science studying the body's functions. However, although cast out of physiology, morphology did not wither away; not at all. Instead, contemporaries of the physicalist physiologists interested in morphological questions simply changed locations in the universities, moving into professorships of anatomy and zoology, where they called themselves "scientific" zoologists. Indeed, one of the reasons the redefinition of physiology succeeded was because it occurred during a time when states were willing and able to expand the number of university professorships. This allowed a level of institutional differentiation among disciplines that had hitherto been lacking: not only did anatomy and physiology gain independence from one another in the medical faculties, but as part of a broader expansion that lasted through the 1860s zoology, too, gained independent professorships in the university philosophical faculties. Although morphology still did not become an independent discipline, in this new setting it became identified with a distinctive program of research, a clear orientation within the newly autonomous disciplines of anatomy and zoology.

In part 2 I turn to the reception of Darwin's theory among German morphologists and the development of evolutionary morphology up to about 1880. Chapter 4 presents the initial reception of the theory in a generational context; here I argue that the experiences (or lack thereof) of members of generations 2, 3, and 4 help us understand the central underlying issues that led them to read Darwin's theory in different ways. Chapter 5 focuses on the development of Haeckel and Gegenbaur's evolutionary research program while both taught at Jena up to 1873 and argues that an unusual combination of personalities, skills, and an institutional setting where disciplinary boundaries remained low gave great impetus to their research program in the first flush of Darwin's theory.

But disciplinary boundaries were higher elsewhere. In chapters 6 and 7 I analyze the reception of Haeckel's and Gegenbaur's ideas within their respective disciplines of zoology and anatomy, considering the intellectual values, assumptions about teaching and research, and competing research programs in each discipline. Most significantly, these chapters show that Haeckel and Gegenbaur's program did not dominate German anatomy and zoology between the 1860s and 1880s. To be sure, while they were together at Jena the two men attracted numerous students, among whom would be some of the leaders of the next generation of morphological researchers (generation 5). But within anatomy more broadly, Gegenbaur's evolutionary comparative anatomy faced competitive pressure from other orientations that were perceived to be more medically relevant, while within zoology Haeckel's ideas about embryological recapitulation were met with sharp criticism. German zoology, far from becoming uniformly Haeckelian in its features, continued to be characterized by a broad range of questions, including not only the construction of phylogenetic trees but also problems relating to functional morphology, generation, and adaptation pursued by the scientific zoologists.

Part 3 gives a new shape to the morphologists' debates of the late nineteenth and early twentieth centuries. Whereas these debates have usually been described exclusively in terms of the philosophical and methodological opposition between "causal" (i.e., experimental) and "evolutionary" morphology, I argue that in Germany a variety of factors made this opposition less important to the history of anatomy and zoology than is suggested by the polarizing rhetoric of the 1890s (which was very real). To begin with, as my analysis in chapter 8 shows, the Haeckel-Gegenbaur program came to suffer from methodological conflicts that threatened to render it incoherent; it no longer provided as strong a framework for inquiry as some had believed it to in the early years. Chapters 9 and 10 argue that on the other side of the dichotomy, the self-consciously experimental and antievolutionary approach to morphology represented by Hans Driesch did not come into ascendance either: although causal-experimental techniques were increasingly used, they seem most often to have been regarded as supplementing rather than supplanting existing methods and questions. Furthermore, in the absence of further expansion of the university system, experimentalists were unable to gain new professorships devoted to their specialty, and (for reasons differing in anatomy and zoology) were not widely successful in their claims to be able to represent the broader disciplines of anatomy and zoology as full professors. Thus neither the Haeckel-Gegenbaur program of evolutionary morphology nor a self-consciously experimental morphology held a dominating position in German animal biology at the turn of the century. Instead, a more diffuse pattern of problem areas and approaches appeared in both disciplines, with anatomists placing greater emphasis on the description of cell and tissue development and zoologists connecting descriptive studies of organisms' life cycles and morphology with theoretical questions about evolution, adaptation, and generation. By presenting this fuller picture of events, I hope to offer some corrective to the existing "revolt from morphology" literature that has been based on the American scene.

Perhaps the most important modification of Russell's story to emerge from my overall analysis concerns the role of scientific zoology, a movement that spanned all three of Russell's phases. The intellectual aspect of this movement has been called "functional morphology," although within a more restricted context than I present here;[31] curiously enough, Russell himself, who divided morphology into "functional" and "pure" types, completely ignored the existence of a functionalist tradition in Germany, thereby framing the problematic of German morphology in terms of the continuity or discontinuity between idealistic morphology, Haeckelian evolutionary morphology, and causal-experimental morphology. In the story I present here, each of these programs from the 1840s on evolved in interaction with the functionalist concerns of the scientific zoologists. In the concluding chapter, I draw out some of the implications of this revision and suggest some directions for further research.

Although it is analytically useful to situate morphologists according to the three axes of their intellectual, institutional, and generational locations, the above sketch suggests that these were most often intimately related, and I hope to demonstrate this more fully in the course of my analysis. Arguments over what made morphology *wissenschaftlich* (or not) were often asserted in the context of competition for institutional resources among researchers with different philosophical outlooks, which in turn were sometimes related to the age of the contenders. It is not my argument, however, that institutional conditions in some way determined the beliefs of the participants, or that their arguments over ideas were always advanced with an eye toward disciplinary autonomy. How much easier it was to colonize existing disciplines, to infuse anatomy and zoology with morphologists, than to create entirely new chairs and institutes! Nor should we assume, as generation theorists might, that generation gaps inevitably cause conflict—that younger scientists always rebel against the ideas

31. Lenoir, *Strategy of Life,* esp. chaps. 4 and 5.

of the older generation. Moments when such rebellion appears evident
to some historians, such as the late nineteenth-century "revolt from
[evolutionary] morphology" undergone by men such as Wilhelm
Roux and Hans Driesch, often appear, upon closer examination, to
be far more equivocal.[32]

The task, as I see it, is to try to discover what was at stake for the
morphologists and their peers at any moment and then set out to
understand the arguments in terms of these stakes. Sometimes the
stakes were institutional, sometimes they were purely intellectual,
sometimes they were different for people of different age cohorts.
Sometimes intellectual arguments supplied the means toward an institutional end; sometimes the reverse was the case. Often the two were
so closely intertwined that it is impossible to separate means and
ends, but sometimes, as when Darwin's theory profoundly affected
morphologists' thinking in the 1860s and early 1870s without immediate institutional or disciplinary consequences, intellectual changes
proceeded independently of institutional change. We simply cannot
know a priori what the relationship between intellectual, institutional,
and generational change will be. Part of the historian's job is to find
out.

32. For the strong version of this argument, see G. Allen, *Life Science* (1975). In
fact, Driesch appears to have been far more of a revolutionary than Roux; one should
also note that they belonged to two different intellectual generations: Roux was born
in 1850, Driesch in 1867. On the relation between evolutionary and experimental
morphology in Germany more generally, see pt. 3 below. There is a growing literature
on this topic as it affected the American scene, although it tends not to emphasize the
generational component especially strongly (see n. 9). Of course, sometimes a persuasive
case can be made for generational revolts; Timothy Lenoir has developed such a case
for the "1847 Group" of physicalist physiologists. See Lenoir, "Social Interests" (1988),
and "Laboratories, Medicine, and Public Life" (1992).

# PART ONE

*Morphology and Physiology*

## CHAPTER TWO

## The Study of Form before 1850

H istorians desiring to characterize the burgeoning interest in animal form in the early nineteenth century have frequently run into difficulties, for the intellectual currents of the first half of the century have proven notoriously hard to sort out. Vitalism, mechanism, teleomechanism, Kantianism, romanticism, *Naturphilosophie*—all of these philosophical stances have been connected to the new fascination for form, and with few clear results. Nor does an institutional or disciplinary approach help us out much, for the institutional arrangements under which researchers pursued the study of form are even less securely charted than the philosophical issues. Here, too, attempts to sort students of form into clearly defined categories—this time associated with particular disciplines—founder. This does not mean that the study of form was "interdisciplinary"; rather, the modern notion of academic disciplines, as they came to be defined by the late nineteenth century, was itself just being molded during this period, and the institutional features that came to mark disciplinary status had not yet been firmly established.

In seeking a way through the murkiness, one might hope that the invention and use of the new term "morphology" would provide some useful guide to the meaning and significance of studying animal form. As early as the turn of the nineteenth century, the term was in use in some circles: the philosopher-poet Johann Wolfgang von Goethe mentioned it in a letter to Schiller in the late 1790s, and in 1800 the anatomist and physiologist Karl Friedrich Burdach introduced it into print, along with a welter of other neologisms in his *Propädeutik zum*

*Studium der gesammten Heilkunst.*[1] By "morphology" Goethe and Burdach meant the study or doctrine of form, and both sought to denote something intellectually more elevated than "anatomy," which, although it dealt with organic structures, was still more closely associated with cutting up plant and animal bodies than with philosophizing about them. Each also sought to convey the importance of change over time for understanding form, a concern that would continue to be one of the central features of morphology and would distinguish it from both traditional human anatomy and the comparative anatomy of the eighteenth century.

Both men continued to make use of the word, Goethe in private and Burdach in print. No one else took it up. Then in 1817, each used it for the first time in the title of a published work, marking this year as another possible launching date for the term. But the impact upon animal researchers seems to have been minimal: Burdach's essay appears to have caused few ripples, while Goethe's writing was far more influential among botanists than among zoologists, anatomists, or physiologists. (Even into the 1860s nonspecialist German dictionaries and encyclopedias defined "morphology" as "the doctrine of plant form.")[2] Indeed, the utility of the term "morphology" as a guide to the topic in the first part of the century turns out to be illusory, for despite some sporadic appearances after 1817, it was only in the late 1840s and 1850s that scholars began to use it with frequency.[3]

1. On Goethe's and Burdach's uses of "Morphologie," see G. Schmid, "Über die Herkunft." In his introduction to *Goethe's Botanical Writings* (1952/1989), p. 9, n. 17, C. J. Engard notes that Schmid's analysis ignores Goethe's use of the word in letters to Schiller.

2. See, e.g., *Allgemeine deutsche Real-Encyclopädie* (Brockhaus Conversations-Lexikon), 10th ed., s.v. "Morphologie": "Morphology or the doctrine of plant form [*Pflanzengestaltlehre*] . . . comprises the depiction of the external forms of plants and their organs, and is divided into a general part, in which the forms of plants and plant organs are discussed, and a special part, which concerns the forms of plants according to their main groups along with their organs. The description of the internal formation of plants and their organs belongs to plant anatomy. Among all parts of pure botany, morphology is the most important, both for the concept of genus and species as well as for the systematic ordering of the plant kingdom, and thus has always been the foundation for doing botany. To morphology also belongs the description of the transformation (see 'Metamorphosis') of plant organs." *Meyers Neues Konversations-Lexikon*, 2d ed., vol. 11 (1865), also defines morphology as "the doctrine of plant form" and attributes it to Goethe.

3. Wilhelm Engelmann's *Bibliotheca Historico-naturalis* (1846), the most comprehensive bibliography for early nineteenth-century German writings on animals, lists only eight authors in its index entry, "Morphologie," and among books and articles by these authors only three titles actually have the word "morphology" in them. The

Since neither the intellectual nor the institutional issues are readily clarified by trying to lump and split researchers into philosophical or disciplinary camps, and since "morphology" itself fails us as a guide, it may be more useful to contemplate the lack of pattern (and perhaps even the lack of a term) itself. This perspective suggests that something new was going on, and that the lack of intellectual, institutional, and linguistic uniformity results from different local attempts to deal with it. What was new?

The study of form was taken up as one leading edge of a larger project, namely, the pursuit of a *Wissenschaft* of life. University scholars sought to incorporate this pursuit into their professorial lives both through the kind of writing they did and through the ways they organized their teaching responsibilities. Writers on form played with a set of conceptual elements that included law, cause, life, development, force, and form. Scholars often disagreed over the relationships between these elements, but all shared the conviction that a true *Wissenschaft* of life would encompass them. At the same time, professors interested in animal life and form took up a variety of teaching arrangements whose common feature is that they joined a theoretical subject they sought to enhance through research—physiology—with one that their states might be more readily persuaded to support— usually anatomy or natural history. Both intellectually and institutionally, then, the study of animal form before 1850 was intimately involved with efforts to develop a broader *Wissenschaft* of life.

That university professors should be engaged in developing the sort

---

three works are J. W. von Goethe, *Zur Naturwissenschaft überhaupt, besonders zur Morphologie* (1817–23; reprinted under the title *Morphologische Hefte*, 1954); M. H. Rathke, *Zur Morphologie: Reisebemerkungen aus Taurien* (1837); and J. H. Schmidt, *Zwölf Bücher über Morphologie* (1831); the other titles contain words such as "Formenlehre," "Grundlagegebilde," "Form," and "Entwicklungsgeschichte," which clearly justify their presence under the morphology entry, but still it is remarkable how few people chose to use the term "morphology" as the overall programmatic term for what they were doing. By the second edition of Engelmann's bibliography, which covered the period 1846–60, the word had gained considerable ground, appearing in more than two dozen titles. Remarkably, German-language titles comprised only the second largest group; over a dozen titles were in English! Although the relationship between the English and German uses certainly bears further investigation, there is little to suggest that the English use was the main inspiration for Germans who used the term. In fact, although almost never used in titles until the late 1840s, it had been gradually seeping into the texts of research and review articles (rarely in textbooks) since the 1820s. In these textual uses, "morphology" usually referred simply to considerations of the arrangement of body parts in an organism without appearing to carry much philosophical or programmatic baggage.

of synthetic understanding of the laws and causes of nature embodied in the word *Wissenschaft* was not entirely new with the nineteenth century, but the early nineteenth-century approach to *Wissenschaft* did have some new features. One was that the *Wissenschaftsideologie* now had state sanction, at least rhetorically; the famous university reforms of the early part of the century codified a shift already under way from a situation in which university professors were allowed to conduct original research to one in which they were expected to do so. At the same time, *Wissenschaft* meant for many scholars and administrators a kind of knowledge separate from practical knowledge. The sanctioning of this "pure" knowledge led to the rapid growth in the study of languages, history, and the natural sciences in university philosophical faculties, along with a marked rise in their status. The "higher" faculties of law, medicine, and theology were also affected by the *Wissenschaftsideologie;* in fact, most academic proponents of a *Wissenschaft* of life were located in university medical faculties. However, there was an important difference between the medical and philosophical faculties. In the latter, "knowledge for its own sake" became a watchword covering both research and teaching. Not only were professors supposed to press the frontiers of pure knowledge, but they were also supposed to impart this ethos (and its methods) to their students—thus would the students acquire *Bildung,* the development of their inner selves. In the medical faculties, however, the notion of *Bildung* was met by a strong belief that knowledge for its own sake was not enough: medical students should receive a training aimed at an eventual medical practice. As we shall see, the distinction between teaching for the ends of *Wissenschaft* and *Bildung* and teaching for practical ends quickly became a source of tension in medical education.[4]

This chapter uses Karl Friedrich Burdach's 1817 essay *Über die Aufgabe der Morphologie* (On the tasks of morphology) as a touchstone for understanding the role of animal form in early nineteenth-century scholars' efforts to create a new *Wissenschaft* of life. Burdach's approach in this essay was atypical in emphasizing the study of form as a *Wissenschaft* in itself, rather than as part of a larger agenda for physiology, but the more general properties of a *Wissenschaft* he envisioned were quite representative of his generation. For the purpose of understanding the elements of such a *Wissenschaft,*

4. R. S. Turner, "Growth of Professorial Research" (1971), "Prussian Universities"; on the late eighteenth-century origins of the tension between teaching for *wissenschaftlich* ends and for practical ones, see T. H. Broman, "Transformation of Academic Medicine" (1987).

Burdach offers an unusually transparent glimpse into a scientific world whose language and principles are quite foreign to our own, one in which the study of form was viewed as a central element in a new, wide-ranging science of physiology. Furthermore, we can also read the essay as a response to the institutional conditions confronting would-be *Wissenschaftler* of life in the early part of the century: it suggests the circumstances under which they tried to institutionalize this new science and the strategies they adopted to do so. The chapter concludes by contrasting the situation confronting the first generation of these university *Wissenschaftler* with the generation that followed in the 1830s.

## WISSENSCHAFT AND THE STUDY OF FORM

When Burdach published his essay *Über die Aufgabe der Morphologie* late in 1817, he was able to capitalize on a title that had captured considerable attention a few months earlier. The first installment of Goethe's latest book, *Zur Naturwissenschaft überhaupt, besonders zur Morphologie* (On natural science in general, in particular on morphology), had appeared in July, introducing the term "morphology" to a broad public, and in his pamphlet Burdach made a bow in the direction of the great poet for calling attention to the term (which, however, had first been put into print by Burdach himself). Despite its title, however, Goethe's book had had very little to say about what made morphology a *Wissenschaft,* or indeed, about what this new science should do.[5]

Burdach's pamphlet laid out much more carefully and explicitly than Goethe's book a program for morphological research. The aim of morphology, Burdach wrote, was "to investigate the meaning and significance of organic forms. . . . Firstly it falls to it to discern the laws of formation [*Bildungsgesetze*], and to grasp the process by which creative Nature brings forth organization [*Gestaltung*], the mode of origin and progress of a structure [*Bildung*], the influence that it exercises and under which it stands in return."[6] According to

5. K. F. Burdach, *Über die Aufgabe der Morphologie* (1817); Goethe, *Zur Natur-wissenschaft.* I have been unable to find this book in this form, but see n. 3 above for reprint title; on its publication history and on earlier uses of the word, see n. 1 above. Goethe's work was a collection of essays, the longest of which was a reprint of his 1790 work on the metamorphosis of plants. Although he also published work on animal form, his greatest scientific impact seems to have been on botanists, coming through this reprinted essay. See G. Schmid, "Über die Herkunft," pp. 602–3, on botanists using the term in connection with Goethe.

6. Burdach, *Über die Aufgabe,* p. 31.

Burdach, this meant studying the ways that the organism's external environment affects it, the relation of the individual to its genus, and of the genus in its turn to the great organism of which all particular organisms are a part. It also fell to the morphologist to consider the degree to which present forms should be understood as the remnants of extinct ones, and in turn the "seed" of future ones. Finally, the morphologist would study the mutual interactions of form and the activities of life, trying to discover how "that which springs from activity [i.e., form] in its turn determines the expression of activity."[7] Thus Burdach outlined a clear program for research under the rubric of morphology. He presented a broad goal—insight into form—and outlined a specific set of problems; elsewhere in the essay, he suggested a method—comparison—by which these problems should be attacked. Perhaps most usefully, he also took considerable care to explain how his program as a whole fulfilled the intellectual requisites of a *Wissenschaft*.

As was typical for someone influenced by *Naturphilosophie*, Burdach's argument for why one should study form was preceded by a lengthy discourse on the nature of knowledge, in which he outlined a model of three kinds of knowledge: practical, descriptive, and *wissenschaftlich*. Practical knowledge, of the sort that allowed physicians to heal people, existed in its own realm, while descriptive and *wissenschaftlich* knowledge formed a two-layered hierarchy. "From descriptive knowledge [*Kenntniß*] we strive towards comprehensive knowledge [*Erkenntniß*], from information-gathering [*Kunde*] to science [*Wissenschaft*]." In contrast to mere *Kenntniß*, the collection of particulars, *Wissenschaft* is the project of finding the fundamental laws or principles that revealed the order beneath the apparent chaos of particulars.

Burdach posited two kinds of *Wissenschaft*. "Pure" *Wissenschaft* takes place entirely in the abstract realm of reason, that is, within the "interior world" of the mind; *Erfahrungswissenschaft*, by contrast, is a "science of experience" that takes into account our very real encounters with the phenomenal world. Of these two sorts of *Wissenschaft* only the latter is satisfactory, according to Burdach, because it alone "occupies all sides of our being," uniting the two sides of our nature, our reason and our senses.[8] Only an *Erfahrungswissenschaft* allows us to discover the ways in which the laws of the interior world are

7. Ibid., pp. 31–32.
8. Ibid., p. 21.

played out in the external world and to recognize the inner unity among the diverse particulars of the external world.

This is hardly a radical claim for what science should do, and it represents the mainstream line on the kind of knowledge a *Wissenschaft* should seek. Even the most extreme *Naturphilosoph,* such as Lorenz Oken, who was accused by many of his contemporaries of trying to create a purely rational system without sufficient attention to the empirical world, protested that he, too, aimed at bringing the two realms together in his writing.[9] The conflicts between the *Naturphilosophen* and various other early nineteenth century scientists revolved not around the basic idea of these two aspects of human nature and their union in science, but rather around the relation of the realm of reason to the truths of nature, the way the union of the two realms was to be worked out, and the status of the knowledge gained in each.[10] While these conflicts were certainly important, at one level they mattered less than the common commitment to the pursuit of *Wissenschaft* through empirical research itself. For example, to the objection of Kantians that there is no way to know if the way we think actually reflects the underlying, real principles of nature or if it reflects only the structure of our mind, Burdach replied, "Even though we can contemplate nature only in a form that corresponds to the organization of our material nature, and cannot know its innermost self, independent of sensory perception [*sinnliche Anschauung*], nevertheless we persuade ourselves that at the foundation of these sensory perceptions there lies unity, order and lawfulness, that the laws of which we are aware in our reason, are realized in [the sensory perceptions]."[11] Without that faith, or something like it, why pursue empirical research at all?

Such general arguments over metaphysics and epistemology were extremely important at the time, but they do not bring us much closer to understanding the connections people made between the idea of *Wissenschaft* and the study of organic form. These connections may be clarified by focusing on the conceptual links that could be forged between law and cause on the one hand, and form, force, and development on the other. Burdach provided one version of these connections in his pamphlet.

"All knowledge" of a comprehensive sort, Burdach wrote, "is in-

9. L. Oken, *Lehrbuch der Naturphilosophie* vol. 2 (1810), p. xvi.
10. See B. Lohff, *Der Suche nach der Wissenschaftlichkeit* (1990).
11. K. F. Burdach, *Die Physiologie* (1810), p. 8.

sight into the causal interconnections among phenomena." That is, we really know a thing in this deepest way when we have grasped "how it emerges from its origin [*Grund*] and reveals itself in its effects," and when we have apprehended "the inner bond that joins things, the unity that lies at the base [*Grund*] of apparent diversity."[12] With this conflation of temporality and logic, Burdach has already plunged us into an unfamiliar realm, for the word *Grund* conveys multiple meanings, the creative ambiguity of which is central to much early nineteenth-century scientific writing. A thing's *Grund* is its foundation, its basis; it is also the thing's origin and its cause. Thus when a writer speaks of "*Grund und Wirkung*" he seems to be talking straightforwardly about "cause and effect," but the phrase may also carry further connotations, as when Burdach writes of how a being "emerges from its *Grund* and reveals itself in its effects"—here a word that means "cause" also means "source": a thing's cause is its origin. Thus the relation of "cause and effect" occurs not just at the level of mechanical causes, of physical contacts between bodies; when seeking the causes of something, you necessarily look backward in time to its origin, and also from its particular existence in the phenomenal world to its source in the ideal world. Most important, a *Grund* is something more general than its effect, something philosophically both higher and deeper. These deep causes or *Gründe* are expressed as the laws of nature; both law and cause at this level reflect universally operating Ideas. As Burdach put it in a later work, "Law is that which is permanent, unchanging, ideal in the ephemeral, changing, phenomenal; the rule of law means nothing other than that an abiding Thought rules over the particular. . . . The Thought . . . of the whole is the *Grund* of the world."[13]

In this usage, then, law is inextricable from cause. The laws that govern nature also provide the common root from which particular phenomena derive; if you trace a set of phenomena back to this common root, you have found at once their cause and the law that governs them. The task of searching for the laws of nature is, in this meaning, the same as the search for causes. Now, of course, Burdach recognized that one could find mechanical causes, events that directly effected subsequent events in space and time, but for him, these were not satisfactory as causal explanations because any mechanical cause had to be the effect of a previous one, which in turn had its own mechani-

12. Burdach, *Über die Aufgabe*, p. 19.

13. Burdach, ed., *Die Physiologie als Erfahrungswissenschaft*, vol. 6 (1840), p. 596.

cal cause, and so on; pursuing this causal chain would leave the researcher always at the same level of explanation and never allow him to reach the higher level he was seeking.[14] Thus he acknowledged the approach of Kant and Blumenbach, which focused on mechanical causes within the organism and took organization itself as a given, but judged it an insufficient goal for science.[15] Burdach's aim was more ambitious: to discover both the laws of form and their causes, the ultimate *Gründe* underlying the laws.

For Burdach, form exists only in the phenomenal world, while the *Grund* of form must reside in the ideal world, as all ultimate *Gründe* do; in his view it is Life as an idea that is the *Grund*. But form does not derive directly from life, for forms are bounded, particular, space-filling entities, while life is an unbounded force. Now forces, according to Burdach, are comprehensible to us only as Thought, that is, within the realm of reason. What connects the interior world of the Idea or force of life to its external, particular manifestation as form is a middle term, formation, which is the *process* or activity through which life expresses itself in form. As he put it, form is the result of formation (*Gestaltung*), and "formation is one with life, inseparably intertwined with it and at the same time a revelation [*Offenbarwerden*] of it; thus [form] can only be derived and explained by [life]."[16] Here again, life, an idea or a force, is presented as the *Grund* (cause, explanation, foundation) of the forms seen in the phenomenal world. But life is manifested through development, and development itself displays certain lawful characteristics. Both philosophically and materially, development signifies the movement from the general to the particular: it shows us at one and the same time how we get from general, ideal *Grund* to specific, phenomenal effect, and also how we get from the general and poorly defined features of the egg to the individuated adult. This logical elision means that when people talked about seeking the *Grund* of a particular form, they could simultaneously mean its temporal origin in the egg and its more general foundational cause in the ideal world. (The connection between development and explanation may have been reinforced by the use of the word "*entwickeln*," which translates as "to develop" but had once meant "to explain.")[17] Despite its strangeness to us, this

14. Ibid., p. 592.
15. This approach, which James Larson identified in 1979, has subsequently been given the felicitous name of "teleomechanism" by Timothy Lenoir. See Larson, "Vital Forces"; Lenoir, *Strategy of Life*.
16. Burdach, *Über die Aufgabe*, pp. 21–22.
17. K. Spalding, *Historical Dictionary* (1952–), p. 661.

view of development as the crucial link between the underlying, ideal order of living nature and the particular, real-world form of the individual organism provided a powerful set of assumptions for many life scientists in the first third or so of the nineteenth century, and continued to undergird the work of morphologists for decades thereafter.

But how was one actually to conduct research on animal form and development? The crucial method is comparison. "Whence stems all our knowledge," Burdach wrote, "if not from comparison?" Given that comparison is still seen as the method of choice in morphology, this might seem almost natural to the modern reader. But it turns out that Burdach's notion of comparison differs somewhat from the type of comparison that we are accustomed to in post-Darwinian comparative anatomy. First one compares the forms of particular anatomical structures with their life activity to try to establish the connection between their structure and their functions. Then one compares various parts to one another to see how the different parts are characteristically associated with different activities and, conversely, how given life activities typically correspond to particular forms. Finally, once these correlations are established, one can compare a given functioning part within the entire animal series to discover the internal links that join these parts to a common origin, "and so recognize them in their true relation to one another."[18]

Comparison here serves a different function than it would later in the century. The sort of comparison Burdach is talking about is not designed primarily to yield some sort of Ur-form; rather, it is aimed, on the one hand, at understanding the pairing of particular structures and functions and, on the other hand, at seeking the common basis or *Grund* of these pairings. More generally, Burdach's "comparison" describes the process of finding correlations and similarities among the manifold variety of living creatures. It is nothing other than the application of reason to the chaos of fact.

Over the next two decades it became increasingly unfashionable to lay out one's metaphysics, epistemology, and even methodology as thoroughly as Burdach had, and certain of his views, such as his open commitment to a teleological life force, were no longer accepted by rising scholars. *Naturphilosophie* became a code word meaning speculation and was regularly used as a convenient label of derogation. But the overall set of linked commitments he laid out—to the unity of nature's laws, the necessity of studying an organism's form to understand its life, the causal role of development, and the role of compari-

18. Burdach, *Über die Aufgabe,* p. 35.

son in achieving the sort of scientific knowledge that was a union of reason and sensory experience—continued to dominate thinking about animal form in the first half of the nineteenth century.

This linked network of assumptions helps us to understand more fully the relationship of morphology to physiology as it is represented in research publications of the early decades of the century. As should be evident by now, morphology was essential to physiology because organization was one of life's defining characteristics.[19] The organism's form and function were codetermining: the organization of the animal as a whole regulated the order of its parts, and both the parts and the whole were designed to fulfill the functional needs of the animal in its conditions of existence. This teleological approach was expanded into a new dimension, time, by the priority given to development as an explanatory device.

Despite the intimate relationships among form, function, and development, however, for most writers of 1820s and the 1830s, the comparative study of anatomical structures remained philosophically subordinated to physiology. In contrast to Burdach's assumption that anatomy itself could yield satisfying philosophical answers when approached with the proper morphological spirit, writings labeled "morphological" (or, more commonly, "anatomical") generally encompassed empirical descriptions of anatomical structures, while those labeled "physiological" involved the explanation or derivation of those structures. Discussions of development could come under either label, but because they tended to be viewed as answers to the question "where did this structure come from?"—a question considered explanatory rather than descriptive, such discussions were more often considered physiological rather than morphological.

An article published by Johannes Müller in 1828, one among the very few that actually used the word "morphology" in this period, suggests how this perceived relationship between morphology, physiology, and development could serve to structure research accounts.[20] In this early essay, Müller compared the nervous systems of invertebrates with those of vertebrates. Following an introduction setting out the problem and the disagreements among previous scholars, Müller divides his analysis into a short "morphological part" and a longer

19. Although not put in exactly these terms, the point about organization and life has been made before; see esp. K. Figlio, "Metaphor of Organization" (1976); Lenoir, *Strategy of Life.*

20. J. Müller, "Über die Metamorphose des Nervensystems" (1828); for a similar use of morphological, physiological, and developmental categories, see K. E. von Baer, "Über das äussere und innere Skelet" (1826).

"physiological part." The former describes the typical arrangement of the nervous system in the three Cuvierian Types constituted by the radiates, the molluscs, and the articulates. Müller argued that the radiates displayed the simplest form of nervous system, lacking both ganglia and a spinal cord, whereas the mollusks displayed ganglia, and the articulates showed a range of differentiation from a network of ganglia to a true spinal cord. This section is entirely descriptive in its aim, though it does make a few generalizations about the pattern of the anatomical structure in each type.

The physiological part is aimed at explaining the physiological significance of these three arrangements, and does so in terms of development; in fact, the two subsections of the physiological part are titled "Development in the Animal World" and "Development in the Embryo." In the first of these subsections Müller shows that "the nervous system (*actu*) develops through isolation of the factors that are contained, at rest and unisolated [*unvereinzelt*] (*potentia*), in the whole. This is everywhere the true physiological concept of development, which appears unchangeably in the whole of nature, in the animal and mental energies, in the generation of organisms, as in the generation of thoughts."[21] In the second subsection, on the development of the nervous system in the embryo, Müller argues that just as the nervous system develops from a more general state to a more differentiated one in the animal world as a whole, so must it do in the development of the individual. "In the fetus as well, nature cannot form the nervous system differently. . . . The history of the fetus's formation must repeat the law of formation in the animal world."[22] Müller was quick to point out that he did not mean that the individual followed in its development a linear "series that doesn't even exist," but rather that the movement from the general (*potentia*) to the individuated (*actu*) was one that existed at all levels of living nature.

In this article, "morphology" clearly denotes the study of form in a straightforward sense: it refers to concrete structures within the body and concerns their size, placement, and spatial relation to one another. "Physiology," by contrast, supplies the answers to the question "why are these structures arranged the way they are?" either in terms of their purposes within the animal organism or in terms of their development. Physiological knowledge, then, was impossible without knowledge of form, but the latter was a lower kind of knowledge than the former.

21. J. Müller, "Über die Metamorphose," p. 16.
22. Ibid., p. 18.

Despite a range of beliefs and rationales that often put early nine-teenth-century writers on form at odds with one another, there is no question that all accepted an intimate association between form, life (whether considered to be an Idea or a force), and development. Well into the 1840s, students of form continued to view scientific knowledge of their subject as consisting in a union of reason and sensory experience that was not entirely inductive; the classic and much-cited expression of this came in the title of Karl Ernst von Baer's treatise on the development of the chick, which was subtitled "observation and reflection." Following the same sort of reasoning that Burdach used, younger scholars such as Müller still justified studying develop-ment in terms of getting back to a more general state undisturbed by later individuation, something that would at the same time allow the researcher access to general laws—only by the 1840s they were more apt to cite von Baer than Burdach as the source for this. Most impor-tantly, they continued to view the study of the developing animal form as one of several elements crucial for creating a broader *Wis-senschaft* of life.

## THE STUDY OF LIFE AND FORM IN THE UNIVERSITIES

Burdach's attempt to elevate anatomy into a morphology that carried the higher status of a *Wissenschaft* was emblematic of the efforts of a much larger group of scholars to incorporate a *Wissenschaft* of life into the mission of the university. These men sought not only to gain time and sanction for their research, but also to integrate it into their teaching. The well-known (and well-worn) story of the transforma-tion of the German universities into institutions that brought together teaching and research has tended to focus on the faculties of philoso-phy. But of particular relevance for the study of animal form in the early nineteenth century is the fact that with Burdach's generation medical education, too, changed to incorporate the ethos of *Wis-senschaft* and to increase research on living nature. The men who launched this change are, strikingly, members of a single generation, which in addition to Burdach (1776–1847), spans Ignaz Döllinger (1770–1841) and Karl A. Rudolphi (1771–1832) at its older end and includes the *Naturphilosophen* Franz Josef Schelver (1778–1832), Lo-renz Oken (1779–1851), J. B. Wilbrand (1779–1846), Joseph Reubel (1779–1852), and Georg August Goldfuss (1782–1848), as well as the non-*Naturphilosophen* Johann Friedrich Meckel the Younger (1781–1833) and Friedrich Tiedemann (1781–1849). Beginning with this generation, there gradually emerged a new career pattern organized around nonclinical research and teaching within the medical faculties.

Burdach's 1817 pamphlet provides a window not only into the conceptual elements of *Wissenschaft* but also into the struggles involved in incorporating such a *Wissenschaft* into the practical business of medical school teaching. The occasion for his essay was the opening of the new anatomy building at the University of Königsberg, a fact not unimportant for understanding the thrust of his remarks. Burdach had requested a new building as a condition of his acceptance of the professorship of anatomy three years before, and now the state had come through on its promise.[23] The purpose of the essay, therefore, was not only to celebrate the opening of the institute but also to tell the ministry overseeing the funding that its money had been well spent and would continue to be used in its interests.[24] At the same time, Burdach sought to justify the still-recent notion that basic research into the fundamentals of life and teaching on the subject were deserving of support even though they brought no direct benefits to the state. This concern continually reappears as a leitmotif in the essay, popping up even when he is in the midst of explaining the most difficult philosophical principles upon which his morphology is based. This essay thus stands near the head of a long line of justificatory statements about morphology's value, the content of which changed little in the course of the nineteenth century.

*Über die Aufgabe der Morphologie* may be read as the published continuation of an unpublished argument Burdach had begun broaching to the Prussian minister of the interior some three years earlier,[25] which stressed the recent rise of anatomy to a *Wissenschaft* and tried to tap into the broader material support the state had al-

23. GStA Merseburg, Königsberg Professoren, Bll. 203–10, 214–16, 341, 343–53; Königsberg Anat. Theater, unpaginated (for complete archival citations, see list of archival sources preceding bibliography).

24. Although Lenoir refers to Burdach's 1817 essay as a speech, in fact it was not the inaugural lecture given at the opening of the new anatomy institute. Burdach's inaugural lecture on the history of the anatomical institute remained unpublished until 1951 (see W. Bargmann, "K. F. Burdachs Rede" [1951]). According to von Baer, though, the 1817 essay *Über die Aufgabe* "was to serve as an outline of the institute's program" (K. E. von Baer, *Autobiography* [1986], p. 159).

25. Higher education in Prussia was overseen by the ministry of the interior until late 1817, when the bureaucracy was reorganized and the ministry for "geistlichen, Unterrichts- und Medizinal-Angelegenheiten" (the so-called cultural ministry, hereafter GUMA) was founded. Burdach appears to have been unaware of the reorganization when he wrote his pamphlet, which he sent to the recently removed Interior Minister Schuckmann. It was replied to not by Schuckmann but by the first cultural minister Altenstein. See GStA Merseburg, Königsberg Anat. Theater, Bll. 105–6.

ready begun to provide for other areas of science. In June 1814, shortly after he had arrived in Königsberg, he wrote a lengthy proposal for a new anatomy institute, outlining the needs of an institute that would meet the standards of the time and requesting that Prussia turn some of its generosity toward anatomy as a science and not just as a practical subject. To the possible objection that Königsberg did not have sufficient numbers of medical students to justify an expensive new institute for anatomy, he responded that more would come if the proper tools for teaching were in place; he then observed that the ministry had not seemed particularly worried about the lack of astronomy students when it built the observatory or the irrelevance to students of the large section of the botanical garden containing medically unimportant plants. If money was available for what was clearly research in the sciences of astronomy and botany, Burdach asked, why not consider anatomy as a research science as well, and outfit it accordingly?[26]

Whether Burdach's arguments were the deciding ones or not is unclear, but the Prussian ministry did see its way to outfitting Königsberg with a new anatomy institute with the most up-to-date features. These included a lecture and demonstration hall with windows on three sides, a mechanical device to raise cadavers for demonstration from the storage cellar directly beneath the lecture hall to its dissection table, and three tiers of seats to accommodate sixty observers (optimistic, given the university's contemporary annual enrollment of less than ten medical students!). In the next room over, a laboratory for students to practice dissections featured wheeled tables for easy transportation of cadavers from the lecture hall. In addition to various offices and rooms for the anatomist and his prosector to prepare specimens in, an anatomical museum upstairs allowed closer comparative study of healthy and diseased body parts in both animals and humans at all ages. The museum's four rooms allowed space for expansion of the initially quite small collection, and the substantial, newly instituted annual budget from the state for materials and books guaranteed that such expansion would take place. All in all, it would seem that Burdach had little to complain about.[27]

26. K. F. Burdach, "Über die Einrichtung eines anatomischen Instituts zu Königsberg. Erster Aufsatz, enthaltend allgemeine Gesichtspuncte, und Übersicht der dem Institute zu ertheilende Mittel" (8 June 1814), in GStA Merseburg, Königsberg Anat. Theater, unpaginated.

27. Burdach, Über die Aufgabe, pp. 53–60; Von Baer recalls the Königsberg anatomy institute in his autobiography, Nachrichten über Leben und Schriften (1866), pp. 221ff.

Thus his argument in 1817 was no longer about material resources but about what one was to do with them. One hardly needed to justify the value of pursuing anatomy on a scientific plane, Burdach asserted: an *Erfahrungswissenschaft* "justifies itself: we are drawn irresistibly toward it, and in order not to waver restlessly between the two worlds [of reason and the senses], or through the attention to one to offend against the rights of the other, we are pressed to move forward from the gathering of individual phenomena to the knowledge of them in relation to the whole."

Moving on to the presumed objections of his audience, he continued, "One should not damn morphology as a *Wissenschaft* from the outset just because it appears to be useless for practical life." To begin with, we don't know what we will find that will be useful, and furthermore, we should not discount the value of the intellectual satisfaction that comes from increasing one's knowledge of nature.

Burdach recognized, however, that this might not be a sufficient goal for an institute built for "the ends of the state." How was a morphologist to justify teaching this arcane set of doctrines about law, cause, life, development, and form to students? He acknowledged that "The state machine requires cogs that will fit into it" smoothly, and "to prepare such useful members of society is the first aim of the teacher"; he should undoubtedly not confuse them or "lessen their usefulness" with too much science. Nevertheless, Burdach hoped that the state would agree that as long as such useful cogs were produced, the teacher should also encourage and foster the life of the mind among his students. Indeed, the state should protect the teacher in this aim and should understand that he is not trying to turn everyone into a scholar but simply to allow those students who felt strongly the internal urge to knowledge to be able to pursue it.[28]

The structure of the courses Burdach proposed in an appendix to this essay explicitly followed the dual aim of training the uncurious practitioner in those aspects of anatomy necessary for practice and providing a more elevated discussion of morphology as a *Wissenschaft*. In his plan, medical students would learn the basics of anatomy in their first year of study, with the subject matter strictly confined to practically applicable aspects. In his own lectures, Burdach would go through the body topographically, covering, for example, the limbs in one semester and the head and trunk in the other, while his prosector, the young Karl Ernst von Baer, would go through organ system by organ system. These two different approaches to the body

---

28. Burdach, *Über die Aufgabe*, pp. 21–23.

would reinforce the important anatomical information. More advanced students could then return to anatomy in their fifth or seventh semester, this time to focus on the dissected body in connection with pathology and surgery. Separate from all this practical anatomy would be a parallel set of courses on morphology as a higher science, aimed at anyone who might be interested. Von Baer would lecture on "zootomy" or animal anatomy, rising through the main taxonomic groups from infusoria to mammals and considering each animal as a whole; simultaneously he would offer an introduction to zootomical research. Burdach, by contrast, would consider the general laws of development and show the various forms appearing in the animal kingdom.[29]

Perhaps predictably, it was this latter part that caught the attention of the recently appointed Prussian minister of culture, Karl Freiherr vom Stein zum Altenstein. In a letter thanking Burdach for the essay, Altenstein expressed his disapproval of dividing the teaching into a set of practical lectures and a set of more theoretical (or as he put it, "zoological") ones. With so much ground to be covered, Altenstein argued, medical students would get lost, and might too easily become accustomed to a "pasison for hypotheses and vague reasoning." He reminded Burdach that the aim of the provincial university was to train good medical practitioners, and requested him to focus his second set of lectures on the more directly applicable subjects of pathological and comparative anatomy rather than indulging his "otherwise estimable efforts to expand his subject."[30]

This slap on the wrist reminds us that even a ministry supportive of *Wissenschaft* in general and *Naturphilosophie* in particular, as evidenced by the appointments made at Bonn and Berlin in the same years, believed that the pursuit of philosophical knowledge at the university (at least at a provincial university) needed to be tempered with the sort of training that made good servants of the state, in this case medical practitioners. What would become a perennial tension between the practical needs of future practitioners and the research interests of professional university professors of medicine is already evident near the beginning of the modern research university.[31]

The strong association of anatomy with practical knowledge and the unquestioned necessity for future practitioners to learn the organi-

29. Ibid., pp. 60–64.

30. Kultusministerium to Burdach, 23 December 1817, GStA Merseburg, Königsberg Anat. Theater.

31. On the tussle over this problem at the founding of the University of Berlin, see Broman, "University Reform," pp. 44–48.

zation of the human body meant that few scholars attempted to make anatomy into a more consciously theoretical subject of teaching, as Burdach attempted to do.[32] But there was another spot in the curriculum available for presenting theory, one indeed designed just for that purpose: the lectures on physiology. In fact, Burdach's aims for morphology were unusual in that they were attached primarily to a *Wissenschaft* of anatomy; more typical of this period was a different approach that solved the problem of combining *Wissenschaft* and anatomy by placing the teaching of the practical subject of anatomy and the theoretical subject of physiology into a single set of hands, while keeping the aims of the two distinct. And here it is essential to understand the meanings of "physiology" in this period, for they hold the key to understanding the organization of both research and teaching in the animal sciences.

Until the mid-nineteenth century, physiology had two distinct meanings. It could be understood narrowly as the science of human (and by extension, animal) organic function, describing how the various organs and systems of the body operated to maintain the life of the individual. But it could also be understood much more broadly as a set of doctrines concerning the nature of life in general. In the late eighteenth and early nineteenth century, this latter aspect of physiology flowered as professors came to value *Wissenschaft* ever more in comparison to teaching future physicians what they needed to know to practice medicine. Physiology, writers in abundance now declared, was the science of life—its goal was to understand the nature of living things in all their aspects.[33] It was a *Wissenschaft* par excellence, aimed at uncovering the most general laws of life.

Thomas Broman has pointed out that at the turn of the century physiology was catapulted into a position of new prominence as professors, especially those under the influence of Schelling's *Naturphilosophie,* made bold claims about the possibility of discovering the fundamental nature of life.[34] The situation reportedly holding at the University of Leipzig in the duchy of Saxony in 1801 probably represents an extreme case: although one of the five full professors in the medical faculty held the official professorship of physiology, the pro-

32. Here I don't mean to say that anatomy is somehow essentially atheoretical, but rather that practical anatomy—that covering dissection and the memorization of parts of the body—was considered a fully known body of knowledge, and as such, a collection of facts rather than theories.

33. See Lohff, *Die Suche,* esp. pp. 26–34.

34. See Broman, "University Reform," "Transformation of Academic Medicine," chap. 5, pp. 225–68, esp. pp. 237–46.

fessors of therapy, anatomy and surgery, and chemistry also offered lectures on the subject.[35] The Prussian University of Bonn, where many professors with *naturphilosophisch* inclinations were hired at its 1818 founding, presents a similar story: during the 1820s and 1830s, lecture courses on physiology were announced not only by A. F. Mayer, who held the chair of anatomy and physiology, but also by the professor of clinical medicine, the professor of gynecology, the professor of pharmacology, and the botanist![36]

As Broman has shown, the new emphasis on physiology as a *Wissenschaft* coincided with three shifts in the institutional and professional setting of professors in medical faculties in the German-speaking states. First, university medical faculties expanded in the first two decades of the nineteenth century, from a typical faculty size of three to an enlarged faculty of six or more full professors. This was accompanied by the final abolition of the eighteenth-century tradition of *Aufrücken,* in which a medical professor advanced over the course of his career from a chair controlling the cluster of theoretical sciences known as the *Institutiones* (physiology, pathology, therapeutics, semiotics, and dietetics) to one controlling other medical sciences, such as pathology, therapy, and chemistry, and finally achieved the chair overseeing clinial instruction. This system was replaced by one in which professors began to concentrate on a single, if broadly defined, group of subjects throughout their careers. Finally, the early nineteenth century saw a split between careers oriented toward clinical instruction and those oriented toward nonclinical sciences. This split was possible partly because of new funding that allowed professors of theoretical subjects to teach full time without having a clinical practice on the side and partly because of the end of the *Aufrücken* system; it was spurred on by a group of professors seeking to pursue *Wissenschaft* without being fettered by the demands of clinical practice. Each of these factors—the size of the universities, the end of the *Aufrücken* system, and the split between clinical and basic science subjects—is relevant for understanding the career structures that developed around the teaching subjects concerned with animal life and form in early nineteenth-century medical faculties.

Although the universities of the early nineteenth century repre-

35. W. His, "Über Entwicklungsverhältnisse des academischen Unterrichts" (1882), p. 26.

36. U. Ebbecke, "Das Physiologische Institut" (1933), p. 72. It was written into the statutes that more than one professor was permitted to lecture on the same subject (see "Statuten; III: Von den Professuren an der Universität," para. 39, reprinted in K. T. Schäfer, *Verfassungsgeschichte* [1968], p. 434).

sented an expansion over the typical sizes of the eighteenth century, they were still quite small.[37] Most medical faculties had between six and ten full professors, whose teaching duties were divided among anatomy and physiology, clinical medicine, pathology and therapy, surgery, obstetrics, "medical encyclopedia" (introduction to the topics and methods of medicine) and history of medicine, pharmacy, medical botany, and often chemistry.[38] Given the small number of medical professors relative to the wide range of subjects considered worthy of offering, it remained common for a single professor to teach a number of different subjects, although these were no longer tied to the *Afrücken* system. The life-science subjects of anatomy, physiology, comparative anatomy, and zoology, all of which shared a concern for animal form, could be parcelled out in a strikingly wide variety of ways.

One approach was to have the same person teach all of them. J. B. Wilbrand, hired in 1808 into the medical faculty at the small and impoverished University of Giessen in the grand duchy of Hesse-

37. All had at least nominal faculties of law, medicine, theology, and philosophy. A few had two theology faculties, one for Protestant theology and the other for Catholic theology. A twentieth university, in Strassburg, was founded after the unification of the empire. Of the nineteen German-speaking universities that would eventually be united under the German Empire, the very largest was Prussia's University of Berlin, rivaled in German-speaking central Europe only by the Austrian University of Vienna. With 256 students enrolled in its founding year 1810, it already had some 1,400 students by 1828; by 1853 that number would rise to over 2,100. In 1810 they were taught by 25 full professors (in 1853, 51) of whom 6 (in 1853, 11) had appointments in the medical faculty. The medical professors were supplemented by 8 (in 1853, 41) *ausserordentliche* professors, a teaching force without faculty voting rights (and sometimes without salary) that became especially prominent in Berlin; in addition, some 13 men (by 1853, 56) worked as medical *Privatdozenten,* whose university income derived entirely from student fees; figures on Berlin teaching faculty are from R. Köpke, *Die Gründung der königlichen Friedrich-Wilhelms-Universität zu Berlin* (1860). At the other end of the scale was the University of Rostock in the small northern grand duchy of Mecklenburg-Schwerin, which, in 1828, had 33 full and extraordinary professors distributed among the four faculties, with student numbers fluctuating between 100 and 150 in the first half of the century. The medical faculty had a teaching staff of 6; in 1836 the number of medical students was reported to be only 8; for 1828, see R. Kilian, *Die Universitäten Deutschlands* (1828/1966), pp. 329–35. According to K. G. von Raumer (*Contributions* [1859], app. 14), the total number of students at Rostock in 1853 had sunk to 108, of whom 10 were enrolled in the philosophical faculty and 24 in medicine; on 1836 numbers, see J. J. Sachs, "Flüchtige Reiseblicke" (1837), p. 63. These two universities marked the extremes in both student and faculty numbers.

38. When the new Prussian University of Bonn was founded in 1818, for example, 6 full professorships were mandated in the medical faculty, as compared with 18 in the faculty of philosophy (see Schäfer, *Verfassungsgeschichte,* p. 433).

Darmstadt to teach physiology, comparative anatomy, and general natural history, also offered the main lectures on anatomy, botany, and zoology, and after 1817 directed the botanical gardens and zoo. He retained control over all of these subjects until his death in 1846.[39] At other universities, not all of them larger than Giessen, these same subjects were taught by two or more professors; however, few university faculties insisted on a hard and fast division of teaching areas. Between 1820 and 1860 at Baden's University of Freiburg, for example, where the responsibilities for teaching comparative anatomy, anatomy, physiology, and zoology were split among two or three professors, the arrangements of who taught what were shuffled nearly every time a professor died or left the university.[40]

One might suppose states with more than one or two universities would have more stable associations among subjects, but in fact they still varied from university to university. Each of the three Bavarian universities, for example, developed its own arrangement.[41] The Prus-

39. Günter Risse, "Wilbrand, Johann Bernhard"; W. J. Schmidt, "Einiges aus der Geschichte" (1938), p. 18; in a similarly broad vein, a set of regulations at the University of Würzburg in Bavaria announced in 1825, "The professor of anatomy and physiology is the head of all establishments of the university that are connected with the teaching subjects of physiological, pathological, and surgical anatomy of humans, human physiology, and zootomy and 'zoonomie' [the study of animal life]" ("Instruction für die anatomischen Anstalten an der Universität zu Würzburg" [Würzburg: Carl Wilhelm Becker, 1825], BHStA München MK 11415). At Tübingen, Wilhelm Rapp covered anatomy, physiology, pathological and comparative anatomy, and zoology from 1818, when he started there as an *ausserordentliche* professor, through his nomination as full professor in 1827, until 1844, when he gave up all but zoology to a younger professor (see K. Mörike, *Geschichte der Tübinger Anatomie* [1988], pp. 43–44; Universität Tübingen, *Vorlesungsverzeichnisse* [1830–47]).

40. Until the mid-1850s, comparative anatomy and physiology were always taught by the same person, but different professors added to that either anatomy or zoology or both. Then in 1855, comparative anatomy was split from physiology when Alexander Ecker, who had previously been teaching comparative anatomy, physiology, and zoology, traded his responsibility for the latter two subjects for the professorship in anatomy, taking comparative anatomy with him. Physiology and zoology became the official province of a new professor and were not separated until the mid-1860s (W. Neuland, *Geschichte des anatomischen Instituts* [1941], pp. 112–38; O. Koehler, "Die Zoologie an der Universität Freiburg im Breisgau" [1957]).

41. At Erlangen (the smallest of the three), anatomy and physiology were taught by one professor, comparative anatomy and zoology by another; at Würzburg, zoology was taught by a professor other than the professor of anatomy, physiology, and comparative anatomy (Universität Würzburg, *Vorlesungsverzeichnisse* [1820–44]); and at Munich (the largest Bavarian university after its move from Landshut in 1826), anatomy, zoology, and physiology each had separate professors, though, as was frequently the case, there was some flexibility in teaching duties (Universität München, Vorlesungsverzeichnisse [1830–45]).

sian universities were somewhat more uniform: at Berlin, Bonn, Greifswald, and Halle, anatomy and physiology were united under one professor, while at Königsberg, Burdach initially taught both subjects but gave up anatomy to his former prosector von Baer in 1826; the sixth Prussian university, Breslau, mandated a separate professorship for physiology upon its founding in 1811. (It should be noted, however, that the second holder of the Breslau chair, Jan Evangelista Purkyně, also taught pathological anatomy and internal medicine.)[42]

The importance of this variety of teaching structures for our story is twofold. First, in all the teaching arrangements discussed by historians, university chroniclers, and biographers, morphology never appears as a separate, regularly taught course. It was simply not considered a subject for teaching, a *Fach*. Instead, the study of animal form was integrated into a preexisting set of teaching subjects. Second and more generally, the variety and flexibility of academic organization in the early nineteenth century shows that when the *Aufrücken* system died out and the clinical and nonclinical subjects began to split apart into separate career tracks, no agreed-upon new way of ordering knowledge and careers around the nonclinical subjects emerged right away. In trying to understand the groupings that did exist, and morphology's place in them, we cannot invoke the shift to a discipline-based intellectual ecology, for academic disciplines in the modern sense, with self-perpetuating communities, distinctive sets of problems, stable institutional loci and career patterns organized around university professorships, were just coming into being then. The emergence of modern disciplines is a problem for the history of this period and cannot be used as an explanation for the emergent patterns of ordering knowledge and careers. The specifics of how the nonclinical subjects relevant to the study of form, including anatomy, physiology, comparative anatomy, and pathological anatomy, were divided up and associated with other subjects such as zoology or therapeutics cannot be accounted for by simply pointing to "nascent disciplines." Instead, we must ask what factors made certain clusterings of subjects seem appropriate.

At one level the answer is straightforward: in the absence of some kind of disciplinary pressure to create a more uniform division of scholarly labor across universities, local circumstances and combinations of personalities account for different ways of organizing teaching

42. Universität Berlin, *Index Lectionum* (1833–54); Universität Bonn, *Verzeichnis der Professoren* (1826–42); Universität Halle, *Catalogus Professorum* (1822–45); G. Kaufmann, ed., *Festschrift* (1911), p. 243.

coverage of these subjects.[43] Thus at Königsberg after 1826, Burdach, who ran the anatomy institute and also taught physiology, gave up the directorship of the institute to von Baer, who was already professor of zoology and director of the zoology museum. Why would a senior professor just reaching the peak of his career relinquish to a junior colleague the directorship of the anatomy institute and the lucrative professorship of anatomy? Because, according to von Baer, Burdach "wanted his son to become prosector [laboratory instructor], and thought that in such an event it was somewhat inappropriate to remain director of the anatomical institute"[44] (a scruple not widely shared by his contemporaries). At the impoverished University of Marburg in the early 1820s, J. M. D. Herold had been neglecting his duties as prosector in anatomy to conduct his research on the development of insects; when the professorship of zoology opened up in 1823, offering it to Herold was viewed as an appropriate way to reward his diligence as a researcher while at the same time liberating the labor-intensive prosectorship for someone who would pay more attention to the job—this despite the fact that Herold admitted he knew next to nothing about the broad subject of zoology.[45]

These two examples illustrate not only the fluidity of teaching arrangements but also another important feature of professorial careers in the first part of the century: the large majority of professors teaching the animal sciences rose through the ranks at a single university, which was frequently the same university where they received their medical degree.[46] Of the forty-one full professorships with control over anatomy, physiology, zoology, and comparative anatomy taken up between 1818 and 1840, only a quarter were given to men from outside the university offering the professorship, and most of those moves (seven out of ten) were lateral ones involving people who had already risen through the ranks at a single university. (Johannes Müller, for example, had followed a typical vertical path at the University of Bonn, from student to *Ordinarius*, with time out for a *Studienreise* right after he finished his medical degree, before his call

43. R. S. Turner ("Growth of Professional Research") has emphasized the changing balance between local and disciplinary ties in the development of the Prussian universities.

44. von Baer, *Autobiography*, p. 192.

45. U. Runge, *Johann Moritz David Herold* (1983), pp. 44–49.

46. A typical example is Emil Huschke (1797–1858), who received his M.D. at Jena in 1818, then spent a year or so traveling and studying in Berlin, Vienna, and Paris, returning to Jena to spend the next eighteen years rising through the ranks from *Privatdozent* through *Extraordinarius* and honorary full professor, finally to *Ordinarius* of anatomy and physiology.

to Berlin in 1833.)[47] In this period, therefore, there was very little competition among states for full professors in the sciences of animal life; such movement as did take place was most common at the time of a first hire as prosector, *Privatdozent,* or possibly *ausserordentlicher* professor.[48] Local considerations, it appears, remained the most important in structuring the careers of academic life scientists.

Despite the continuing fluidity of teaching arrangements and the predominance of local factors in the selection of professors, certain patterns in career paths and teaching arrangements for these theoretical sciences nevertheless emerged that transcended local circumstances. More and more, a scholar would move up through the academic ranks while staying within a relatively narrow range of teaching responsibilities. There were still cases in the 1830s of people who moved into academic positions in these fields from medical practice—the physiologists Carl Theodor von Siebold and Friedrich Stannius, for example, both practiced medicine for some years before gaining academic employment.[49] But this kind of horizontal move was becoming ever more unusual; a standard academic career path for studying animal life was being forged, even if the particular subjects an individual taught might vary over time. This academic path lent some stability to professional and professorial career expectations, even within a predominantly local framework.

At the same time, the various local teaching and administrative arrangements settled upon tended to share a common feature: professors who wanted to pursue empirical research on animal life and form almost always sought a combination of teaching duties and director-

47. On Müller, see W. Haberling, *Johannes Müller* (1924).

48. Of the lateral moves, four were associated with the establishment of new universities—Bonn in 1818 and Munich in 1826—which obviously couldn't draw on their own students. The two largest states, Prussia and Bavaria, were willing to hire professors from other states but did so infrequently, at least for these subjects; there, as in Baden, a more likely source of an external appointment was another university within the same state. States with just a single university rarely hired laterally from outside for these subjects, and when they did, the hires were often people who already had personal ties to the university. This finding contrasts with that of Turner who, in his dissertation, "Prussian Universities," pp. 359–61, argued there was greatly increased career mobility during the *Vormärz* period in Prussia.

49. On Stannius, see G.-H. Schumacher and H.-G. Wichhusen, *Anatomia Rostochiensis* (1970), pp. 104–5; on Siebold, see H. Körner, *Die Würzburger Siebold* (1967), pp. 291–355; Körner points out that Siebold was active as a researcher and had been seeking some kind of academic appointment from the early 1830s on; indeed, he taught at the obstetrical school at Danzig before moving on to a university professorship (Körner, pp. 294–302).

ships that gave them access to the resources necessary for that research. The most common of these was to join anatomy to physiology, a combination that had not been traditional but that, by the early nineteenth century, held sway at more than half the universities under study here (Erlangen, Würzburg, Bonn, Berlin, Halle, Greifswald, Leipzig, Giessen, Heidelberg, Jena, and Tübingen). Both anatomy and physiology had been teaching subjects in the eighteenth century and earlier, but they had differed in their manner of presentation and were not usually taught by the same professor.[50] In the reorganizations of the early nineteenth century, however, as the ideal of *Wissenschaft* swept through university medical faculties, professorships of anatomy and physiology came increasingly to be held by a single person.

Paralleling the relationship expressed in research papers, the public rationales for this pairing were most often put in terms of anatomy serving physiology. Although anatomy could be studied independently of physiology, few of the professors who controlled both subjects would claim that the sort of knowledge gained by purely anatomical study was *wissenschaftlich* in its highest sense. Anatomical generalizations were meaningful only to the extent that they told one something about either the laws or the *Gründe* of life—questions that were considered by their nature to be physiological. As one textbook put it, "Whereas anatomy concerns itself with the presentation of the form and structure of the individual organs, or the material carriers of the life-processes, . . . physiology . . . places the anatomical detail, which it assimilates, under general points of view."[51] As late as 1841, the Munich professor of physiology Joseph Reubel argued that anatomy was merely the means to a physiological end—this was why at most universities the physiologist also held the chair of anatomy (which, it will be noted, was not the case for Reubel himself; this perhaps accounts for the characteristically pugnacious tone in which he wrote.)[52]

As much as it might reflect the real convictions of their authors, this rhetoric should not blind us to the possibility that anatomy served physiology in a more concrete way as well. Although physiology,

50. Thus E. T. Nauck notes that in the eighteenth century, physiology and anatomy were taught by separate professors at the universities of Leipzig, Freiburg, Halle, Heidelberg, and at the Collegium medico-chirurgicum in Berlin (Nauck, "Bemerkung zur Geschichte" [1950], p. 157, n. 6). That the two could be held by one person is testified to by the case of Albrecht von Haller, who taught anatomy, physiology, botany, and certain clinical subjects, but this was by no means the dominant tradition.

51. R. Wagner, *Grundriß der Encyclopädie* (1838), p. 58.

52. Separatvotum des Prof. Dr. Reubel, 30 August 1841, UA München, Senatsakten, "Anatomie als Disziplin."

understood as a set of doctrines, had a claim to being considered more important theoretically, research-oriented professors such as Burdach who wanted access to a budget, building space, and research materials would be more likely to find them provided under the rubric of anatomy.[53] While the concept of supporting either anatomy or physiology as a research science was novel in the early nineteenth century, there was already a difference in the material bases available for teaching the two. Physiology had traditionally been taught by lecturing, as befitted its theoretical status. Anatomical teaching, by contrast, was demonstrative in nature, with lectures supplemented by demonstrations and exercises in dissection. For these pedagogical purposes, all states paid some attention to the provision of human cadavers and animal specimens—which could be useful for research as well. For someone interested in physiological research, which at the time included microscopic anatomy and the study of organic development, building up collections in macroscopic, microscopic, and comparative anatomy—justifiable in terms of teaching—may simultaneously have offered a chance to conduct research along the way.

Direct evidence for this dual significance of collections is difficult to find, but hints are available. In his letter nominating himself for the professorship of anatomy and physiology at Berlin in 1833, for example, Johannes Müller pointed to the recently deceased Meckel as a model anatomist, one who both built a museum and conducted superb physiological research. Substantial anatomical collections had been built up at Berlin, too; Müller proposed they be placed under the control of "a man who understands how to unite interest in human, comparative, and pathological anatomy, and who, through successful activity in the foundation of all medicine, physiology, understands how to animate all of medical instruction." Indeed, Müller's ambitions soared even higher. He saw in Berlin "the sources for an activity that may be compared with Cuvier's marvelous work, one that will achieve through the cultivation [*Betreibung*] of anatomical materials for physiology what Cuvier once won for zoology through the application of anatomy."[54] If one could only find a man who knew how

53. In 1839 at Breslau, Jan Evangelista Purkyně, in arguing for a physically separate physiological institute, explicitly used the structure provided by anatomical institutes as a model (see Kremer, "Building Institutes for Physiology," p. 84).

54. Johannes Müller to Minister Altenstein, 7 January 1833, quoted in E. Du Bois-Reymond, "Gedächtnisrede auf Johannes Müller" (1912), pp. 178–79; it should be noted that the medical faculty at Berlin emphasized Müller's prowess as an experimenter and not primarily as a comparative anatomist (Du Bois-Reymond, pp. 174–75).

to attract and inspire students with this task, Berlin could become for the new age the sort of center that previously existed only in Paris under Cuvier (who had recently died). In Müller's undisguised presentation of himself as the new (German) Cuvier, he himself would animate anatomical teaching and at the same time develop a physiological research school centered around the collections. Although few collections could compare with Berlin's, the number of research articles in this period based on fortuitously received specimens does suggest that at many universities, collections provided important sources of material for professors conducting anatomical-physiological research.

Holding a professorship of anatomy was useful for another reason besides control over space and specimens, too: all medical students could be expected to take anatomy lectures and labs, and thus receipts from student fees would be high. This source of income, which professors often funneled into developing privately owned collections for research and teaching, was especially important for professors whose states offered little in the way of regular annual support for equipment, supplies, or collection development. Thus even as the internal demands of the scholarly community to provide *wissenschaftlich* generalizations increased the prestige of physiology, the external demands of the state for the training of physicians in anatomy provided material support that could be used for the twin ends of teaching and research.

At those universities where the professor of anatomy and physiology also ran the zoology museum (the case at Giessen, Tübingen, and at Heidelberg during Tiedemann's time) and at those where physiology was united with zoology and comparative anatomy but not human anatomy (Göttingen and Marburg), control over the zoological collection may have provided a similar source of material. Although few students studied zoology, and therefore revenues and prestige deriving from zoological teaching were low, control of the museum could once again provide an important resource for research and public support. Despite regular complaints from professors that the space and equipment provided for these collections was inadequate, in fact at many universities they were well financed relative to other administrative units.[55]

55. See, e.g., the comparison of funds devoted to zoology versus physics in C. Jungnickel and R. McCormmach, *Intellectual Mastery of Nature* (1986), vol. 1, p. 12, n. 36. For a fuller comparison, see K. F. W. Dieterici (*Geschichtliche und statistische Nachrichten* [1836/1982]), who reported on the comparative costs for the Prussian

It is not clear yet how far this frequent link between teachers of physiology and the control of collections may be pushed. At this point one can only speculate, but the existence of this union may ultimately help us understand why anatomical, comparative anatomical, and developmental studies overshadowed other ongoing orientations of physiological research, such as vivisection and chemical studies, in early nineteenth-century Germany. Philosophical considerations about the nature of life should not be underestimated, of course, but neither should such practical considerations as the availability of specimens and equipment. At a few universities, such as Göttingen and Halle, the originally private comparative anatomy collections of Blumenbach and Johann Friedrich Meckel the Elder, already well developed in the eighteenth century, provided a foundation for later scholars. But elsewhere the possible dual function of a collection for both research and teaching may have encouraged scholars to make efficient decisions about the sort of research they did. The biographer of Johann Friedrich Meckel the Younger suggests a sense of this, referring to Meckel as "the born possessor and creator of a scientific collection of such profusion in a single area, comparative anatomy, that it practically made it a natural duty for him to investigate it."[56]

In the first forty years of the century, then, several new aspects of the careers open to scholars interested in animal life helped to support the study of animal form. First, research became not only an accepted part of a professorial career in the sciences concerned with animal life, but a sine qua non. Second, teaching arrangements remained flexible, allowing people interested in animal form to claim expertise for a variety of related teaching subjects, despite the fact that morphology itself was not one of them. Third, it seems likely that the relatively generous material support provided by some states for anatomy and zoology helped reinforce the morphological approach to physiology by making it especially appealing to attach physiological teaching to teaching in these subjects.[57]

---

universities' medical and scientific institutes in 1834; at these universities, the collections for anatomy and zoology (sometimes combined, as at Halle) were among the best-financed institutes, often second only to the library and in the same pecuniary neighborhood as the surgical clinic and botanical garden.

56. R. Beneke, *Johann Friedrich Meckel der Jüngere* (1934), pp. 21–22.

57. I do not mean to suggest that the association of physiology with anatomy or zoology rather than, for example, chemistry, meant that an anatomist/zoologist/physiologist could not develop a chemical approach. Heidelberg's professor of anatomy, physiology, and zoology Friedrich Tiedemann, in addition to building the anatomy collection into a major one, conducted research on both chemical and vivisectional

## WISSENSCHAFT: THE NEXT GENERATION

The study of animal form in the first four decades of the nineteenth century was not distinct from a larger project of developing a science of animal life; indeed, its significance must be understood in the context of the gradual institutionalization of the *Wissenschaftsideologie* in the universities. Despite the occasional feeble protest, few researchers argued that the study of form directly served practical medical knowledge. Rather, the study of form came into the universities on the coattails of physiology—or perhaps more accurately, *as* the coattails of physiology.

As the *Wissenschaftsideologie* became a firm part of university life, however, the character of scientific writing on form and life changed. In the first couple of decades of the century many writers, like Burdach, devoted considerable effort to expressing the epistemological basis of research, stressing the role of reasoning and its relationship to empirical research in the attainment of knowledge. But by the late 1820s and 1830s younger writers, like Johannes Müller, were apt to be more reticent about the philosophical foundations of their research and laid greater weight on empirical findings for their own sake. Certainly this shift reflects in part the decline of *Naturphilosophie* as a scientific fashion—but *naturphilosophisch* writers did not die off that quickly: Oken, Carus, and many of their philosophically inclined contemporaries continued writing well into the 1840s. There may, however, be another way of accounting for the decline of this sort of philosophical discussion concerning the nature of *Wissenschaft* in these decades: perhaps such arguments no longer spoke to what was at stake for the researchers involved. When the incorporation of *Wissenschaft* into the universities was at issue, it was of central importance to define what sort of *Wissenschaft* it would be. Despite the many philosophical differences among writers on form and life, there was certainly consensus that this new *Wissenschaft* would have to include empirical research. Stepping back from the details of the philosophical debates, it may be appropriate to view the turn-of-the-century generation broadly as research-oriented professors working to gain sanction for their *Wissenschaft* in the university context, while arguing about the nature of that *Wissenschaft*.

By the early 1830s a new generation of men was beginning to achieve the level of full professor with the assumption that carrying

---

problems in the 1820s. The chemical orientation appears to have been less well developed than the morphological one, however, until the mid-1840s.

out research was a part of the job. For this generation, which included the professors of anatomy and physiology Carl August Sigismund Schultze (1795–1877), Johannes Müller (1801–58), Friedrich Arnold (1803–90), Rudolph Wagner (1805–64), Carl Theodor von Siebold (1804–85), and the professors of zoology H. G. Bronn (1800–1862) and A. A. Berthold (1803–61), the task was not to persuade their government ministries that *Wissenschaft* was worth pursuing at all, but rather that ever more material support—usually in the form of funding for collections and equipment—was necessary to carry out their research and teaching. With a firm, if not always copious, institutional base, this generation appears to have been more occupied in the 1830s and early 1840s with consolidating and expanding state support than with fighting battles over metaphysics and epistemology. Although numerous ideas were already being floated about increasing the material resources devoted to physiology (especially experimental physiology), these did not challenge the dominant consensus about the intimate intellectual relationship that had been constructed between physiology and the other sciences of animal form.[58]

By the 1840s, the study of form was firmly embedded within the matrix of teaching and research on animal life in the German-speaking universities—but it did not have a distinct, programmatic identity. That would come only in the turbulent period of the late 1840s and 1850s, when a new conception of physiology helped to rearrange the disciplinary order among the sciences associated with animal life and form. As we will see in the following chapter, it was through this adjustment (and expansion) of the disciplinary landscape that morphology took on a new, more clearly defined shape.

58. See Kremer ("Building Institutes for Physiology") for some Prussian proposals for independent physiology institutes; on C. A. S. Schultze and experimental physiology at Freiburg, see Nauck, "Bemerkungen zur Geschichte."

## CHAPTER THREE

## Rearranging the Sciences of Animal Life, 1845–1870

In 1855, the comparative anatomist Oscar Schmidt wrote that scholars had recently begun using the term "animal morphology" to signify a particular orientation of research concerned with the laws of form. Since this rubric covered just the same questions, aims, and methods that scientists had been applying for several decades, Schmidt continued, "one can legitimately ask, why apply a new name to an old thing?" Why, indeed? Schmidt's own answer was that a number of younger scholars (himself included) were using the term to designate a project that would rise above the masses of anatomical and zoological detail collected over the previous few decades and organize them under a few "governing ideas" to establish the "norms of form and the relationships among species and higher groups."[1] Waving the banner of animal morphology thus meant reasserting the higher purpose of studying form and underlining its status as a true *Wissenschaft*.

However much it might capture his views about the nature and significance of animal morphology, Schmidt's assessment of what was new about his subject is glaringly incomplete. Schmidt was certainly correct that animal morphology first came to be viewed as a particular orientation in the late 1840s and early 1850s. But rather than representing simply another periodic effort to stand back and make sense of accumulated detail, the orientation that emerged from the intellectual crucible of the late 1840s and 1850s was one of numerous elements

1. O. Schmidt, *Entwicklung der vergleichenden Anatomie* (1855), pp. 115–16.

that separated out of the old amalgam of life sciences, and that was part of a far-reaching reorganization of knowledge that would extend over the next few decades.

In fact, I shall argue, animal morphology was defined in contradistinction to a new "physicalist" program in physiology led by Carl Ludwig (1816–95), Emil Du Bois-Reymond (1818–98), Hermann von Helmholtz (1821–94), and Ernst Brücke (1819–92), who sought to redefine physiology away from what they saw as its focus on organisms as teleologically defined wholes and toward the study of those aspects of organic function measurable by physical instruments.[2] The physicalist physiologists are important to the history of morphology because part of their aim was to eject from physiology the study of form, generation, and development—aspects of life that seemed at the time essentially unmeasurable. It is through their rhetoric and the responses to it from their opponents that morphology came to be defined as a project distinct from physiology, and it is no accident that the word "morphology" appears with increasing frequency in programmatic writings from the late 1840s on.

The construction of morphology and physicalist physiology as mutually defining orientations was just one part of a broader trend in the animal sciences, however: in the late 1840s and 1850s, there emerged a range of distinct programs of research and teaching that took different views on what made good *Wissenschaft*. Virtually simultaneously with morphology and physicalist physiology there appeared a vigorous program of "scientific zoology," which took aim at classification in much the same way that the physicalist physiologists attacked the older physiology; programs of mechanical anatomy and microscopic anatomy appeared as well. All of these programs were led by men of the generation born between 1815 and 1822, who were establishing their careers in the late 1840s and 1850s. Nearly all of them had something to say about the proper scientific way of studying animal or human form.

The simultaneous emergence of these orientations indicates a profound intellectual splintering around 1850, one that was institutionalized over the next twenty years as the universities continued to expand in ways that accommodated these differences. Whereas before midcentury few universities could afford the luxury of separate professors of physiology, anatomy, and zoology, by the 1870s only the smallest and poorest of the German-speaking universities could do without

2. Cf. Lenoir, *Strategy of Life*, esp. chap. 5; K. Rothschuh, *History of Physiology* (1973).

any one of the three. This specialization gave scholars in these disciplines an opportunity to articulate new goals and to seek new relationships among them. The study of form figured in both the disciplines of anatomy and zoology, but its programmatic significance differed. Few anatomists used the word "morphology" in identifying their research programs, whereas among zoologists, morphology became an explicit part of a new program for scientific zoology. The subject of this chapter, then, is this intellectual and institutional fanning out, for the meaning of morphology in the middle decades of the century can only be grasped by means of its relations to the competing intellectual currents and expanding institutional possibilities of the time.

## THE CONSTRUCTION OF "OLD" MORPHOLOGY AND "NEW" PHYSIOLOGY

In 1877 the physiologist Emil Du Bois-Reymond wrote, "After the immeasurable efforts in morphology, which had occupied nearly all the powers of those conducting research into organic nature in the first half of the century, there came the day of physiology."[3] This statement would have been incomprehensible to a physiologist like Karl Friedrich Burdach, for whom morphology was a leading route to physiological knowledge. Before the late 1840s, it would have seemed peculiar indeed for anyone to argue that physiology was dominated by morphology, for since the turn of the century morphological problems had comprised only one among several different respected lines of research within physiology. Johann Friedrich Blumenbach, for example, taught his students in 1804 that the acquisition of physiological knowledge was aided by the study of anatomy, comparative anatomy, vivisection, experiment, chemistry, and physics.[4] Jan Evangelista Purkýně (1787–1869), who is credited with directing the first autonomous German physiological institute at the Prussian University of Breslau in the late 1830s, regarded physiology as divisible into six "doctrines," just one of which was "physiological morphology" or "physiological anatomy," comprising the study of development and the general laws of organic form.[5]

---

3. Quoted in Rothschuh, "Hyrtl contra Brücke" (1974), p. 163.
4. J. Brunner, lecture notes, p. 1, Medhist. Inst. Zürich.
5. The others were physiological physics; physiological chemistry; physiological dynamics (doctrines of the various forces, including the life force); physiological psychology; and physiological anthropology, which viewed the human race as a total organism (Purkýně to Altenstein, 1 June 1836, in J. E. Purkýně, *Opera*, vol. 12 (1973), pp. 219–34; this division, which he articulated at least as early as 1836, continued to

By the same token, although the generation of physiologists that included Johannes Müller, Friedrich Arnold, Rudolph Wagner, and Carl Theodor von Siebold emphasized physiology's links to anatomy, comparative anatomy, zoology, and embryology, none of them would have claimed that these avenues of research constituted the whole of physiology. Müller had done his share of research in chemical physiology in the 1830s, and Siebold had a chemical assistant when he was professor of physiology at Breslau.[6] Wagner, too, supported chemical research in his institute at Göttingen, reminding his government in 1858 that in his physiological institute he sought to link physiology not just to medicine but also to the physicochemical sciences. Under his leadership, the institute sought to place the human body and mind in the context of a more "general theory of organic nature [*allgemeine organische Naturlehre*]," a project that included support for an entire section of physiological chemistry with its own director. Consonant with this view of physiology as an all-embracing science of life, Wagner proudly noted that researchers who had studied under him had gone on to diverse professorships in clinical medicine, chemistry, zoology and comparative anatomy, and physiology.[7] In addition to these signs of interest in a range of physiological approaches, in 1860 Wagner, who was in ill health, actively recruited for his successor one of the leading "physicalist" physiologists, Hermann Helmholtz![8] This was not a group fundamentally opposed to physical or chemical research in physiology per se.

Nevertheless, by the early 1850s physiologists of various orientations were using the term "morphological" to describe the approach of men such as Müller, Wagner, and Siebold to physiological research, because it was recognized by proponents of both the "new" and the "old" physiology as descriptive of the chief difference between the two. Siebold, for example, bemoaned the fact that his theory of parthenogenesis, one of the most important advances in the understanding of reproduction, was ignored by the physicalists, but he accepted

---

structure his discussions of physiology as late as the 1850s (see Purkyně, "Über den Begriff der Physiologie" [1851]).

6. Historians have not focused a great deal of attention on Müller's brief excursion into chemistry; at least some of it was conducted with his student Theodor Schwann (see J. Müller and T. Schwann, "Versuche über die künstliche Verdauung" [1836], and references to earlier work therein); Siebold to Wagner, 27 July 1853, UB Göttingen, Cod. MS R. Wagner.

7. Wagner, "Denkschrift" (6 August 1858), p. 4 (unpaginated in original), UA Göttingen, Personalakte Keferstein.

8. Wagner to Henle, 23 January 1860, in H.-H. Eulner and H. Hoepke, *Briefwechsel* (1979), p. 103.

the label of morphologist, accounting for their reaction in a letter to Wagner by saying, "indeed, it must make this school very uncomfortable to have to acknowledge that here the morphological, histological school has contributed something substantial to animal biology."[9] This acknowledgment by a general physiologist of the splintering of his field into schools suggests that an earlier common dream, for a physiology to which researchers with various special interests would pool their contributions, was breaking down into hostile, competing camps.[10]

Whereas the morphologists generally accepted physical and chemical research as worthwhile aspects of a general physiology, the physicalists attacked the very legitimacy of morphological research, claiming that the study of form had nothing to offer physiology. In the 1850s, the most vocal physicalist on this score was Carl Ludwig.[11] In his famous 1852 textbook he provided a new definition of physiology that stood in sharp contrast to the definition provided by the eminent teacher, Johannes Müller. According to Müller, "Physiology is the science of the properties and appearances of organic bodies, animals and plants, and of the laws according to which their actions [Wirkungen] ensue."[12] Ludwig rejoined, "Scientific physiology has the task of determining the functions of the animal body, and of deriving them consequentially from its elementary conditions."[13] By shifting the emphasis from the general properties and appearances of living bodies to the determination of function, Ludwig's definition called for shifting physiology's task and also for abandoning its existing relationship to the study of form. Instead of focusing on the organization of the material substrate that made function possible, with all the implicit associations that this notion contained, physiologists would concentrate on the physicochemical basis of function. In particular, they would seek ways to measure exactly the physical and chemical processes taking place in the body.

With this change of aim, the old interlacing of physiological and anatomical research would be swept away. It was still necessary for the physiologist to know some anatomical facts, but anatomy could

9. Siebold to Wagner, 26 April 1857, UB Göttingen, Cod. MS R. Wagner.

10. Lenoir notes that "from the late 1840s through the late 1860s physiology in Germany was not dominated by a single approach," citing the physicalists, the morphologists, and a chemical orientation as the leading competing camps ("Science for the Clinic," pp. 144–45).

11. It should be noted that later, Ludwig became considerably more broad-minded.

12. J. Müller, Handbuch (1844), p. 1.

13. Ludwig, Lehrbuch, vol. 1 (1852), translated in M. H. Frank and J. J. Weiss, "'Introduction,'" (1966), p. 77.

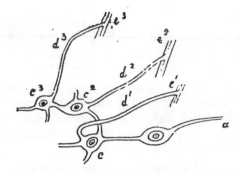

Fig. 3.1. Ganglia and their connections (illustrating Wagner's views): *a,* nerve; *c,* ganglion-cells; *d,* nerves; *e,* muscles (from Carl Ludwig, *Lehrbuch der Physiologie des Menschen* [1852], vol. 1, p. 125).

be treated as a closed body of descriptive knowledge, nothing more than a *Hilfswissenschaft* upon which the explanatory science of physiology would draw. According to Ludwig, anatomical researchers had made little contribution to solving physiological problems because they neglected to analyze anatomical structures in terms of their ultimate chemical and physical properties. He finished off his comments with the following words of comfort to the beginning student "who finds physicalist physiology harder than the comfortable discussions of a morphologist. . . . We answer you: the small effort that you have put there [into morphology] is lost, because the result doesn't lead you to the desired insight, whereas the larger effort that you put in here is fully rewarded."[14]

Ludwig's withering opinion of the morphological approach to physiology was not confined to textbook pronouncements about physiology's aims, but also extended to ongoing physiological research problems and methods. This can also be seen in the criticisms he heaped on Rudolph Wagner concerning the transmission of reflex reactions in the spinal cord. In 1852, Ludwig had printed a criticism of Wagner's views in his *Lehrbuch.*[15] When Wagner arrived at the definitive statement of his theory of reflex transmission two years later, Ludwig once again attacked his position.

According to Wagner, ganglion cells acted as nodal points or junctions that linked seemingly unconnected nerves to one another (fig. 3.1). Reflex impulses were transmitted through a continuous network of nerves and ganglion cells, from the initial point of stimulus along

14. Ludwig, *Lehrbuch,* vol. 2 (1856), pp. iv–viii, quotation on pp. vii–viii.
15. Ibid., vol. 1 (1852), pp. 124–25, 139ff.

one nerve to another, eventually reaching the muscles. Wagner agreed with Ludwig that physical and chemical factors were critical for understanding the transmission of nerve impulses, but he stressed that the physiological phenomena could be completely explained only by reference to this anatomical network.[16]

Ludwig responded to this conception of nervous anatomy and physiology in a biting article in late 1854. An anatomically based theory of reflex transmission was indefensible: the work of Du Bois-Reymond and Conrad Eckhard proved "in the strictest sense of the word" that nervous reactions were independent of the anatomical or chemical structure of the nerves and corresponded to electrical impulses. The physiology of the reflex reactions, in other words, was not dependent on the organism's anatomical organization. Ludwig further claimed that Wagner's research was itself of dubious value. He questioned the validity of Wagner's anatomical observations and said that even if they were correct, they were insufficient to support his theory. He had already noted in his textbook that the particular anatomical configurations did not bear a unique relation to particular physiological events. Of course, ganglion cells were present where reflex actions occurred and absent where they did not—but they were not the only such structures. Even more damning, sometimes when nerves were connected to ganglion cells, not all of the physiological events took place that were ascribed to the connection. In some cases none of them occurred. How could one then possibly claim that the structures determined the function? Indeed, Ludwig concluded, what use did Wagner's study have for physiology at all? "If confidence in microscopic research has not already been shaken a thousandfold; and even if we others had not, from years of experience, acquired a profound insight into the limits of the capabilities of all existing methods of microscopic preparation; even so, it would still be irresponsible to accept [a theory of] the structure of the spinal cord pieced together from these unconnected facts. . . . The facts that Herr Wagner uses as the base of his . . . hypothesis are undoubtedly praiseworthy beginnings toward insight into the structure of the spinal cord, but they are no more."[17] Ludwig thus rejected Wagner's anatomical methodology and results as even a conceivable basis for explaining the physiological phenomenon of reflex impulses. In Ludwig's presentation, the inadequacies of Wagner's morphological explanation combined with the merits of Du Bois-Reymond and Eckhard's theory to send a pow-

16. R. Wagner, "Neurologische Untersuchungen" (1854), pp. 89–104, esp. p. 95n.
17. Ludwig, "Zur Ablehnung der Anmuthungen" (1854), p. 273.

erful double message: physiological phenomena were by their nature distinct from anatomical structures, and physiological problems could not be answered with morphological methods.[18]

If Ludwig clearly expressed the usefulness of anatomical methods and the morphological viewpoint from the perspective of the physicalists, the opposing point of view also had an articulate spokesman in the physiologist and anatomist Karl Reichert (1811–83), another former student of Johannes Müller of a slightly older vintage than the physicalists. During the 1850s he published a series of "Reports on the Progress of Microscopic Anatomy" in Müller's *Archiv*. His rendering of the differences between the physicalists and the morphologists defined the camps in terms of their attitude toward organization and development. In his report for 1852 he acknowledged that important progress had been made through physical and chemical studies of function but argued that physiology properly had a larger task, "[f]or although morphological phenomena and their conformity with natural law [*Gesetzmäßigkeit*] cannot yet be clothed in mathematical formulae, they still offer the sole basis for the conception of the most important life phenomena, of organization, of generation, of development, and their lawful behavior. . . . Similarly, physics and chemistry offer us no starting point for interpreting the lawful agitation within organic matter from which Form arises as the effect."[19]

Reichert elaborated the differences between the two views, as well as their consequences, in his reports for 1854 and 1855. The morphologists, Reichert wrote, started out from the axiom that the organism's fundamental characteristic was its organization into a system in which the whole regulated the parts.[20] This he called the "systematic" view of the organism. Ludwig and his group took an "atomistic" stance, considering the organism as simply an aggregation of atoms that were built up to form, first, cells, then tissues, then organs, and ultimately the organism. For Reichert, this view denied the evidence of development, in which the process of coordinated differentiation clearly demonstrated the historical and ontological priority of the whole over the parts.[21]

18. For his part, Wagner responded that he was "truly sorry to have to declare Ludwig and Eckhard, in judgments on the application of microscopical . . . research to physiology, completely incompetent" (Wagner, "Begründung meiner . . . 'Anmuthungen'" [1854], pp. 308–9n).

19. K. Reichert, "Bericht 1852" (1853), p. 1.

20. As Reichert himself indicated, this is a Kantian position.

21. K. Reichert, "Bericht 1855" (1856), esp. pp. 1–2, 10–11, 19; for other discussions of this essay and his similar 1854 report, see E. S. Russell, *Form and Function* (1916), pp. 192–94, and Lenoir, *Strategy of Life,* pp. 219–24. Reichert was quite right

Because of their fundamentally differing premises, Reichert argued, the two positions could not be reconciled. The difference in approach admitted of no possibility for compromise, either for pursuing an understanding of form or for learning about the nature of organic function. Indeed, the very definition of function differed according to which view one took: "In the atomistic sense the concept of function is equivalent to a simple output [*Leistung*]; . . . by contrast, for the systematist function is the output of one of the parts determined within an organized system."[22] Particular physiological functions also held different meanings: "The condition of irritation signifies for the atomist a simple cause and effect relation," whereas in the systematic view of nature, one must consider the effect of the main irritating factor and subsidiary stimuli on the other parts of the system, as well as the results on the system as a whole. Only out of this understanding can one develop a general concept of "irritability."[23] Reichert's underlying message was that anatomy was more than a description of the material in which functions took place: functions simply could not be understood without direct reference to the structural organization of the animal, and understanding structural organization meant studying development.

When we balance the physicalist's perspective against that of the morphologist, three important phenomena emerge. First, we see both sides agreeing that their differences centered on the legitimacy of considering the phenomena of organization, development, and reproduction as part of physiology. But second, we also see an asymmetry in their views. Whereas the picture created by the 1847 group contrasted their new, physicalist orientation with the outdated morphological one (thereby incidentally collapsing all of physiological research into this opposition), their opponents viewed the contrast as one between, on the one hand, their own broadly based physiology, which included the study of form along with the study of chemical and physicochemical aspects of life, and, on the other hand, the narrowly one-sided approach of the physicalists. In addition, we see that the characterization by later historians of the controversy as one between physiologists and morphologists projects onto the late 1840s and 1850s a

---

about the physicalists' ignoring development and generation: indeed, Ludwig explicitly defined the study of development out of physiology, saying that it was impossible to treat it as a physiological phenomenon until the anatomists had developed "more or less clear suggestions" concerning the ways that the primitive structures gave rise to tissues and organs (Ludwig, *Lehrbuch*, vol. 2, p. vi).

22. Reichert, "Bericht 1854" (1855), p. 8.

23. Ibid.

situation that did not hold until later. At the time, the difference was seen as one between two approaches within physiology; it was only in the wake of the institutional divisions following the mid-1850s that the story began to be rewritten into one between physiologists and morphologists, that is, between people inside and outside physiology.[24]

Alongside their arguments over the methods and goals of physiology, the physicalists also confronted the morphologists at the institutional level. Beginning in the 1850s, adherents to the physicalist approach increased their efforts to gain institutional autonomy from anatomy and comparative anatomy and to attain independent chairs of physiology. For example, in arguing for separate chairs for anatomy and physiology, the physicalist Carl Vierordt (1818–84) contrasted the descriptive science of anatomy with the "explanatory" science of physiology, asserting that each had its own methods and aims, and that the "principles and approaches of a descriptive science are not applicable to solving physiological problems."[25] He admitted that anatomy had its own scientific task, saying it was not simply the collection of facts but that embryology and comparative anatomy provided the route to a higher, philosophical understanding of form. These two pursuits, however, could contribute little to understanding processes, which by their nature had to be investigated experimentally. Why, then, were anatomy and physiology so often taught by a single person? Only because the true, distinct tasks of the two had not been recognized. Physiologists especially were to be blamed for this: "The needs on both sides, but especially the interests of physiology, make it indispensible that the . . . demand for their separation be recognized and implemented."[26] Other physicalist physiologists

24. In *Strategy of Life,* Lenoir argued that "Helmholtz and Du Bois-Reymond were not defending a position against physiologists," but having already conquered physiology, faced opposition from "*zoologists* and *morphologists*" (pp. 232–33, emphasis in original). More recently, however, Lenoir has acknowledged the existence of a morphological camp among physiologists (see "Science for the Clinic," pp. 144–45). Karl Rothschuh captures a sense of the historical shift in progress in an article's subhead, "The General Controversy between Morphologists and Physiologists, Microscopists and Physicalist Physiologists" ("Hyrtl contra Brücke" [1974], p. 163). It appears that Du Bois-Reymond was instrumental in constructing this history and projecting it onto the earlier part of the century, although he acknowledged that there was a time (before the "truth" was evident to all) when the true physiology was viewed merely as the "physicalist school" (Du Bois-Reymond, "Der physiologische Unterricht" [1877], esp. pp. 636–37).

25. C. Vierordt, "Über die gegenwärtigen Standpunkt" (1849), p. 298.

26. Ibid., p. 299; Vierordt himself reaped the advantages of this change when he received the first independent chair for physiology at Tübingen in 1855 (see H.-H. Eulner, *Entwicklung der medizinischen Specialfächer* [1970], p. 57).

shared his view, pressing for independent chairs to benefit the discipline and, of course, to better their own situations.[27]

As Vierordt's example suggests, the physicalist physiologists could be explicit about linking the drive for institutional independence to their own programmatic aims for physiology. We should not be too hasty, however, in assuming that physiology's successful institutional "liberation from anatomy" resulted from universal acceptance of the intellectual claims and goals of the physicalist program. As Arleen Tuchman has pointed out, in the 1850s the young physicalists were not in a particularly strong bargaining position when demanding independent chairs for physiology.[28] Why, then, did they succeed?

A full answer to this question would be a separate project in itself, but recent studies offer some provisional suggestions. In a recent essay, Richard Kremer has drawn attention to three Prussian proposals for physiological institutes presented between 1836 and 1846, each of which had a different aim and organization. Kremer argues that

27. When Ludwig published the first volume of his polemical textbook, he was still officially teaching both anatomy and physiology at the University of Zurich. Unable to negotiate a better position for himself, and unwelcome in Prussia because of his radicalism and atheism, he moved to the Josephs-Akademie in Vienna, which trained Austria's military surgeons. It was not as prestigious a position as an *Ordinarius* at a German university, but at least he was next door to his good friend Brücke (at the University of Vienna), and he did not have to teach anatomy. When Du Bois-Reymond was invited to consider a post in anatomy and physiology at Bonn in 1854 (the same one that Helmholtz eventually received), he pressed hard to be released from the obligation to teach comparative anatomy. Du Bois-Reymond finally turned down Bonn in hopes of an eventual professorship of physiology in Berlin; meanwhile he would remain in Berlin as *ausserordentliche* professor and lobby to better his position. Helmholtz received the position in Bonn, where at first he shared his teaching duties there with another professor. He apparently managed to avoid teaching comparative anatomy in his first year, and when the other professor left in the fall of 1856, Helmholtz wrote to his father, "Now I am the only official representative of physiology, and the ministry can therefore make no more demands on me regarding comparative and microscopic anatomy, for which I would otherwise have had to be responsible. For if I read human anatomy in the winter and physiology in the summer as the main courses, my time is occupied, and one cannot reasonably demand more of me" (L. Koenigsberger, *Hermann von Helmholtz* [1902], vol. 2, p. 283). Helmholtz thus managed to hold his anatomy teaching to a minimum. Two years later he was offered the new physiology chair at Heidelberg, his first position free of any teaching obligations other than physiology.

28. A. Tuchman, "Science, Medicine and the State," p. 171. Tuchman attributes this outside impetus to government officials in the state of Baden who, she argues, appear to have believed that the "exact" approach would ultimately lead to improved health care; Lenoir makes a similar point for governmental support of Ludwig's physiology institute by the state of Saxony in "Science for the Clinic." While the argument is intriguing, and certainly the role of the state in decisions concerning support for science and medicine was important, I am not sure that such top-down policy considerations were responsible for the broader success of the physicalists.

the men making these proposals helped to establish a "discourse of experimental physiology" that became the "currency" for the physiological market. Significantly, these earlier entrepreneurs were not physicalist physiologists, nor were they seeking disciplinary autonomy in the form of separate chairs.[29] As Kremer hints, however, their efforts to carve out separate spaces for teaching and research meant that by the late 1850s, when the physicalists did begin to argue for distinct professorships, much necessary groundwork had been laid. Thus at the disciplinary level, there was already a sense that physiology was something distinct from anatomy, marked in part by its experimental nature. The physicalists drew out certain aspects of that existing discourse at the expense of others, but they did not create the field of discourse itself.[30]

Although Kremer's argument helps us understand the circumstances enabling the physicalists to get a hearing, it remains unclear why states were willing to support their bid for independence. Arleen Tuchman and Timothy Lenoir have argued that Baden and Saxony, respectively, supported physiology as part of a larger movement to modernize the state and its citizenry. For Prussia, by contrast, Kremer argues that there is little evidence in favor of this thesis. No one has yet analyzed Bavaria and the smaller German principalities for their relationships between the state, university professorships, medicine, and modernization, but the available data certainly do indicate increasing state funding for science and medicine in the 1850s and 1860s.[31] At this point, the precise reasons for this new interest on the part of the state must remain murky.

29. R. L. Kremer, "Between *Wissenschaft* and Praxis" (1991), p. 168.

30. Kenneth Caneva, whose recent work, *Robert Mayer* (1993), includes a detailed exposition of the status of vital forces in the physiological and medical literature of the 1830s and early 1840s, reads the situation slightly differently; he argues that already by the late 1830s "the clear tendency within physiology was to seek as far as possible to account for all vital phenomena in terms of underlying chemical and physical processes. . . . Processes such as embryological development . . . , which could not be plausibly so treated, increasingly fell outside the competency of physiology" (p. 101). Thus Caneva suggests that the split that I have seen emerging around 1850 was already apparent a decade earlier. Although it does seem likely that there was increasing tension through the 1840s over the relationship of morphological problems to other ones in physiology, I see little evidence that most physiologists had written these out of the purview of physiology until the end of the decade. Nevertheless, such an emerging division, even if still within physiology, might have provided a body of support for the more extreme stance of the physicalists in the later 1840s, and in this way also contributed to the creation of a new physiological discourse.

31. Tuchman, "From the Lecture to the Laboratory"; Lenoir, "Science for the Clinic"; Kremer, "Building Institutes for Physiology"; H. W. Kupka, "Ausgaben der süddeutschen Länder" (1970); the financial support for medicine and the natural sci-

One possible source of support for physiology's independence that has gone largely unnoticed came from those younger professors of anatomy and physiology who leaned toward anatomy, and who appear to have been ready to get rid of physiology. Despite the common historical wisdom that professors were always seeking to preserve and increase their student numbers because of their financial dependence on lecture fees, this seems not to have been the main consideration for teachers of anatomy and physiology at mid-century. Few professors argued against what would effectively be a lightening of their teaching load and the enhancement of the overall importance of their subjects that would result from adding another professor. The sheer number of teaching subjects that a professor of the animal sciences expected to cover seemed to mandate some sort of division of labor.

In this situation, the particular orientation of the new physiologist seems not to have been the primary issue. Tuchman has shown that for Jacob Henle (1809–85) at Heidelberg, it seemed more important that he find a leading scholar to help relieve his load than that he get someone from a particular school. Henle, who had been appointed in 1843 to teach anatomy and physiology and who by 1849 was also teaching comparative anatomy and pathological anatomy, clearly felt himself to be overburdened and wanted to be rid of physiology.[32] When he started agitating for a new physiologist, he first proposed the physicalist Carl Ludwig; when negotiations with the politically radical Ludwig fell through in the aftermath of the revolutions of 1848 Henle, with the backing of the faculty, suggested Carl Theodor von Siebold to take over comparative anatomy and physiology—a man who epitomized the morphological approach to physiology. That effort did not work out either, and Henle and the medical faculty returned to candidates from the physicalist school. (For various reasons it took until 1858, six years after Henle gave up and left for Göttingen, for Heidelberg to hire a physiologist; when they did so it was Helmholtz.)[33]

We might also consider the case of Albert von Kölliker (1817–1905). From the moment of his appointment as professor of anatomy

ences in these decades is small compared to what it would become in the last third of the century, the period of the so-called institutional revolution of German science, but it was nonetheless significant in the terms of the time.

32. See Henle to Wagner, 1 July 1852, in Eulner and Hoepke, *Briefwechsel*, p. 48.

33. For a detailed exposition of Henle's efforts and the relationship of physiology to other subjects at Heidelberg, from which this account is borrowed, see Tuchman, "Science, Medicine and the State," chap. 6.

and physiology at the large and vigorous medical faculty at the Bavarian University of Würzburg in 1849, he began building a highly successful empire that embraced human and comparative anatomy, physiology and physiological anatomy, embryology, and histology.[34] Reminiscing on his teaching efforts Kölliker wrote, "Obviously it was impossible for me to teach all of human anatomy, physiology, embryology, comparative anatomy, and microscopy alone, although in some semesters I taught 14–16 hours a week, all the more so since I had also introduced topographic anatomy, comparative histology, comparative embryology, and comparative physiology as optional courses [Nebenfächer]." For some fifteen years, Kölliker would rely on junior associates to help him cover the courses, but finally it became too much, and in 1865 he gave up physiology to a new professor, Albert von Bezold.[35] As the examples of Kölliker and Henle suggest, the separation of anatomy and physiology might be viewed as a mutual liberation for anatomists and physiologists. And the fact that physicalist physiologists were offered those chairs may have as much to do with their availability and their evident willingness to remain on their own intellectual territory as with the reductionist or materialist content of their program.

The institutional separation of anatomy and physiology could be occasioned in a variety of ways. As we have seen with Kölliker and Henle, sometimes the professor himself was actively involved in promoting a separate professorship, at least in part to reduce his teaching load. More often, though, it took the death or departure of the professor controlling anatomy and physiology (or physiology, comparative anatomy, and zoology) to precipitate a division, sometimes by first providing for a new ausserordentliche professor. In 1852, for example, when Friedrich Arnold left Tübingen for Heidelberg he was replaced by two ausserordentliche professors, one to teach anatomy and pathology and one to teach physiology. Both were promoted to full professorships three years later.[36] A similar cluster of openings provided the occasion for anatomy and physiology to split at Jena, Berlin, and Bonn within a year of each other, helping to bring momentum

34. There is no full-length work on Kölliker that considers his place in nineteenth-century German biology and medicine. The most useful sources are A. von Kölliker, *Erinnerungen* (1899); and G. Feser, "Das anatomische Institut" (1977), which provides considerable information about Kölliker, his students and assistants, and the work conducted at Würzburg's anatomy institute.

35. Kölliker, *Erinnerungen*, pp. 32–34, quotation on p. 32.

36. Mörike, *Geschichte der Tübinger Anatomie*, pp. 55–57.

to the establishment of new chairs. At Jena in 1858, the professor of anatomy and physiology Emil Huschke died, and Carl Gegenbaur, who had originally been hired to teach zoology, was asked to take over Huschke's chair. He refused to take on physiology, and so it was given to an *ausserordentliche* professor; two years later the position was made a full professorship. At Berlin, Müller's death in 1858 occasioned the creation of two distinct chairs: one for physiology, which was given to his former student Emil Du Bois-Reymond; and one for anatomy, which was given over to another of his students, Karl Reichert, after Jacob Henle had turned it down because it would have entailed too much administrative work. The same year, Hermann Helmholtz was appionted as Heidelberg's first full professor in physiology, and the professorship of anatomy and physiology he left at Bonn was divided the next year.[37]

At most universities, physiology and anatomy were bound in one professorship, but at a few, physiology was tied to comparative anatomy and zoology instead. At these universities, which in the 1850s included Munich, Göttingen, Freiburg, and Rostock, a parallel set of shifts took place: the union was dissolved through a professor's death, departure, or decision to narrow his area of concentration. This trend began in the early 1860s. By then, Carl Siebold at Munich had already turned physiology over to the new anatomist T. L. W. Bischoff (in 1854); following the pattern of dividing anatomy from physiology, Bischoff gave up physiology to a new professorship in 1863. In 1860 at Göttingen, Rudolph Wagner turned over physiology to his former student Georg Meissner, who had been professor of physiology, zoology, and histology at Freiburg since 1857. In this case, both men contracted their official teaching area: Wagner to zoology and comparative anatomy, Meissner to physiology. Shortly after Meissner's departure from Freiburg, physiology and zoology split there as well. Although Meissner's successor Otto Funke was initially named professor of both subjects, he appears to have been quite happy to hand over the teaching of zoology to the young *ausserordentliche* professor August Weismann in 1867. At Rostock, Carl Bergmann's empire of anatomy, physiology, and comparative anatomy was divided upon his death in 1865. Anatomy was given to one professor, physiology to a second, and comparative anatomy was eventually united with zoology

---

37. On Jena, see G. Uschmann, *Geschichte der Zoologie* (1959), pp. 31–32; Henle to Minister for GUMA, 6 June 1858, GStA Merseburg, Berlin Med. Fak., p. 18; Eulner, *Entwicklung der medizinischen Spezialfächer*, pp. 49–61.

in 1871, when the university established its first independent chair of zoology.[38]

For the history of morphology, the institutional shifts are important because in most cases the new physiology professors concentrated on problems approached with physical or chemical methods rather than morphological ones. Physiological researchers whose interests had little to do with physics or chemistry—those concerned with generation, development, and comparative anatomy, problems ever more frequently viewed as morphological—found themselves less and less successful (or even interested) in filling professorships in physiology. Increasingly they found their niches in anatomy and zoology. Both the latter disciplines were affected, therefore, by the new institutional autonomy of physiology and the concurrent redefining of its appropriate problems and approaches, because they provided the new homes for those researchers defined out of "modern" physiology. However, as we will see in the next two sections, the circumstances of anatomy and zoology differed; the study of form took on correspondingly different attributes and programmatic links in each discipline.

## MAKING ANATOMY WISSENSCHAFTLICH

It would have been logical and tidy if the institutional split between anatomy and physiology meant that physiologists took for their territory the study of function and anatomists adopted the study of form, but this happened only to a degree. It is certainly the case that the younger group of physicalist physiologists in the 1850s decided that problems of development, generation, and "laws of form" were not in their province, but not all anatomists chose to define physiological concerns out of their work. A number of anatomists agreed with physiologists that anatomy was purely descriptive and physiology was explanatory; these men could hardly afford to drop their claims to physiological significance if they hoped to be *wissenschaftlich*. A further complicating issue is that, just as there were different views among physiologists about what constituted real or good physiology, so, too, did there exist differences among anatomists about what would make anatomy "scientific." For the 1850s and for much of the 1860s one can identify two broad orientations that represented differ-

---

38. Eulner, *Entwicklung der medizinischen Spezialfächer*, p. 58; E. Ehlers, "Göttinger Zoologen" (1901), pp. 443–44; E. T. Nauck, *Zur Vorgeschichte* (1954), pp. 36–37; Schumacher and Wischhusen, *Anatomia Rostochiensis*, p. 112.

ent views of scientific anatomy. One, called "mechanical" or some-times "physiological" anatomy, had relatively few adherents but was intellectually quite coherent. The other, microscopic anatomy, domi-nated the discipline but was much more diffuse intellectually. Though both orientations were concerned with form, none of the mechanical anatomists and only a few of the microscopic anatomists identified their work in programmatic terms as primarily morphological. Most chose to avoid the word, which, after all, had been given a rather bad odor by the younger generation of physiologists.

Mechanical anatomy, which can be traced back to the 1820s with the work of the brothers Ernst Heinrich, Wilhelm, and Eduard Weber, focused on the statics and mechanics of the human body. Concentrat-ing on the human locomotor apparatus and the circulation of the blood, the Weber brothers applied the concepts of physical mechanics to explain the form of anatomical structures.[39] Few other anatomists appear to have adopted this approach before the 1850s, when Georg Hermann von Meyer (1815–92) took it up, and he remained the leader of the mechanical orientation for the next thirty years. Begin-ning in 1856 with his *Textbook of Human Physiological Anatomy* and extending to an 1883 essay titled "Stellung und Aufgabe der Anatomie in der Gegenwart" (The position and task of anatomy in the present), Meyer worked steadily to gain recognition for the me-chanical approach.

In the introduction to his textbook Meyer stated that "the investi-gation of form as such should be regarded as nearly completed by existing works; and really, very little new of substance has been added in recent decades to the earlier knowledge in this area. However, anatomy is in no way therefore a finished science, as one so often hears," because this was just the first step in anatomical knowledge. Now that it had been taken, he continued, anatomy could move on to the next task, which was "to conceive of the body as a complex of physiological apparatuses, and to derive an understanding of its form from the functional character of the individual parts."[40] As he would put it later, the human body was to be examined in the same way that a student of mechanics would study the construction of a machine, so that a mechanism such as the arm would be considered

39. On E. H. Weber, see V. Kruta, "Weber, Ernst Heinrich"; on the other Webers, see A. E. Woodruff, "Weber, Wilhelm Eduard"; *Biogr. Lex.* I, s.v. "Weber, Eduard Friedrich Wilhelm"; E. and W. Weber, *Die Mechanik der menschlichen Gehwerkzeuge* (1836); E. H. Weber, *Über die Anwendung der Wellenlehre* (1889).

40. G. H. von Meyer, *Lehrbuch* (1856), pp. v–vi; also quoted in F. Koch, *Der Anatom Georg Heinrich von Meyer* (1979), pp. 50–51.

in terms of the functional coordination of bones, muscles, ligaments, and nerves.[41] The anatomist further has as his task "to set out the general laws of the arrangement and order of the apparatuses more precisely and more consciously than has been done before."[42]

Meyer's approach encompassed two steps: first, form could not be understood without reference to function—the appropriate way to analyze the body was by functional groupings, not structural ones. To understand function, one could then further break down the work done by these groupings into the mechanical contributions of each part, which could be reduced to simple physical cause-and-effect laws. The two aspects of this approach were expressed by the two terms used interchangeably to describe it: it was both physiological and mechanical. As this description suggests, Meyer was one of those who did not agree that anatomy and physiology should move into completely separate spheres. First as professor of anatomy and physiology at Zürich under Carl Ludwig's direction and then as professor of anatomy there after 1856, he continued to maintain that the structure of the human body could be understood only by reference to its function.[43]

In addition to Meyer, mechanical anatomy inspired the interest of several other German-speaking anatomists between the 1850s and 1870s: his near contemporaries Ludwig Fick (1813–58) of Marburg and Carl Langer (1819–87) of Vienna, and a younger cohort consisting of Christian Wilhelm Braune (1831–92) of Leipzig, Wilhelm Henke (1834–96) of Tübingen, Nikolaus Rüdinger (1832–96) of Munich, and Christoph Theodor Aeby (1835–85) of Bern. These men shared two core interests that may be traced back to the Webers: the mechanics of the joints, a topic that was particularly suited to the kind of physical analysis of stress and pressure that they viewed as providing an explanation of form; and the structure of the vascular system, in which the mechanics and dynamics of blood flow played a formative role.[44] They also joined in adopting from physics the methods of quantitative measurement and the expression of their findings in mathematical terms. As to the end goal of these measurements,

41. [G.] H. von Meyer, "Stellung und Aufgabe der Anatomie" (1883), p. 359.
42. G. H. von Meyer, *Lehrbuch* (1856), pp. v–vi; also quoted in F. Koch, *Der Anatom Georg Heinrich von Meyer* (1979), pp. 50–51.
43. K. Bardeleben, "Georg Hermann von Meyer" (1892).
44. On Fick and Aeby, see the listings under their names in *Biogr. Lex.* I; for the others, see C. Rabl, "Carl Langer" (1888); K. Bardeleben, "Wilhelm Braune" (1892); A. Froriep, "Wilhelm Henke" (1896); K. von Kupffer, "Nikolaus Rüdinger" (1897).

the mechanical anatomists appear to have shared the view of the physiologist Carl Ludwig, who said that the task of anatomy was to serve physiology by giving "the constant and, where possible, mathematically expressible conditions" of organic form.[45]

The approach of this group was eminently acceptable to the physicalist physiologists, and it is perhaps not entirely coincidental that both took off in the 1850s. But the mechanical anatomists shared another characteristic besides their close link to physicalist physiology, which was also important for their institutional location in medical schools: they tended to have connections to practical or surgical anatomy.[46] All of them were known as skilled practical anatomists

45. Ludwig, "Einleitung," *Lehrbuch,* vol. 1, p. 11. Indeed, the interaction between mechanical anatomists and physicalist physiology, as represented by Carl Ludwig, was very close: four leading mechanical anatomists had direct contact with him during their careers. Already in the early 1840s he formed a close friendship with his near contemporary Ludwig Fick, when the latter was first prosector in anatomy at Marburg and he was the second prosector. When Fick advanced to professor of anatomy, he saw to it that Ludwig was promoted, first, to prosector and then to *ausserordentliche* professor, and gave him space in the anatomy institute for his physiological experiments (H. Schröer, *Ludwig* [1967], pp. 33–34); Ludwig also became a mentor for Fick's youngest brother, Adolf, who conducted early work in physiological anatomy and eventually was employed as a physiologist (see R. Bezel, *Der Physiologe Adolf Fick* [1979]). During the period when Ludwig became professor of anatomy and physiology at Zurich, from 1849 until 1855, Hermann von Meyer worked as his prosector. That year, Ludwig left for a professorship in physiology of the Kaiserliche Josephs-Akademie in Vienna, the training center for Austrian military surgeons; the mechanical anatomist Joseph Langer also taught there, from 1856 to 1869. Although Langer had already begun developing his own physiological approach in the early 1850s, his interests could only have been reinforced by Ludwig. When Ludwig moved to Leipzig in 1865, he had an opportunity to encourage a third mechanical anatomist, Christian Wilhelm Braune. Braune was already working as an assistant in anatomy at Leipzig, under his father-in-law E. H. Weber, so he had a chance to imbibe the approach from two directions. Although Weber undoubtedly exerted the most important influence on Braune before Ludwig's arrival, Braune had ample opportunity for close contact with Ludwig thereafter. He made public his intellectual debt to Ludwig by dedicating the *Atlas of Topographic Anatomy* to him in 1872 (C. W. Braune, *Topographisch-anatomischer Atlas* [1872]).

46. As professor of anatomy at the Josephs-Akademie in Vienna, Langer had to be most concerned to teach the macroscopic features important for surgery. Braune had originally habilitated for surgery, and even after he switched to anatomy, he took time out for stints as a military surgeon in the campaigns in Schleswig, Bohemia, and France. Similarly, Henke had a strong interest in surgery. Born with a club foot, he originally directed his medical and surgical interests toward the pathological development of the foot, but gradually expanded to include the general mechanics of the joints. Finally, Nikolaus Rüdinger, who began his healing career as a bathkeeper (*Bader*), was educated as a medical surgeon before rising to the post of prosector of macroscopic anatomy at Munich (for biographical sources see n. 44).

whose major contributions to teaching took place in the dissection room.[47] Thus the mechanical anatomists could make claims both to serving the higher scientific goals of physiology and to expertly teaching the practical anatomy essential to the medical student.

Mechanical anatomy is a readily identifiable orientation in anatomical research, associated with a fairly clear-cut set of problems, approaches, and people. But its members were distinctly in the minority. The other readily identifiable realm of anatomical research in the period does not succumb to such neat description, in part because it has received so little historical attention.[48] Known variously as microscopic anatomy, histology, or general anatomy, this area of research concerned the microscopic study of cells and tissues. Improvements in the microscope in the 1820s and 1830s, the rapidly developing cell theory after 1838, the spread of that doctrine by Johannes Müller's circle of students, and new staining and fixing techniques in the succeeding decades all contributed to making the study of cells and tissues one of the hottest areas of all biological research in the 1840s and after.

In the late 1830s and the 1840s, those who conducted research on cells and tissues tended not to distinguish sharply between its anatomical and physiological aspects, consistent with both the dominant intellectual paradigm linking the two and the institutional union of the subjects.[49] Theodor Schwann's cell theory, for example, invited researchers to consider the cell as the critical unit both for such functions as nutrition and respiration and for development. If cells really were the fundamental units of life, Schwann proposed, then one might well expect to find at the cellular level the essential organic activities and the laws by which they operated. At the same time, he sought to explore the relationship between cells as the basic organic elements

47. On this score, one might also add the name of Friedrich Schlemm (1797–1858), who was made second professor of anatomy at Berlin in 1833 when Johannes Müller was appointed director of the anatomy institute. Schlemm, who had worked as a military surgeon before gaining a university education, published at least one article on mechanical anatomy: "Über die Verstärkungsbänder" (1853).

48. Bracegirdle reviews what little literature there is in *History of Microtechnique*, 2d ed. (1978).

49. Researchers concerned with structure and function at the microscopic level also claimed for their work new possibilities for pathological inquiry, that is, a new basis of medicine. Virchow's cellular pathology of the 1850s is the most prominent historically, but even before Virchow, Henle had included the study of tissues in his program for a "rational medicine," a program he pursued as coeditor with Carl Pfeufer of the *Zeitschrift für rationelle Medizin* beginning in 1844.

and the "secondary" elements, tissues and organs—a project concerned with development.[50]

Despite the broad excitement generated by the cell theory and its apparent promise for both anatomical and physiological research, however, anatomical and physiological approaches to the study of cells and tissues were already becoming distinct by the early 1850s. One can see this by comparing Jakob Henle's pathbreaking histological handbook of 1841, *Allgemeine Anatomie*, with the 1852 textbook of his former student Albert von Kölliker, which became the new standard for the field. Although Henle emphasized anatomy in the title, he still found it necessary to devote considerable space not only to the forms and general physiological function of tissues but also to their chemical composition and interactions. The first hundred pages of this thousand-page textbook, on chemical aspects of tissues, Henle derived from the handbooks of the chemists J. J. Berzelius, C. J. Loewig, and F. Simon, reflecting both his sense that it was necessary to cover this subject and his lack of expertise in the area.[51]

By the time Albert von Kölliker wrote his textbook of histology in 1852, the perceived obstacles to understanding the chemistry and physiological functioning of cells and tissues had grown more complex. He argued that in the current state of knowledge it was possible to construct a coherent account only of tissue anatomy; the time was not yet ripe for a comprehensive doctrine of tissues that would include their chemical composition and functioning as well. In his textbook, therefore, the latter would "come into the question only insofar as they are connected to the origin of the forms and their variety."[52] As this comment hints, the orientation of Kölliker's textbook, like that of his research, was developmental; for him the central question was to understand how the different tissues emerged out of the fertilized egg cell.

Kölliker's approach suggests a divergence between anatomical and physiological approaches to the study of tissues that would emerge more clearly in the two decades after 1850. As anatomists and physiologists gradually gained disciplinary autonomy and separate physical locations, both groups supported research on tissues, but increasingly microscopic anatomists confined their studies to developmental questions—questions that the physicalist physiologists viewed as outside

50. T. Schwann, *Mikroskopische Untersuchungen* (1839); on Schwann, see M. Florkin, "Schwann, Theodor," and literature cited in that article.

51. J. Henle, *Allgemeine Anatomie* (1841).

52. A. von Kölliker, *Handbuch* (1852), p. 3.

the scope of their own research. Research on development, long considered a physiological category as one of the life activities of the organism, became transferred almost entirely over to anatomists.

This transfer took place because it served the interests of both physicalist physiologists and microscopic anatomists. The physicalist physiologists rejected laws of development as, at best, lacking in explanatory power; at worst, these laws might open the door through which the despised *Bildungstrieb,* the developmental force associated with vitalism, might enter science. In addition, as Ludwig argued in 1856, physiologists lacked the empirical basis for an appropriately physicochemical account of development. "Until more or less clear suggestions exist as to how the primitive forms of the developing animal or organ contribute to the emergence of the secondary structures," Ludwig asserted, physiologists would leave the study of development to the "pure" anatomists.[53]

Anatomists accepted this task partly because it offered them a new arena for their traditional mission of discovering and describing new anatomical structures. The old task of identifying the macroscopic structures of the human body seemed pretty well complete; now microscopic anatomy offered a vast new territory, for tissues needed identification at every stage of development, and the emergence of tissues from undifferentiated cells provided the prime vehicle for pursuing the age-old problem of individuation. This explorer's motivation for delving into microscopic anatomy and development is not insignificant, for by a long and powerful tradition, anatomy was a science of empirical discovery, and a claim to immortality could be made by getting one's name associated with a new anatomical structure.

Important as it was for individual anatomists' careers and their day-to-day research, however, this sort of empirical investigation was not usually considered the aim of microscopic anatomy as a *Wissenschaft.* Most anatomists viewed their studies of cell and tissue development as contributing to a broader set of developmental laws concerning anatomical structures. As Kölliker argued in his 1852 histology textbook, to bring their study to the level of a *Wissenschaft* its practitioners needed to aim at deriving the laws by which the tissues "originated, developed and finally reached their permanent form." To do this, one would seek to follow all the usual rules of an *Erfahrungswissenschaft,* that is, "to separate ever further from the whole sum of

53. Ludwig, *Lehrbuch,* vol. 2, p. vi.

individual facts and phenomena the accidental from the ever-present, the unsubstantial from the substantial, until gradually a series of more general and most general statements of experience [*Erfahrungssätze*] emerge, from which then finally mathematical expressions or formulae will be derived, at which point the laws have been found." Unfortunately, Kölliker continued, so far histology has yet to derive "one single law"; the empirical knowledge from which the law would emerge was still so scanty that at the present one could not even attain the level of general statements. While he was optimistic that the basic information for human microscopic anatomy and development would soon be found, a more general set of laws would require comparison with other organisms, and this would require "not years but decades" of new empirical research.[54] Thus both from the physicalist physiologist's perspective and that of the microscopic anatomist, the immediate goal for microscopic anatomy in the 1850s was to establish better the "facts" of tissue development and differentiation, from which someday more general laws of development (whether physicochemical or morphological) would be derived.

As represented by the two leading orientations of mechanical and microscopic anatomy, then, anatomists differed in their positions concerning the relationship of anatomy to physiology and the sort of work anatomists should be doing. The mechanical anatomists, dealing mainly with structures whose basic topographical description was well established, were concerned to work out in detail the effects of physical forces on the shapes and functioning of human organs; they were following a physiological approach whose spirit was consonant with physicalist physiology. Microscopic anatomists in the long run were seeking general developmental laws of form, but they were ambiguous about whether or not they viewed development as a physiological process or as a set of anatomical stages, the laws of which were distinct from physiological function. In either case, in the short run they were most concerned to identify the relationship between earlier cells and later tissues.

Where did morphology fit into all this? Few anatomists in the 1850s employed the term, perhaps because in the context of the wrangle with the physicalist physiologists, using it may have been considered too polemical. Yet one can certainly find a few professors of anatomy conducting research that was morphological in the sense of seeking broad laws of form that were developmental in character—

54. Kölliker, *Handbuch*, pp. 3–4.

research that was"physiological" in the older sense of physiology. Friedrich Arnold, professor of anatomy and physiology at Heidelberg from 1852 until 1858, when he gave up physiology, described his approach as "genetic-physiological" in 1844 and as "physiological" in 1873. But he clearly adhered to the older, morphological meaning of physiology shared by his contemporaries Wagner, Müller, and Siebold, claiming that the physiological approach strives "to view the human body and its parts from a higher morphological standpoint."[55] Karl Reichert, who adhered more closely to the approach of his older colleagues than his younger ones, refocused his work onto the nature of the protoplasm, in opposition to the cell theory. Both of these men opposed the view that reduced the whole organism to an aggregation of activities at the cellular level, claiming instead that the organism as a whole governed the activities of its parts, and both saw in development an overriding causal explanation of form.[56]

The most influential anatomist to call his approach "morphological" was Albert von Kölliker, who, as we saw in the previous section, ran a broadly based empire in anatomy and physiology in Würzburg's prominent medical faculty. In 1861, he was teaching students in his physiology course that the "physiology of the individual" was divided into "morphology" and "physiology in the narrower sense, the doctrine of the activities and functions of the human organism." "Morphology," he taught, was the same as "*wissenschaftliche* anatomy, or the doctrine of the conditions of form of the human body; that is, . . . not the usual descriptive anatomy, rather formation should be understood in its laws; embryology, histology and the so-called anatomy of organs and systems belong to it; it is to be distinguished from the usual systematic anatomy, which only gives us a synopsis [*Zusammenstellung*] of the facts . . .; *wissenschaftliche* anatomy [that is, morphology] is a true *Wissenschaft,* which is more than the usual anatomy, which is only so to speak a dictionary."[57]

Long after he stopped teaching physiology in 1865, Kölliker would continue to hold "morphology" synonymous with *wissenschaftliche*

55. Arnold, *Handbuch* (1844), pp. iii–iv; "Aeußerung über den Commissions-bericht betr. die Besetzung der an der Universität Heidelberg erledigten Lehrkanzel der Anatomie" (8 May 1873), p. 3 (unpaginated in original), BGLA, file 235/29858.

56. See, e.g., Reichert, "Über die neueren Reformen" (1863); in addition, Alexander Ecker, in his 1886 memoirs, recalled that at Freiburg in 1857 he was pleased to trade his responsibilities for physiology and zoology for those of anatomy and comparative anatomy so that he could return to morphological studies, but it is not clear whether he would have described his research in those categories at the time (A. Ecker, *Hundert Jahre* [1886], p. 115).

57. F. Sidler, physiology lecture notes (Würzburg, 1861), Medhist. Inst. Zürich.

anatomy. As late as 1883 he would argue that, in contrast to systematic or purely descriptive anatomy, *wissenschaftliche* anatomy had as its task the understanding of the lawful origin and transformation of living forms, and "this anatomy . . . should be the end goal of every morphologist."[58]

The advocates of both the major orientations described here, mechanical anatomy and microscopic anatomy (whether explicitly tied to morphology or not), set out the same goal for making anatomy *wissenschaftlich:* to find and describe the natural laws underlying the formation of the body. But in doing so, they proceeded from varying assumptions about the nature of those laws and the level at which one should look for them. Meyer and the mechanical anatomists rejected the notion that development provided a true explanation of organic form. Instead, they believed that the laws of form were causal-mechanical ones operating on the body as it functioned. The "laws" of form they were seeking were couched in terms of physical causes, such as the effect of muscle torsion on the shape of bones. The microscopic anatomists generally agreed that development provided a source of explanation, but among them there was a range of opinion about the extent to which the cell should be considered an autonomous unit, and the extent to which its activities were controlled by the whole organism.

Even among those few microscopic anatomists whose work was identifiably aimed at searching for broad, nonreductionist laws of form, there were differences. Although Kölliker's view of morphology resembled that of Reichert and Arnold insofar as its orientation was developmental and it was seeking "higher" laws of form, it differed in its claims. As represented by his discussion of laws in his histology text, Kölliker held a much more modest notion of natural law than did Reichert and Arnold, seeing it as the ultimate level of descriptive generalization rather than the ultimate cause. Reichert and Arnold's view of cause differed in turn from that of the mechanical anatomists in that it did not rely solely on physicomechanical causes but also included a vital cause, the Idea that provided organization for the organism as a whole.

As the brief survey presented in this section suggests, the separation of physiology from anatomy did not immediately lead to a clear division of anatomical and physiological problems or to a unified vision of anatomy's mission. Mechanical anatomists continued to view their work as speaking to both structure and function and to both anatomi-

58. Kölliker, *Die Aufgaben der anatomischen Institute* (1884), p. 15.

cal specificity and physiological generalization. And a few older morphologists refused to separate their task from their conception of
physiology. But on the one issue of development, there was consensus:
this was now in the realm of anatomy, not physiology as it was conceived of by the physicalist physiologists. The majority of the discipline was engaged in the project Kölliker outlined in his textbook,
namely, the study of the development of tissues and organs out of
cells. Despite Kölliker's own linguistic preferences, however, in the
1850s and 1860s "morphology" did not become the catchword for a
newly autonomous discipline of anatomy.

## ZOOLOGY: FROM NATURAL HISTORY TO MORPHOLOGY

Anatomy was not the only discipline whose intellectual and institutional features were renegotiated as physiology gained institutional
independence and came to be centered around the problems of the
physicalist school. Simultaneous with the physiological "revolution"
in university medical faculties, a parallel transformation was taking
place in the philosophical faculty: zoology, too, was gaining institutional autonomy and developing a new scientific program. As with
the case of embryology in anatomy, certain problems and approaches
that had been common to zoology and physiology now acquired the
label "zoological." The morphological areas of comparative anatomy,
development and generation, particularly of invertebrates, were
touted as the foundation of a new program of "scientific" zoology,
appropriate to a newly independent discipline.

In the structure of scientific knowledge institutionalized by German-speaking universities in the first half of the nineteenth century,
zoology had rarely been considered sufficiently autonomous to merit
a separate professorship. Of the nineteen universities under study
here, only five—the Prussian universities of Berlin, Breslau, and Halle,
the Bavarian university at Würzburg, and the University of Heidelberg
in Baden—established full professorships of zoology in the philosophical faculty before 1840 (see table 3.1). Elsewhere it was either appended to comparative anatomy and physiology in the medical faculty
or considered part of general natural history in the philosophical faculty. Full professors might teach zoology as an auxiliary subject, or
*ausserordentliche* professors might take it over as their main subject.
Only beginning in the late 1850s did the German universities as a
group move toward establishing independent full professorships in
zoology in their philosophical faculties. This shift may best be seen

TABLE 3.1 Foundation of Full Professorships in Zoology at German Universities

| YEAR | UNIVERSITY | NEW FULL PROFESSOR |
|------|-----------|-------------------|
| Full Professorship Created in Philosophical Faculty | | |
| 1811 | Berlin | Lichtenstein |
| 1811 | Breslau | Gravenhorst |
| 1816 | Halle | Nitzsch |
| 1830 | Würzburg | Leiblein |
| 1837 | Heidelberg | Bronn |
| 1846 | Leipzig | Pöppig |
| 1851 | Bonn | Troschel |
| 1855 | Giessen | Leuckart |
| 1860 | Königsberg | Zaddach |
| 1864 | Rostock | F. E. Schulze |
| 1865 | Jena | Haeckel |
| 1868 | Göttingen | Keferstein |
| 1868 | Kiel | Möbius |
| 1872 | Strassburg[a] | Schmidt |
| 1873 | Freiburg | Weismann |
| 1876 | Greifswald | Gerstaecker |
| Existing Full Professorship Transferred from Medical Faculty to Philosophical Faculty | | |
| 1863 | Marburg | Claus |
| 1874 | Erlangen | Selenka |
| 1885 | Munich | R. Hertwig |
| Existing Full Professorship Transferred from Medical Faculty to Faculty of Natural Sciences and Mathematics | | |
| 1863 | Tübingen | Leydig |

Source: Lynn Nyhart, "The Disciplinary Breakdown of German Morphology, 1870–1900," Isis 78 (1987):371.
[a]University itself founded in this year.

as part of the same splintering into disciplines that brought autonomy to physiology—but in zoology the separation brought researchers from the morphological tradition into collision with the zoologists allied with natural history.

The "Humboldtian" tradition of natural history that provided the main avenue for zoological research before mid-century did not separate zoology from other areas of inquiry but instead viewed animals, plants, and the earth itself as interacting pieces of a self-sustaining whole. Practitioners of this style of science sought to collect, name, and classify as many of these entities as possible, and to discover the laws underlying the geographical and historical patterns they presented.[59] This tradition was by no means exclusively tied to the universities; the early Humboldtian explorer-naturalists were usually men of means sufficient to finance their own collecting expeditions and were as likely to be affiliated with an academy of science as with a university. Thus Alexander von Humboldt himself never held a university position, although he lectured at the University of Berlin in 1827–28 and played an active role in sponsoring less financially fortunate naturalists in their professional pursuits.

The universities did offer some support for natural history, although research in this area was more oriented toward classifying the specimens brought home than toward the more expensive proposition of sending out expeditions. One can find formal appointments made in the first third of the century at some universities that named men "professor of natural history," and at others, individuals hired to teach botany, mineralogy, or zoology would offer courses in "general natural history."[60] Even at the universities where zoology gained independent professorships before 1850—Berlin, Breslau, Halle, Würzburg, and Leipzig—the natural history approach dominated. Eduard Pöppig (1798–1868), for example, spent most of the 1820s traveling in the Americas, where he recorded quantities of meteorological, hydrological, and temperature measurements while shipping thousands of plant and animal specimens back to his alma mater, the University of Leipzig. He returned to Germany to found the zoological museum at Leipzig, where he was named *ausserordentliche* professor of natural

59. The term "Humboldtian science" was defined by Susan Faye Cannon, who applied it mainly to British and American explorers, and who emphasized the physical sciences aspects of this style of science (S. F. Cannon, *Science in Culture* [1978], pp. 73–110); for a brief description of the natural history side and the German scientists who took part, see I. Jahn, R. Löther, and K. Senglaub, *Geschichte der Biologie* (1982), pp. 374–75. For a more detailed discussion of the dynamic approach to natural history, see T. F. DeJager, "Treviranus" (1991).

60. J. M. D. Herold was professor of natural history at Marburg from 1824 until 1862 (A. Geus, "Zoologie" [1978], pp. 167–73); Leipzig's professor of natural history, C. F. Schwägrichen, gave courses in "natural history of the three kingdoms" and botany from 1802 into the 1840s, retiring only in 1852; information on Schwägrichen is taken from Leipzig's course catalogues from the 1830s and 1840s and from *ADB*, s.v. "Schwägrichen, Christian Friedrich Gottfried." At Freiburg the medical doctor Karl Julius Perleb became full professor of natural history in the philosophical faculty in 1828; he ran both the *Naturalienkabinett* and the botanical garden until his death in 1846 (on Perleb, see Nauck, *Zur Vorgeschichte*, pp. 18, 30); H. G. Bronn was originally hired at Heidelberg as professor of applied natural history and zoology; he also held courses on geology and plant physiology (on Bronn, see Jahn, Löther, and Senglaub, *Geschichte der Biologie*, p. 642).

history in 1833 and full professor in 1846, and where he spent his career sorting specimens and building the collection.[61]

Zoology's subordination to natural history both reflected and perpetuated a lack of career opportunities for specialists in zoology. Within the university, to be sure, the introductory zoology course was required for medical students and in some states for teaching candidates. (The philosophical faculty, it will be recalled, existed mostly to train future *Gymnasium* teachers.) But few students in philosophical faculties concentrated on zoology for their advanced work or dissertation. On the one hand, one could study much the same material by concentrating on comparative anatomy in the medical faculty, and with a medical degree one had a guaranteed future as a practitioner if it proved impossible to find a position as a zoologist. On the other hand, if one was headed for *Gymnasium* teaching, neither zoology nor natural history qualified as major examination subjects: one had to concentrate in Greek, Latin, or math and physics. In Prussia, neither zoology nor even natural history was included as an auxiliary subject in the qualifying examinations for teaching candidates. This paralleled the situation in the schools, where in the entire *Gymnasium* curriculum too few hours of instruction existed to merit a full-time specialist in natural history. In fact, before 1850, natural history at the *Gymnasium* level was frequently taught by someone trained in theology or philology.[62] Generally, then, zoology was a low-status, low-priority subject, surviving in the university mainly as an educational requirement but lacking recognition as a true *Wissenschaft,* or even a distinct discipline.

Beginning in the late 1840s, however, new efforts were made to change zoology's lowly position. Just as the physicalist physiologists began by promoting their approach at the expense of the one then dominating, a group of zoological reformers, too, claimed to provide new, higher standards of science for their discipline. Significantly, they were mainly men from the morphological tradition within the medical faculties, and what they were bringing with them was the very same

61. See *ADB,* s.v. "Pöppig, Eduard;" and Jahn, Löther, and Senglaub, *Geschichte der Biologie,* p. 717.

62. Scheele, *Von Lüben bis Schmeil* (1981), pp. 43–45. Marsha Richmond has recently conducted some important research on the development of teacher education in Bavaria and its relation to the development of zoology. She points out that beginning in the mid-1850s, demand increased for science teachers, including those educated in the natural historical sciences (M. Richmond, "From Natural History Cabinet to Zoological Institute" [1991]).

broad physiological program rejected by the physicalist physiologists.

True to German academic tradition, the self-proclaimed vanguard of the "new" zoology mobilized by founding a journal. The *Zeitschrift für wissenschaftliche Zoologie* was designed to challenge the collect-and-classify orientation of traditional zoology. As initially planned in 1847, the journal was to include both "scientific" botany and zoology, and was to be coedited by Carl Theodor von Siebold, Albert von Kölliker, and the botanists Alexander Braun and Carl Naegeli. Siebold and Braun were colleagues at Freiburg at the time; Kölliker and Naegeli were at Zürich. In early 1848, Siebold and Braun drafted a prospectus for the new journal that so clearly states their intentions that it is worth translating at length.

We desire to give our journal the most scientific character possible, . . . in the objective sense of real scientific research, through the most comprehensive and pure [*geläuterte*] presentation of the facts, of their lawful determination and of their causal connection. To this purpose we exclude all announcements of new genera and species that do not relate to this task, unless these offer us a more thorough-going insight into plant and animal construction [*Bau*], into the life-history of animals and plants, or into the lawful organization of the organic realms. For the same reason we will exclude any kind of simple notes and natural history news, as well as everything concerning medical, economic, agricultural, forestry, and gardening practice, to the extent that it does not also offer a particular scientific point of contact, for example to anatomy and physiology. On the other hand, from the truly scientific side of botany and zoology nothing will be excluded; but we will especially make a point of enriching and promoting those parts of the science that are currently in need of cultivation, namely, the disciplines known as morphology, comparative anatomy and histology; and the developmental history of the organic structure of the plants and animals, which binds together and gives rise to all of the others; as well as the investigation, bordering on these, of the physical laws of life phenomena, of the physiology of plants and animals.[63]

63. Quoted in E. Ehlers, "Siebold" (1885), p. xiii. The prospectus seems to have been modeled on that of the *Archiv für Naturgeschichte*, originally published under the editorship of A. F. A. Wiegmann, which was founded in 1835 and continued publication until 1926. The editors of the *Zeitschrift* merely inverted the language of the *Archiv*'s prospectus to describe their own emphasis. Compare the passage quoted above with this one from the *Archiv*: "Essays in descriptive zoology, descriptions of new genera and species, reports on the mental abilities, mode of life, and geographical distribution of already-known animal species, even anatomical [*zootomische*] reports, insofar as these justify or secure the systematic position of an animal or a whole group, will find a suitable place here" ("Prospectus" [1835]).

The botanical half of the journal never appeared, but the zoological side did, and quickly became one of the leading publishing outlets for zoology. The contents of the first issue are fairly representative of what the journal would look like: it contained two articles by Kölliker, one considering the scientific basis on which a group of single-celled creatures might be classed as animals, and the other concerning smooth muscle tissue; an article by Freiburg's physiologist Alexander Ecker (a close friend of Siebold), on the development of a group of nerves in the torpedo-fish; a study by Johann Joseph Scherer, professor of chemistry at Würzburg, entitled "Chemical researches on the amniotic fluid of man in various periods of its existence"; and a contribution from Siebold on parthenogenesis and alternation of generations in the butterfly.[64] Over the following years, contributions from Kölliker, Siebold, and their many students would form the bulk of the *Zeitschrift,* but it was by no means confined to them. Developmental studies of lower organisms and the histology of more highly organized animals would continue to dominate its pages. Classification was not ignored, but the emphasis was on what they considered the scientific basis of classification, namely, development and comparative anatomy. This was especially important for those forms that underwent changes of form in their life cycle—a topic which Siebold, Kölliker, and their scientific comrade Rudolf Leuckart all investigated in detail.[65]

Now, of course, this claim to novelty in classificatory method was not entirely justified; since the time of Cuvier, comparative anatomy of internal parts had been an important tool for classifiers, and the work of von Baer and others in the 1820s and 1830s had firmly established development as part of the zoological canon. But the number of claims concerning the significance of development for classification shot up dramatically in the 1840s and 1850s, for a number of reasons. In this period, researchers began to focus their attention on marine invertebrates, and as their strange patterns of generation and development became more widely recognized, comparative developmental studies became indispensable. Thus the new material for investigation itself contributed essentially to the increased importance of morphological and developmental studies in classification. In addition, these claims were coming from a group with a new interest in ascribing classificatory significance to morphological subjects. Since

64. *Zeitschrift für wissenschaftliche Zoologie* 1 (1848), Heft 1.

65. Siebold, "Über den Generationswechsel der Cestoden" (1850), esp. pp. 198–99.

medical school physiology was increasingly becoming focused on non-morphological research, those who studied these subjects needed to justify their interest in new terms. It was remarkably convenient for them to claim that their special study provided an ingredient necessary for classification, one of zoology's main missions.

Kölliker and Siebold sought something more, however: they promoted comparative anatomy and embryology not simply as better tools for classification, but as means to understanding the laws of form. Classification was a necessary part of zoology, they acknowledged, but it need not be its only task and should not be its main one. In this view they were joined by a number of zoologists of Kölliker's generation, including Johann Friedrich Will (1815–68), who had worked under Siebold and Rudolph Wagner before becoming full professor of zoology (in the medical faculty) at Erlangen in 1848, and Franz Leydig (1821–1908), a student and colleague of Kölliker who became full professor at Tübingen in 1855. Also important to the development of the new zoology was Julius Victor Carus (1823–1903), who had worked with both Siebold and Kölliker and who, though he was never sufficiently interested in independent research to earn a full professorship, was instrumental in promoting morphology in his capacities as historian, translater, editor, and general observer of the zoological scene.[66]

The most important morphologist to work as a professor of zoology, however, was Rudolf Leuckart (1822–98). Indeed, by the early 1860s, he had become one of zoology's leading scientific lights; anyone who wanted a good introduction to state-of-the-art zoology went to Giessen to study with him.[67] A student of Rudolph Wagner and for several years his prosector at Göttingen, Leuckart made a name for himself in 1848 at the age of 26 with his *Über die Morphologie und Verwandtschaftsverhältnisse der wirbellosen Thiere* (On the morphology and conditions of relationship of the invertebrates), in which he definitively restructured the classification of the invertebrates

66. Unlike the others, however, Carus drew a sharp distinction between the studies of form and function, something Leuckart, at least, objected to strongly (see J. V. Carus, "Einleitung," in *System der thierischen Morphologie* [1853], pp. 3–36); Leuckart, "Bericht 1848–1853" (1854); on Will, see H.-J. Stammer, "Kürzer Abriß" (1956), p. 29; also, *ADB*, s.v. "Will, Friedrich"; on Leydig, see P. Glees, "Leydig, Franz"; and M. Nußbaum, 'Franz Leydig" (1908); on Carus, see G. Robinson, "Carus, Julius Victor."

67. K. Wunderlich, *Leuckart* (1978), an informative, full-length biography, p. 20; see also A. Weismann, "Erinnerungen," typescript autobiography, p. 38. I am grateful to Professor Karl Sander of the University of Freiburg for making a copy available to me.

on the basis of his morphological research. He dissolved Cuvier's grab-bag class of the radiates and replaced it with two classes, the coelenterates and the echinoderms. Its major achievement was a taxonomic one, yet it demonstrated the power of careful comparative studies of internal form and development for understanding the invertebrates.[68]

With this book, Leuckart was drawn into the wrangle between the morphologists and physicalist physiologists. Although hailed by classifiers as a major innovation, the book called down the derision of Carl Ludwig, who published a critique of it in one of the leading review journals of the day. The reorganization of classes interested Ludwig not at all; in his view such a rearrangement could not constitute a substantial contribution to science, since it was based solely on morphology. To Ludwig, before one could reasonably ascertain the relationship of animals to one another, one must know their "elementary functions." Since these were not yet known, any attempt to ask "through what circumstances the animals present on the earth came into being and what the connections are between them, . . . in a word, how they must be systematically ordered, . . . is folly." Earlier in the century, he went on, even the leading scientists had presumed the world to be organized by ideal principles and vital forces, and then it had been legitimate to organize the animal world solely according to structural types. But those days were long gone, and with the present knowledge and understanding of anatomy and physiology, such a pursuit was unjustified, "a scientific or artistic dalliance." He closed the review by saying that it "would be a good sign for German science if the book found no readers."[69]

Leuckart responded with a notice in Kölliker and Siebold's journal entitled *"Ist die Morphologie denn wirklich so ganz unberechtigt?"* (Is morphology then really so totally unjustified?). Enclosing a letter of support from Martin Heinrich Rathke (1793–1860), one of the grand old men of morphology, Leuckart defended morphology on the basis of its history of scientific success, and pointed out that one did not, in fact, need to know the elementary functions in order to distinguish different plans of animal organization. Finally, echoing his elders, Leuckart said Ludwig's main mistake was that he assumed that "the physiological view excludes the morphological view, and viceversa. . . . In any case, it is still very much a question whether morphology and physiology are really different [*verschieden*]; but it is certain

68. R. Leuckart, *Über die Morphologie* (1848).
69. C. Ludwig, Review of Leuckart, *Über die Morphologie* (1849), pp. 342–43.

that both can exist next to each other without infringing on one another, that both are equally justified, yes even necessary, to supply complete insight, a complete understanding of animal form. Through morphology we receive insight into the schema of construction, physiology by contrast teaches us of the purpose [*Zweckmäßigkeit*], of the necessity of the special form."[70]

Leuckart demonstrated his own understanding of what physiology should be a few years later when he coauthored the *Anatomisch-physiologische Übersicht des Thierreichs* with Carl Bergmann (then a coworker with Leuckart in Wagner's institute in Göttingen). In contrast to Leuckart's 1848 book, this work promised to describe anatomical structures only insofar as they were relevant to understanding physiological processes. Nevertheless, each chapter was solidly grounded in anatomy, first describing the relevant structures before moving on to analyzing their functions in the body and how those functions were accomplished. Bergmann and Leuckart explicitly recognized that although the basic life functions were the same in all animals, the varying structural plans of different animal types dictated that the life functions be carried out in different ways.[71] As Leuckart phrased it in an article that appeared while he was working on the *Übersicht,* "manifestations of life and structure are to each other as the two sides of an equation."[72]

The *Übersicht des Thierreichs* and Leuckart's earlier works on morphology demonstrate clearly his understanding of what would constitute scientific zoology. It would unite classification with the studies of animal function and organization in the synthetic framework provided by functional morphology. It was precisely the same program, in fact, as that called for by Siebold and Kölliker in the *Zeitschrift für wissenschaftliche Zoologie.* All of these men saw in morphology the route to a new, higher science of zoology.

The sources and timing of this call for a new zoology are significant. Most of the leading proponents of scientific zoology did not originally hold independent chairs of zoology in university faculties. Instead, they had taught physiology in a medical faculty, together with zoology and comparative anatomy. Their turning to zoology was part of the larger splintering of disciplines in which physiology would be redefined from a general science of life to a science of animal function. The topics of development, generation, and form did not

70. R. Leuckart, "Ist die Morphologie" (1850), p. 273.
71. C. Bergmann and R. Leuckart, *Anatomisch-physiologische Übersicht* (1852), pp. 36–37.
72. R. Leuckart, "Bau der Insekten" (1851), p. 19.

just disappear from science when they dropped out of physiology; they were retained as part of scientific zoology.

The older advocates of scientific zoology taught in medical faculties, the home of physiology. Kölliker and Bergmann continued to teach in medical faculties throughout their careers, Kölliker at Würzburg and Bergmann at Rostock. Siebold remained a professor in both the medical and philosophical faculties at Munich, but from the mid-1850s narrowed his focus to zoology, trading his primary responsibility in physiology (in the medical faculty) for coverage of natural history (in the philosophical faculty).[73] J. F. Will held a full professorship of zoology at Erlangen beginning in 1848, but it, too, was in the medical faculty. Leuckart was the only one of these men to hold an independent professorship in zoology in a philosophical faculty, and he, too, had been trained in the medical school tradition of a broad-based physiology.

The scientific zoologists who made up the next generation were in a different position. With the attack on morphology by the physicalist physiologists and the separation of physiology from anatomy, comparative anatomy, and zoology, the institutional justification for their presence in the medical faculty diminished. Over the course of the 1850s and 1860s, the morphological subjects and approach at many universities were gradually transplanted from medical faculties to philosophical faculties. This occurred as the medically trained students of the early scientific zoologists moved over into faculties of philosophy to take newly independent positions in zoology. Leuckart's career set the pattern: he had received a medical degree under Wagner and had worked as a *Privatdozent* in the medical faculty at Göttingen before being hired into the philosophical faculty at Giessen, initially as *ausserordentliche* professor of zoology and comparative anatomy, and then as full professor in 1855. In the 1860s and 1870s, this pattern was repeated by a younger cohort of medically trained zoologists that included August Weismann (1834–1914), Ernst Haeckel (1834–1919), and Franz Eilhard Schulze (1840–1921).[74]

73. B. Hoppe, "Entwicklung der biologischen Fächer" (1972), p. 371.

74. August Weismann taught zoology and comparative anatomy in the medical faculty at Freiburg from 1865 until 1873 when, upon his promotion to full professor, he and his teaching duties were transferred to the philosophical faculty. At Jena, Ernst Haeckel was named *ausserordentliche* professor of zoology and comparative anatomy in the medical faculty in 1862 and moved to the philosophical faculty as full professor in 1865; at Rostock, Franz Eilhard Schulze first taught comparative anatomy as *ausserordentliche* professor in the medical faculty before being named full professor of zoology in the philosophical faculty in 1871. All of these scientists shared the spirit of the new age of zoology, in which morphological physiology rather than classification would

The movement of medically educated morphologists into zoology professorships was not universal. To be sure, at universities where zoology's first full professorship in the philosophical faculty was created after 1850, the chair was most often given to someone from the medical tradition of morphology rather than that of pure systematics; this helped shift the balance in the profession toward morphology. But at several universities, zoology was already established as an independent professorship in the philosophical faculty before the morphologists called for a new scientific zoology (see table 3.1). At these universities, where the state had frequently already committed considerable resources to developing a museum, there was some resistance to the invasion by morphologists.

This was especially the case in Prussia, where nearly all of the universities faced a turnover in their zoology professorships in the late 1840s or 1850s. At Bonn, when Goldfuss died in 1848, the morphologist Rathke was placed on the list but the first choice was the natural historian Franz Hermann Troschel (1810–82) who "as a zoologist in the narrower sense of the word would better meet the needs of our university."[75] At Berlin, zoology revolved entirely around the museum; this preserved the orientation toward systematics long after other universities had given themselves over to the morphological approach. In fact, when the systematist Martin Heinrich Lichtenstein died in 1857, the faculty recommended hiring Siebold as the best all-around zoologist alive, but the minister chose instead Lichtenstein's right-hand man, the systematist Wilhelm Peters (1815–83).[76] Systematists were also chosen at Breslau in 1856 and at Königsberg in 1860.[77]

---

be the main thrust of the discipline; see Wunderlich, *Leuckart*, pp. 11–20; on Weismann, see Nauck, *Zur Vorgeschichte*, pp. 37ff; on Haeckel, see Uschmann, *Geschichte der Zoologie*, pp. 34–50; on Schulze, see K. Heider, "Gedächtnisrede" (1922), p. lxxxviii, n. 1. This was not a completely new phenomenon in the 1860s; many earlier professors of natural history also had received medical educations, and some of them were also morphologically inclined. Marburg's J. J. Herold, e.g., concentrated on invertebrate development as early as the 1810s. In addition, not all comparative anatomists moved over to zoology: the famous cytologist Max Schultze, e.g., was *ausserordentliche* professor of comparative anatomy at Halle before being named full professor of anatomy at Bonn in 1859.

75. Bonn Philosophische Fakultät to Minister der GUMA, 20 January 1849, GStA Merseburg, Bonn Zool. Prof., pp. 45–46.

76. Berlin Philosophische Fakultät to Raumer (Minister der GUMA), 17 December 1857, pp. 106–13; Raumer to King, 2 February 1858, p. 115, both GStA Merseburg Berlin Zool. Sammlung.

77. Adolf Eduard Grube (1812–80) was appointed at Breslau, Gustav Zaddach (1817–81) at Königsberg.

At Halle, systematic zoology was defended in the strongest possible terms. In 1858 the systematist Gottfried Andreas Giebel (1820–81) was promoted from *Privatdozent* to *ausserordentlicher* professor for zoology partly on the recommendation of the dean of the philosophical faculty, who stated that Giebel's career should be promoted in order that the entire profession of zoology not become filled with "men who know all the changing conditions of the animal cell, but who are little practiced in the determination and description of species."[78] When Halle's full professor of zoology, the Humboldtian naturalist Hermann Burmeister (1807–92), departed for Buenos Aires three years later, he, too, recommended that Giebel succeed him so that the zoological museum that he had built up would be well taken care of. "On these grounds I am against filling the vacancy through a zoologist from the modern physiological-morphological school, as it is represented by the names von Siebold, Leuckart, Leydig, Gegenbaur, Leiberkühn, Meissner, and others. Men of this sort would neglect my singular creation, because they do not perceive its worth. . . . I therefore warn the faculty frankly and emphatically against such recommendations; it runs the risk that the university would lose the valuable thing it already has in exchange for that which is in itself worthless."[79] Giebel was named to succeed Burmeister and held the post until his death in 1881, when he was succeeded by the morphologist Hermann Grenacher.[80]

Despite this last instance, the replacement of systematists by morphologists took place with considerably less rhetoric and dissension than did the ejection of morphology from physiology. Partly this was because the transition was spread out over a longer time and was never universal. Furthermore, by the time most of the new zoology professorships were created in the 1860s and later, the distinction between morphologists and systematists was becoming increasingly blurred. Few zoologists were like Giebel, for whom the morphological researches of the 1830s and 1840s had "passed by without leaving a trace."[81] In fact, a Humboldtian naturalist educated outside the medical tradition of morphology, like H. G. Bronn, could make major contributions to morphology. Nor did the functional morphologists reject classification as a legitimate goal for zoology: Leuckart himself is credited with naming 164 new zoological taxa over the course of

78. Decan Leo, quoted in O. Taschenberg, *Geschichte der Zoologie* (1894), pp. 96–97.

79. Burmeister, quoted in ibid., p. 98.

80. W. Schrader, *Geschichte der Friedrichs-Universität* (1894), p. 287.

81. Taschenberg, *Geschichte der Zoologie*, p. 101.

his long career.[82] Finally, by the time the last full professorships were created, the impact of Darwin's theory was being felt. The theory of evolution would offer a new rationale for merging classification and morphology and would finally join the two major strands of zoological practice into a single, coherent theoretical edifice.

## CONCLUSION: MORPHOLOGY IN THE DISCIPLINES

The institutional and intellectual breakup of the older clusters of subjects centered on physiology led to the widespread acceptance of the label "morphology" to mean the study of form, chiefly based on comparative anatomy, embryology, and sometimes microscopic anatomy, and its excision from the "modern" task of physiology. As one historian has aptly put it, "the opposed pair earlier was 'physiology-pathology'; now it was 'physiology-morphology.'"[83] But the connotations of morphology were slightly different in each of the disciplines newly separated by the breakup. Among physiologists, it signified the older approach of physiology, now rejected as outmoded by the younger generation of physicalists but still pursued by such men as Müller, Wagner, and Siebold. Among a few anatomists, but not the majority, it was associated with areas of anatomical subject matter— embryology, microscopic anatomy, and comparative anatomy—that still promised new empirical discoveries, and in their eyes it could confer *Wissenschaftlichkeit* on their discipline as well, by engaging them in the search for developmental laws of form. Finally, among zoologists, proponents of morphology—here virtually identical to the "older" physiology—touted it as the "new" approach that would bring *Wissenschaftlichkeit* to the hitherto lowly classificatory subject.

The splintering of anatomy, physiology, and zoology into disciplines with distinctive concepts of *Wissenschaft* did not happen instantaneously but took place over a good quarter-century, between 1848 and 1872. This timing is significant, for the programmatic and institutional shifts provided the framework for a third aspect of the history of morphology: the reception of Darwin's theory. As we will see in the following chapters, in the second half of that period, Darwin's theory added a new wrinkle to the claims for *Wissenschaftlichkeit* in the disciplines of anatomy and zoology; at the same time, Darwinism would create a new set of meanings for morphology.

82. Wunderlich, *Leuckart*, p. 59.
83. Koller, *Das Leben Johannes Müller* (1958), p. 138.

# PART TWO

*Evolutionary Morphology, 1860–1880*

## CHAPTER FOUR

## Descent and the Laws of Development

When Darwin's theory arrived in the German-speaking biological community in the early 1860s, it entered a diverse
scene in which philosophical differences over the study of
life were increasingly being channeled along disciplinary
lines. Although one can turn up reactions to the theory from members
of a wide variety of life-science disciplines, Darwin's work did not
engage scholars in all of them with equal intensity. In addition to
botanists, who are outside the scope of this study, the scientists who
became especially quickly and deeply engaged with Darwin's theory
in the first decade or so after its appearance were primarily those
interested in problems of animal form and form change. Most of these
men were professors and advanced students of zoology, although
given the still-changing disciplinary situation, a few were affiliated
with physiology and anatomy.[1]

In characterizing certain morphologically oriented zoologists as
those most closely engaged with Darwin's theory, I seek to avoid
dichotomizing the community into Darwinian and anti-Darwinian
camps. To begin with, few people either rejected or accepted what
Darwin said in its entirety, and even people sharing similar criticisms

---

1. The other group of German scientists to take particular interest in Darwin's
theory was made up of botanists. For most anatomists and physiologists in the 1860s,
by contrast, Darwin's theory was less central to their research concerns (see Montgomery, "Germany" [1974]); for more detail on German botanists' interests in Darwinism,
see E. Cittadino, *Nature as the Laboratory* (1990), and T. Junker, *Darwinismus und
Botanik* (1989).

of the theory might part company on considering themselves Darwinians. Nor is it especially helpful for understanding German preoccupations to take some features of the theory as constituting its transhistorical "essence" and examine the German responses to those particular features. Instead, I propose to examine the responses by those most closely engaged with Darwin's theory primarily to uncover the scientific problems and philosophical presuppositions they found most compelling.

Even among zoologists, the level of engagement with Darwin's theory varied. As we have seen, the community was split into two leading research orientations in the 1850s and 1860s: men following the two-pronged natural historical tradition of collecting and classifying information about life on earth and searching for the historical laws governing it; and those self-avowedly engaged in "scientific zoology," who viewed the study of individual development and generation as central to the task of the zoologist. Scholars of both orientations might have had reason to react to Darwin's theory, which challenged the typological view of species central to the principles of pre-Darwinian classification. It also opened up intriguing possibilities for connecting the apparent progressive development of life on earth with individual development. But as it happened, adherents to the two orientations did not respond to the theory with equal intensity. As we will see in the next chapter, the group that most vigorously took up the issues raised by Darwin's theory were the younger scientific zoologists, who reshaped their work into a variety of Darwinian approaches. Natural historians were less engaged, although there is an important exception to this rule. As far as can be determined, most men who came from the more strictly empirical end of classification—generally men who were already disposed to be hostile to scientific zoology because of its implication that what they were doing was not scientific—tended to dismiss Darwin's theory out of hand and not waste a lot of time on it. But at least one zoologist within the natural history orientation—H. G. Bronn at Heidelberg—was interested in theoretical problems concerning the relation of living things to their physical surroundings and the ways both changed together over time. Like many of the scientific zoologists, he found ample issues for discussion in Darwin's theory.

For understanding the role played by Darwinism in the history of morphology, analysis according to different research orientations is still insufficiently fine, for not all students of form and form change latched onto the same aspects of Darwin's theory or developed it in the same way. To make sense of these differences, a generational analysis is especially germane. Close examination from this perspective suggests that each different generational cohort held a particular bundle of scientific and philosophical commitments—commitments which, to a certain extent, transcended the divide between natural history and scientific zoology. Each of these generations had come into their field at a different moment in its intellectual and disciplinary history, and for each somewhat different issues were at stake.

When zoologists started discussing Darwin's theory in the 1860s, members of three distinct generations took up the problem (generations 2, 3, and 4 in table 1.1). By that time, none of Burdach's generation remained; they had all died by the early 1850s. The oldest important generational cluster, comprising scholars born around 1800, were nearing the end of their careers in the 1860s. For them, Darwin's theory represented only the most recent in a long series of dramatic proposals to redraw the shape of life science. They could look back to the time of their early training, when *Naturphilosophie* was still something to be reckoned with; as mature scholars during the previous two decades they had lived through and been engaged with the rise of the cell theory, the debate over vital forces, and the beginnings of the drastic reshaping of physiology that was just beginning to show permanent institutional effects. This cohort included some men who still held chairs in anatomy and physiology, such as Carl Theodor von Siebold and Rudolph Wagner, as well as representatives of the natural history tradition such as H. G. Bronn and Eduard Pöppig. Relatively few members of this generation invested a great deal of energy in discussing Darwin's theory, but those who did were listened to with respect.

A second cohort, whose births were clustered between 1815 and 1823, included the leading scientific zoologists. By the time these men received their university education (still almost exclusively in medical faculties) between the late 1830s and mid 1840s, *Naturphilosophie* was very much passé, even if a number of *Naturphilosophen* were still writing and teaching. For this middle generation, the cell theory, the rejection of vital forces, and the opening up of the territory of invertebrate generation and development were the most powerful developments shaping their scientific/theoretical worlds, while the gradual shifting of their subject (and themselves) out of university medical faculties and into philosophical faculties was undoubtedly the most significant factor shaping their professional identities.

Finally, there were those born in the first half of the 1830s, who were just establishing their careers when Darwin's theory appeared. Educated mainly by the scientific zoologists, these men concentrated their research on development and generation, especially among invertebrates, but also engaged in taxonomic classification, often using their developmental studies as aids. They comprised the first generation who could identify themselves throughout the duration of their careers as researchers and teachers exclusively of zoology in university philosophical faculties.

This chapter focuses on the reactions to Darwin's theory by one or two key members of each of these generations, with some comparison to others in the same cohort, in order to examine the philosophical assumptions revealed in their reactions to Darwin and to each other. As translator of the first two German editions of the *Origin,* H. G. Bronn shaped the interpretation most German scholars took away from the book. His older contemporary Karl Ernst von Baer (1792–1876) wrote one of the most detailed German-language responses to Darwin's theory, which commanded attention because of the high esteem in which von Baer had long been held. Albert von Kölliker, a leading scientific zoologist of the middle generation, was prominent not only by virtue of his many empirical discoveries and synthetic writings in embryology and histology but also by his position in Würzburg's eminent and large medical faculty, where he taught hundreds of students. His critical view of Darwin's theory was one of the most conservative of his generation; though others shared some of his criticisms, most scientific zoologists approved of Darwinism subject to certain modifications. Nevertheless, his preoccupations give us considerable insight into the leading concerns and the style of reasoning of that generation. Ernst Haeckel was the most visible member of the youngest generation, the scholar who most vocally sought to make himself the apostle of Darwin in Germany. As one of the first writers to interpret Darwin's theory at length, and one who did so in a highly inflammatory manner, Haeckel quickly became a lightning rod for discussions of Darwinism. After the publication of his *Generelle Morphologie* in 1866, almost any German scientific writer responding to Darwin also found himself responding to Haeckel. None of these men is strictly representative of his generation—if so, they would not have been so influential or original—but comparing their views of Darwin's theory can help us tease out some of the underlying assumptions their different cohorts held about their science.

At the outset it seems important to establish what these men took Darwin's theory to be. In reviews and summaries of the theory by German scientists, two basic elements were seen as central and were generally understood the same way. First (and especially significant for anyone concerned with classification), the theory linked the entire organic world through descent by means of generation and inheritance. Following Darwin, the "relatedness" of animal types no longer referred merely to similarities of form; rather, it now meant that genealogy or blood ties connected different animal varieties, species, and higher taxonomic groups to one another. He thereby created the possibility for the long-sought "natural system" of classification by interpreting classificatory relationships in terms of temporal distance from a common ancestor. Varieties of a single species had just begun to split apart; species in the same genus had split off recently in geological time; genera within one family had separated longer ago; and so on. In this way, the hierarchy of classification was transformed into a genealogical tree.

German readers agreed that the second central feature of Darwin's theory was his mechanism of natural selection. In contrast to older ideas about a developmental sequence in the pattern of life on earth, which were usually attributed simply to a "law" of development, Darwin provided a specific, material cause for the genealogical branching. Acting on the variations existing among all individuals, nature selected those individuals that held a slight advantage in the struggle for existence. Those favored individuals would pass on their characteristics to their offspring through inheritance, while less well endowed ones would die with fewer offspring or none at all. Darwin's theory held that as advantageous variations accumulated over time, new, distinct varieties would emerge that would eventually diverge further into new species. This process repeated itself throughout the organic world, so that the living world consisted of an ever-greater diversity of forms that one could (at least in theory) follow back in time to their joint ancestors.[2] Although some German readers exhibited some confusion about how natural selection worked, all agreed that it was the explanatory kernel of Darwin's theory.

Beyond this baseline of common understanding of Darwin's theory, views of what he said, its significance, its correctness, and its implications for zoological science diverged. However, the writers discussed here show certain common concerns, especially in regard to the conceptual categories by which they sought to explain the progressive change of life on earth: the role of conditions internal to the organism versus those external to it; the universal disapproval of accident as the basis for explanation; and, most especially, the use of analogies between individual development and the historical development of life on earth. At the same time, however, the meanings they attached

2. C. Darwin, *Origin of Species* (1964).

to such crucial terms as law, force, and cause differed, indicating a profound change across the generations in epistemological and metaphysical assumptions about what constituted appropriate explanation in their science.

## DEFENDERS OF GOAL-DIRECTEDNESS: HEINRICH GEORG BRONN AND KARL ERNST VON BAER

Heinrich Georg Bronn and Karl Ernst von Baer make an odd pair. Bronn, primarily known as a paleontologist and systematist, appears to have been considerably more sympathetic to *Naturphilosophie*— and to Darwin's ideas—than his older contemporary, von Baer, who is best known for his embryological work and has been considered an exemplar of anti-*naturphilosophisch* thought.[3] Nevertheless, in the context of Darwin's theory, their similarities emerge more strongly than their differences. Both were old men nearing the end of their lives when they wrote about it, and both found themselves defending a teleological view of nature that they saw being threatened.

The professor of natural history and zoology at the University of Heidelberg from 1837 until his death in 1862, H. G. Bronn is perhaps best known historically as the founder of *Bronn's Klassen und Ordnungen des Thierreichs* (Bronn's classes and orders of the animal kingdom), a gigantic project of classifying the entire animal world that is still producing new volumes today. In his own time, he was a leading paleontologist and editor of the important journal *Neues Jahrbuch für Mineralogie, Geognosie, Geologie und Petrefaktenkunde*. His own research included considerable work on classifying fossil animals and plants, and seems to fall very neatly into the categories of pure systematics and paleontology. But from the early 1840s on, he also wrote a number of synthetic works addressing the history of life on earth, works in which "laws of development" played an increasingly prominent role. These culminated in his prizewinning submission to the Paris Academy of Sciences in 1857, which was published in slightly revised form in German the following year under the title *Untersuchungen über die Entwickelungs-Gesetze der organischen Welt während der Bildungs-Zeit unserer Erd-Oberfläche* (Investigations into the developmental laws of the organic world during the period of development of our earth's surface), and a complementary work, also published in 1858, titled *Morphologische Studien über die Gestaltungs-Gesetze der Naturkörper* (Morphological studies on the

3. Lenoir, *Strategy of Life*, pp. 72–95.

laws of formation of natural bodies). Given his vast paleontological knowledge and interest in the history of life in relation to the changing physical environment of the earth, it is perhaps not surprising that he volunteered to supervise the German translation of Darwin's *Origin of Species*.[4]

Bronn's ideas about the laws of development exerted a powerful influence on subsequent German readers through his translation of Darwin's *Origin*. As translator of the first two German editions (based on the second and third English editions), which appeared in 1860 and 1862, he shaped the interpretation given to particular English terms. At the same time, his criticisms, appended in a fifteenth chapter and in text notes, received as wide an audience as the translation did. Both Karl Ernst von Baer and Ernst Haeckel, for example, read Bronn's translation, absorbing his vocabulary and his criticisms.[5]

One of Bronn's primary assumptions was already evident in the title of his translation: *Über die Entstehung der Arten im Thier- und Pflanzenreich durch natürliche Züchtung, oder Erhaltung der vervollkommneten Racen im Kampfe ums Dasein*[6] (On the origin of species in the animal and plant kingdom through natural selection, or the preservation of the more perfect races in the struggle for existence). To the English reader, the subtitle is especially striking. Darwin's subtitle reads "or the preservation of favored races in the struggle for life." The revealing word in Bronn's translation is "vervollkommneten," a term that does not translate as "favored," but rather as "improved," or "perfect."[7] When Darwin wrote of "favored races," he meant those that were slightly better adapted to their conditions of existence than their competitors. Naturally, these could be construed as improved or more perfect. But Bronn's interpretation had a peculiarly continental flavor to it. It smacked of Lamarck, Oken, and all the others who had understood the history of organic forms as an

---

4. Darwin to T. H. Huxley, 2 February 1860, *More Letters*, ed. F. Darwin and A. C. Seward, vol. 1, p. 139; for an excellent study of Bronn's relationship to Darwin and his understanding of Darwin's theory, see T. Junker, "Heinrich Georg Bronn" (1991).

5. On Haeckel, see I. Jahn, "Ernst Haeckel" (1985), p. 75. Von Baer cites Bronn's second edition and discusses some of Bronn's criticisms in "Über Darwins Lehre" (1876), pp. 272, 286–88.

6. Darwin, *Über die Entstehung* (1860).

7. Thomas Junker has also called attention to this odd translation ("Bronn," p. 201). When a new translation by Julius Victor Carus appeared in 1867, after Bronn's death, the word "Züchtung" (cultivation) was replaced by "Züchtwahl" (selection) and "vervollkommneten" (perfected) was replaced by "begünstigsten," a term much closer to Darwin's meaning of "favored."

ascent to perfection. To Bronn, as we will see, "favored races" could *only* be those that were more perfect, as judged by some criteria outside of their success in the struggle for existence. They were more perfect from an absolute point of view, as a consequence of the law of progressive development.

Bronn's theoretical commitments also emerge in the critical essay he appended as chapter 15 of the German translation. His doubts about the theory centered on two major objections. First, like many of his contemporaries, he did not think it reasonable that varieties would branch off into distinct groups while intermediate forms died out. If variations were so important, it would follow far more plausibly from the theory that a chaos of forms would emerge; since this did not appear in nature, the theory was deeply flawed. Second, Darwin assumed only one or a few original creations, without giving a naturalistic explanation for the origin of life. This seemed no more scientific to Bronn than positing many creations, so long as one had to rely on the hand of the Creator—a dependence he thought unscientific. Despite these profound defects, however, Bronn thought the book worth translating because it came closer than any previous theory to providing a single law of progressive development for the entire organic world. In fact, as Thomas Junker has pointed out, although Bronn could not bring himself to accept transformation, "this theme fascinated him," and despite his criticisms he exerted himself strenuously to make the *Origin* accessible, through translation, to German readers.[8] Bronn himself likened Darwin's theory to "the fertilized egg, out of which the truth will gradually develop; it is perhaps the chrysalis, out of which the long-sought law of nature will unfold."[9]

On the face of it, Bronn's objections seem straightforward enough. But he also listed numerous other possible objections to which he assigned less weight.[10] Why these two held particular significance for him becomes more evident in light of his own theory of organic development, which he had presented the year before.

Bronn had summarized his scheme in a speech at the 1858 meeting of German Naturalists and Physicians. Here he explained that there were two great laws of the natural world, encompassing both organic and inorganic nature. First there was the "law of progressive development" (i.e., all things in nature develop over time), a law that was as

8. Ibid., p. 189.

9. Darwin, *Über die Entstehung*, p. 518. Bronn's earlier review of the *Origin* focused primarily on the latter criticism (see D. L. Hull, *Darwin and His Critics* [1983], pp. 120–24).

10. For an analysis of the range of Bronn's objections, see Junker, "Bronn."

true for the planet earth as it was for the plants and animals residing on it; and second, there was a "law of adaptation" that stated that organisms were adapted to their external conditions of existence. As these two laws operated together, the earth's development guided and limited the development of living forms. "Now, since these conditions of existence were continually changing as a consequence of the progressive formation [*Bildung*] of the earth's crust, so too was the character of the population of the earth subject to a continual change. And as the conditions of existence accordingly perfected themselves ever more, gradually some higher animal forms could find a livelihood, which would have been impossible earlier."[11] According to Bronn, species were not transformed through inherited modifications; instead, each species had "its own time of origin, its own duration, and its own life's end." Nevertheless, the natural world as a whole displayed a progressive development toward a "higher" existence, as less perfect species were gradually replaced by more perfect ones over the course of geological time. Furthermore, the climb toward perfection constituted "a progression . . . from one starting-point to a particular goal, following a uniformly adhered-to plan." One aspect of this development to perfection was differentiation, a development from homogeneity to heterogeneity. Here too, the conditions of existence defined the limits to variation. Because the temperature of the early earth was more uniform over its entire surface than it is today, Bronn wrote, the earth's population must also have been more uniform.[12]

Bronn's synopsis of his arguments reveals his basic concerns. As a researcher in paleontology, he sought to account for the broadly progressive appearance of animal forms in the geological record, while as a systematist, it was important for him to retain constancy of species and higher taxonomic categories. His solution solved both his problems fairly neatly: each species had a limited life span, and upon its death it would be replaced by another, more perfect species. As a solution to the more general problem of elucidating the laws of form, it posited two principles in some tension with each other. The law of progressive development could only be understood as a generative principle, a continual push toward the goal of perfection. The law of adaptation was the conservative principle limiting the realization of progress. That is, if organisms had to be adapted to their environment, then the state of the environment acted as a constraint on what forms were possible. The actual result found in the geological record de-

11. Bronn, "Über die Entwicklung der organischen Schöpfung" (1859), p. 30.
12. Ibid.

pended on the dynamic interaction of the two laws. This formulation was typical of Germans concerned with the history of life in the nineteenth century.[13]

The character of these laws offers an important clue to understanding Bronn's reaction to Darwin's theory. At least one historian has argued that Bronn's laws were strictly phenomenological ones, derived ostensibly by generalizing from experience, and indeed, Bronn frequently repeated that his laws were entirely grounded in observation and experience.[14] Nevertheless, it seems clear that these laws had greater explanatory power than one might legitimately adduce to mere descriptive generalizations. Not only do they help explain the regularities and differences among organic forms, but they seem to be able to *act:* "the laws of progression and adaptation *bind* the [taxonomic] circles with one another, *make them vary* internally, *raise* one above the other," and assert themselves throughout the taxonomic levels.[15] The source of this greater explanatory power lies in what Bronn understands these laws, especially the law of progressive development, to stand for: the law of progressive development is the phenomenal expression of an as yet undescribed "creative force" (*Schöpfungs-Kraft*) analogous to other forces in nature such as gravity or the forces of chemical affinity.[16] Whenever Bronn invokes the law of progressive development, then, he is also invoking this underlying force. Bronn did not view himself as using a mystical vital force to cover up his ignorance of the material causes of organic change; he shared the view of most physical scientists and many life scientists of his generation and older that forces *were* the causes of phenomena.[17] The ordered and progressive character of the history of life did not seem to Bronn reducible to existing physical and chemical forces, but to be scientific, one needed something analogous to them, a force that was as real as physical and chemical forces but which accounted for both

13. See DeJager, "Treviranus," esp. chap. 6. The situation in France was quite different: Cuvier agreed that the need to conform to the conditions of existence was a conservative force, but did not oppose to it an inner drive toward perfection; Lamarck saw the conditions of existence as providing opportunities for expression of the drive to perfection, rather than as constraining them (on Cuvier, see W. Coleman, *Georges Cuvier Zoologist* [1964]; on Lamarck, see R. Burkhardt, *Spirit of System* [1977]).

14. M. Rudwick, *Meaning of Fossils* (1985), pp. 225–26; see, e.g., Bronn, *Untersuchungen* pp. 85–86.

15. Bronn, *Morphologische Studien* pp. 81–82, quotation on p. 109 (emphasis added).

16. Bronn, *Untersuchungen*, pp. 77–82.

17. On the equation of forces with causes in physics, see Caneva, *Robert Mayer,* pp. 160–65; on the biological tradition attributing the history of life on earth to interacting forces, see DeJager, "Treviranus," esp. chap. 6.

progressive change and the simultaneous maintenance of species boundaries. Only a "creative" force seemed to be able to provide that cause. Its essential nature might be unknown, but then so was that of gravity.

It is Bronn's commitment to a creative force that accounts for the importance he assigned to his two leading objections to Darwin's theory. For him it was evident that there was order to the organic world, as expressed in the distinctiveness and fixity of species. Not only did Darwin's mechanism not account well for the distinctiveness of species from one another (something believed by most systematists to be empirically observable), but if its logic were followed out consequentially, in Bronn's view, one would have to accept, and indeed look for, taxonomic chaos. The creative force, however, had the power to maintain species while gradually replacing less perfect types with more perfect ones. How, precisely, it did that must remain unknown, but the phenomena seemed to him to require a force with that power.

His second objection shows even more clearly the crucial role played by the creative force in his philosophical framework: the underlying problem was not that Darwin did not adequately justify having one Creation over many, but rather that having recourse to divine intervention at all, even in one case, was profoundly unscientific. Although Bronn certainly believed that God had created the laws of nature,[18] he required an explanation wholly in naturalistic terms: invoking God at any point as a direct cause was unacceptable. But it is clear from his earlier work that he found the frequently suggested materialistic alternative of spontaneous generation equally untenable. The grounds of this latter rejection are professedly empirical—in his view spontaneous generation had been experimentally disproven by Ehrenberg, Schwann, and others. But Bronn also associated spontaneous generation (and/or an initial Creation) with arguments in favor of species transmutation, which he rejected,[19] and it seems likely that his inclination to accept the experimental disproofs of Ehrenberg and

18. "Such a force, although we aren't cognizant of it, would not only agree with all the rest of nature's arrangement, but it also seems to us that a Creator who guides the development of organic nature through such a force placed right in it, just as He guides the inorganic world through the simple combined efforts [Zusammenwirkung] of attraction and affinity, appears much more sublime than if we assume that He has to continually tend to the introduction and changing of the plant and animal world on the surface of the earth" (Bronn, Untersuchungen, pp. 81–82).

19. Ibid., pp. 77–79; see, however, Junker, "Bronn," on the internal tension between Bronn's commitment to fixity of species and his fascination for theories of development.

others (which were certainly contested in his time) was strengthened by this association. From his point of view, the only explanation for progressive, orderly change that fell within both the empirical and metaphysical bounds of science was a creative force.

In holding to this view, Bronn was very much in keeping with other members of his generation, especially with physiologists. With only rare exceptions (notably Carl Theodor von Siebold), the physiologists among his contemporaries, Johannes Müller, Emil Huschke, A. W. Volkmann, Rudolf Wagner, and Friedrich Arnold, had long ago agreed upon the existence of a vital force, although they generally confined their use of it to discussions of individual development and self-maintenance, rather than applying it to the development of the organic world in general. They differed among themselves over the exact nature of this force—whether it was a property of organic matter or something separate from it, what its relationship was to consciousness or the soul—and also over the extent to which it was possible or appropriate for scientists to try to delve more deeply into the philosophical issues raised in invoking it.[20] Such differences, which themselves were points of contention in the 1830s and early 1840s, diminished in importance over the 1840s and 1850s, as the physicalist physiologists successfully tarred them all with the same vitalistic brush. By the time Darwin's theory appeared, the similarities among the members of this generation with respect to the existence of a vital force or principle seem to have become more significant—even to themselves, and certainly within the morphological community as a whole—than their differences.

Most members of Bronn's cohort were fairly reticent in print on the subject of Darwin's theory.[21] Halle's professor of anatomy and physiology A. W. Volkmann (1801–77) gave a public lecture on it in 1866; there he joined Bronn in arguing that science was necessarily limited to a level of explanation that did not reach all the way to

20. For the broad context of German physiology and the role of force in it in the 1830s and 1840s, see Caneva, *Robert Mayer*, chap. 3, which explicitly discusses the views of Müller (pp. 90–97) and Volkmann (pp. 160–65); for Wagner's views, see his *Grundriß der Encyklopädie* (1838), pp. 36–38; Ehlers, "Göttinger Zoologen," pp. 50–55. For Arnold's view, see Arnold, *Handbuch*, Bd. 1, p. 4; on Burmeister's view, see his *Geschichte der Schöpfung* (1848), esp. p. 309.

21. For some this was not a matter of choice: Johannes Müller and Emil Huschke, for example, had died in 1858, the year before the *Origin* was published. Others were losing interest in publishing or had stopped altogether: Pöppig had published nothing original since the 1830s, and Friedrich Arnold had only two publications after 1859, which concerned the anatomy of cranial nerves.

God, and the issue of whether God was the ultimate cause or not was a matter of belief. He closed his discussion of this problem with the slightly equivocal comment, "He who erects a structure of crass materialism and fatalism on the grounds set out by Darwin does so at his own risk."[22] The religiously conservative Rudolph Wagner was more direct, raising explicitly what Darwin had left unsaid in the *Origin:* that humans would, on Darwin's theory, derive from an apelike ancestor. "One confuses here a certain small number of *similarities* [to apes], for example in the Negroes, with *transitions* [*Übergängen*]."[23] For these men, Darwin's theory raised metaphysical issues that were of the utmost significance but were beyond the pale of science.

The only other member of the older generation to respond in detail to Darwin's theory was the eminent embryologist and natural historian Karl Ernst von Baer. Von Baer shared Bronn's criticisms of Darwin's theory but was less hopeful about the direction in which it might develop. Sixty-seven years old when the theory appeared, von Baer had had a rich investigative career, moving from his early, reputation-making studies on embryology and generation of the 1820s and early 1830s to natural historical studies from the late 1830s through the 1840s, to questions of physical anthropology in the 1850s and 1860s. Even though he had been living and working in St. Petersburg since the mid-1830s, he continued to publish in German, and many Germans still considered him the founder of classical embryology and their spiritual guide. He was still very much a figure to be reckoned with in European science.[24]

Even more clearly than Bronn's response to Darwin, von Baer's was part and parcel of a broader critique of the scientific times in which he found himself at the end of his life. Although he wrote several different essays delineating his reaction to Darwinism, the longest and most fully elaborated was "Über Darwins Lehre" (On Darwin's theory) a 250-page critique that appeared in 1876, the year of

22. A. W. Volkmann, "Darwinsche Theorie" (1866), pp. 17–19, quotation on p. 18. Avoiding the issue of the *Lebenskraft* (the existence of which he had come to doubt by this time), Volkmann did nevertheless say that Darwin's theory was only apparently atheistic; on his views concerning the *Lebenskraft* during the same period, see A. W. Volkmann, "Über die Grenzen" (1867), pp. 3–5.

23. R. Wagner, "Die Forschungen über Hirn- und Schädelbildung" (1861) p. 184; Wagner's annual essay reviews on general zoology and the natural history of man in the *Archiv für Naturgeschichte* for 1860 and 1861 devote considerable space to the theory but primarily record the views of others.

24. On Bronn, see B. Raikov, *Karl Ernst von Baer* (1968); von Baer, *Nachrichten,* trans. as *Autobiography.*

his death, as part of his *Studien aus dem Gebiete der Naturwis-senschaften.*[25]

Von Baer saw little novelty in the general idea of transformation of species over time— even the occasional emergence of new species out of varieties was an idea he himself had played with in the 1830s, and he reasserted his support of limited transformation.[26] However, concerning Darwin's particular mechanism of transformation, von Baer was less sanguine. He echoed Bronn's critique of Darwin's a priori assertion that life had only one or a few origins, and of the assumption that divergence from common forms yielded distinct species rather than a chaos of forms. In von Baer's view, both were implausible. Furthermore, the existence of a few intermediate forms by no means proved that species originated through divergence from a common form. On the contrary, the rarity of intermediate forms reinforced what naturalists already knew, that species tended to remain constant. The discovery of transitions meant only that the fixity of species was not absolute; it certainly did not prove that species always derived from other species.[27] To these criticisms von Baer added a lengthy discussion of the relation between humans and apes, arguing on grounds of perfect adaptation to different environments that man could not have derived from an apelike ancestor.[28]

Von Baer's objections rested only in part on questions of evidence and cracks in Darwin's reasoning. In the penultimate chapter, he also confronted the fundamental differences in underlying assumptions between himself and the Darwinians. In von Baer's view, these assumptions weighed at least as much as specific evidence and internal logic, and his analysis revealed a wide gulf between the two views of nature.

Von Baer had already hinted at it earlier in the work, when he wrote that "many anatomists and zoologists suffer so much from teleophobia that they flee even from the word 'goal' [*Ziel*]."[29] But he saved his main attack for the last part of his work. "But we must especially fight against Darwin's view of the entire history of organisms only as a result of material effects, and not as a development. It seems to us unmistakable that the gradual progression [*Ausbildung*] of organisms to higher forms and finally to man was a development,

25. von Baer, "Über Darwins Lehre" (1876). For a recent analysis of von Baer's response to Darwinian evolution, see Lenoir, *Strategy of Life*, pp. 246–75.

26. von Baer, "Über Darwins Lehre" (1876), pp. 358, 383, 418–19.

27. Ibid., pp. 284–85, 302–3.

28. Ibid., pp. 306–45.

29. Ibid., p. 333.

a progress toward a goal."[30] This statement contains two principles held by von Baer to be inextricable: that purposiveness or goal directedness is an inescapable fact of nature, and that any kind of organic development can only be understood as a goal-directed process. The case that he knew best was, of course, individual development or embryogenesis. Anyone who had devoted as many years to studying embryology as he had could not help but confront goal directedness everywhere in the process of development. Structures in the embryo, such as lungs, develop morphologically long before they have any function. How else are they to be understood, except as being formed for a future purpose? For von Baer, individual development could only be interpreted as a process directed toward the future, with the purpose of producing a morphologically and functionally independent being. He applied the same view to his understanding of the historical development of organic forms. Evolution was goal directed in just the same way as embryogenesis.

In comparing the development of the individual to that of the animal world as a whole, von Baer reasserted his long-held opposition to the idea that the animal kingdom formed a linear series recapitulated in individual development. But parallels between the two processes did exist. When an embryo develops, it changes much more rapidly during the early stages of its development than later on. The forms that succeed one another in the earlier stages vary much more widely than those in later stages. As development progresses, the process of change slows down and produces fewer significant differences from stage to stage. Similarly, wrote von Baer, the capacity for variation and change, for the creation and modification of forms, must have been much greater earlier in the history of evolutionary development.[31]

For this reason, he thought, Darwin had committed a serious mistake in contending that one could only judge the processes of the past by those observed to operate in the present. The compelling example of individual development revealed that this need not be true; indeed, the analogy forced one to presume that the form-producing forces were probably much more active early in geological time. This helped explain why one did not see substantial modifications in animal and plant forms occurring in the present: the capacity for modification had diminished. It also explained why when new forms first appear

30. Ibid., p. 425.
31. Ibid., p. 430.

in the fossil record, they very rapidly develop into many different species, of which only a few persist. Finally, viewing the history of life as a kind of development allowed for the existence of intermediate forms without assigning too much weight to their appearance: just as there were structures in embryonic development that served only to make possible the next stage, soon disappearing, creatures like the pterodactyl were only transient stages in the overall development of organic forms.[32] To von Baer, the analogy with the goal-directed embryo provided a more satisfactory account of evolution than did Darwin's theory.

In his closing chapter, he recapitulated his main differences with "the Darwinians," culminating in a thundering defense of purposiveness in nature and of evolutionary history as a development whose goal to this point was the emergence of man. Man was probably not the final goal—he was also only a stage in the development of nature—but it seemed to von Baer that this development was nearing the end, as evidenced by the current very slow rate of form change displayed by the organic world. In any case, evolution had to be understood as a development in the same sense as individual development. Development by definition meant unfolding toward a particular goal, and thus teleology was necessarily a basic part of the intellectual armamentarium of the biologist. But von Baer hastened to stress that acceptance of teleology did not mean a reliance on the direct intervention of God through supernatural miracles, as "the Darwinians" maintained.

> Rather, scientific research goes out from the assumption that the forces of nature and therefore the laws of nature (which are nothing other than the measure for the forces) work eternally and unchangingly. Scientists *believe* in this invariability, without being able to *prove* it completely. . . . This unity is quite probably the same one that man felt and sensed before any research into nature, and he designated this unity and absoluteness with the word God. The idea that this deity would arbitrarily and capriciously abrogate his own laws must be unbelievable to a naturalist, because otherwise he cannot be a naturalist at all.[33]

Scientists, like laymen, had their articles of faith, but that did not mean that they believed in an interventionist Creator. A teleological law of nature was still a law of nature, and that law was the only way to explain nature's unity and harmony.

32. Ibid., pp. 430–31. The argument concerning transitional forms in embryos and the fossil record was not unique to von Baer; the anatomist and physiologist Friedrich Tiedemann had made a similar argument in 1808 (see DeJager, "Treviranus" p. 404).
33. von Baer, "Über Darwins Lehre," pp. 461–62.

Bronn's and von Baer's responses to Darwin's theory were some-what different in tone: von Baer openly denounced it, while Bronn's stance was much more equivocal.[34] And yet their particular objections to it were of like kind. First, the theory was incomplete: it ignored or dealt inadequately with problems that required explanation for it to be fully coherent. It did not explain the origin of life forms, nor did it account satisfactorily for the existence of distinct species. Second, Darwin was mistaken in making adaptation a force for change: both individuals and living nature as a whole were constrained by the phys-ical conditions of existence; adaptation was a necessity that imposed limits on form. Third, Darwin's theory denied the inherently teleologi-cal nature of the organic world. Although von Baer's and Bronn's laws of development are not expressed in identical terms—Bronn lays greater stress on the forces underlying the law, while von Baer places more on its teleological nature—the two men seem to have shared a metaphysics and general style of reasoning. Both men viewed the Cre-ator as the ultimate source of the laws of nature, as indeed of order in the universe in general, while clearly agreeing that discussion of such matters lay beyond the bounds of science. These views are repre-sentative of a generation whose fundamental scientific convictions were formed before the 1830s, and who saw in Darwin's theory a challenge to their most cherished views about the nature of life and science more generally.

## PROCEEDING WITH CAUTION: ALBERT VON KÖLLIKER

A quarter-century younger than von Baer, and seventeen years younger than Bronn, the anatomist and physiologist Albert von Köl-liker understandably had a somewhat different orientation. Kölliker had studied under Müller and Henle in the late 1830s, when both the speculative theories of the *Naturphilosophen* and the teleological thinking of the following generation came under sharp criticism as metaphysical and beyond the appropriate bounds of science.[35] By con-trast, the value placed on empirical discoveries, especially those based on the recently improved microscope, shot up. By the end of that decade, the cell theory provided a new source of excitement in the biological sciences, and Kölliker became a leader in the efforts to refound the study of tissues on the basis of new knowledge about

34. Thomas Junker's fine article "Heinrich Georg Bronn" has helped me see the ambivalence in Bronn's position much more clearly than I originally did.

35. Kenneth Caneva outlines the arguments of this period in great detail (*Robert Mayer*, esp. chap. 3 on physiology).

cells.[36] Like most others who worked in this area in the 1840s and 50s, Kölliker sought to connect cell and tissue studies by following the progressive differentiation of homogeneous cells into different tissue types and thence into distinct organs. Thus tissues were to be understood by their development. Although most of·these studies were carried out on vertebrates, whose tissues were the most differentiated, Kölliker also turned his morphological investigations of tissue development to the invertebrates. Beginning in the mid-1840s and extending through the 1850s, he spent a good deal of time studying the life cycles of various marine invertebrates, investigating the newly discovered phenomena of polymorphism (the appearance of different forms within the same species) and alternation of generations (the successive appearance of forms with different modes of reproduction within a single species). His response to Darwin's theory must be read with this background in mind.

In February 1864, Kölliker gave a speech before the Physical and Medical Society of Würzburg entitled "*Über die Darwin'sche Schöpfungstheorie*" (On the Darwinian theory of creation).[37] There he laid out what he saw as the main strengths and weaknesses of Darwin's theory and proposed his own alternative that avoided the pitfalls he attributed to natural selection. Many of Kölliker's specific objections were familiar enough among Darwin's critics: no one had ever observed the gradual emergence of one species from another; the geological record did not show intermediate forms; and no varieties were known that had diminished fertility, as they should if they were incipient species.[38] But Kölliker's more fundamental objections to the theory of natural selection are startling: he first accused Darwin of being "a teleologist in the fullest sense of the word" and then claimed that Darwin "does not believe in general natural laws."[39]

What an odd way to talk about Darwin! How could Kölliker have misunderstood him so completely? His assertion of Darwin's teleology does rest on a basic misreading of the *Origin*, but at the same time it clearly reveals Kölliker's own concerns and biases. He thought that Darwin had posited an internal "tendency of organisms to produce useful variations," upon which natural selection then worked.[40] Such

36. The most complete account of Kölliker's career remains his own autobiography, *Erinnerungen.*

37. A. von Kölliker, "Über die Darwin'sche Schöpfungstheorie" (1864).

38. Ibid., p. 185.

39. Ibid., p. 175.

40. Ibid., p. 177.

a tendency was teleological because it seemed to assume that utility acted as a principle driving imperfect organisms toward perfection. But according to Kölliker, "every animal suffices for its purposes, is in its way perfect, and needs no further perfecting."[41] Here Kölliker's statements both mirrored and opposed the ideas of von Baer and Bronn. Unlike his older colleagues, he opposed the idea of a goal-directed progression of animal types being driven toward perfection—in this regard, he was what von Baer would call a "teleophobe."[42] At the same time, he shared their idea that an organism's adaptation to its environment was a fact of nature. It is difficult for a modern reader to see this view as anything other than teleological, but Kölliker's distinction is important for understanding the fine changes over time in what the zoological community considered "teleological." Adaptation could be viewed as a fact of nature and not implying any teleology because it was not directed toward the future; the sort of purposiveness with which Bronn and von Baer invested their laws of development, by contrast, looked suspiciously willful to someone of Kölliker's generation. Unfortunately for modern readers seeking to clarify Kölliker's position, he did not examine further his views on adaptation and teleology, perhaps because adaptation and environmental influences held little interest for him.

In an elaboration of his views on Darwin's theory published in 1872, Kölliker dropped the accusation that Darwin's thinking was teleological, presumably because he had discovered that he had misunderstood Darwin's conception of variation. Instead, he now understood Darwinian variation as resulting from accident.[43] But variation through accident was just as unacceptable to Kölliker as purposeful variation, for in his scientific worldview, both operated outside of the laws of nature. And although he shifted his assessment of Darwin's specific failings between 1864 and 1872, his solution was the same: to replace the erroneous principle with a law of nature, specifically, a law of development.

So we return to the second peculiar assertion that Kölliker made in 1864: that Darwin did not believe a general law of nature could account for the observed phenomena. For Kölliker, the crucial phenomenon to be accounted for was "the regular, harmonious series of

41. Ibid.
42. von Baer, "Über Darwins Lehre," p. 333.
43. A. von Kölliker, "Anatomisch-systematische Beschreibung" (1869–70, 1872); the relevant section is "Allgemeine Betrachtungen zur Descendenzlehre" (1872), pp. 206–37.

all organic forms progressing from simpler to more perfect."[44] To explain this, one did not require Darwin's theory. "The existence of general laws of nature explains this harmony, even if one follows the assumption that all beings originated independently and separately from one another."[45] In Kölliker's view, Darwin needed to connect the organic world through inheritance only because he did not trust such general laws to produce a similarly harmonious, regular result.

Kölliker later clarified both his opposition to descent from a single primitive form and the general law of nature he sought to put in its place. Like Bronn and von Baer, he thought the likelihood of a single original creation of life, whether by spontaneous generation or by some other event, extremely small. If more than one individual appeared at the beginning of life, then necessarily unconnected phylogenetic trees would develop over time, as the different original individuals divided, produced offspring, and eventually yielded divergent species. The consequence of this was to undercut the basic premises of Darwin's theory.[46] But even without descent from a single ancestral type, a certain harmony existed in organic nature that needed to be accounted for. The particular law of nature Kölliker called on to explain the patterns of organic existence was a law of development "that drives the simpler forms to ever more diverse developments [*Entfaltungen*]"[47]—a law that bears a striking resemblance to those of Bronn and von Baer, despite Kölliker's objection to the idea of goal directedness. Among other things, this law easily explained the parallelism between individual development and the development of the animal series, without recourse to a single, genetically connected family. "If the development of a given creature follows immutable rules, it is impossible at the outset that the animal kingdom could obey other laws."[48] Thus in his basic premises about the law of development, Kölliker followed the precept, well established among pre-Darwinian zoological thinkers, that a single law of nature ruled over all levels of the organic world, and that that law assumed within it a principle of development.

In his 1872 work, Kölliker spelled out in greater detail what he meant by this law of development. First, he asserted that it was just like any other law of nature, with no properties special to organisms.

44. Kölliker, "Über die Darwin'sche Schöpfungstheorie" (1864), p. 177; note that this is not what Darwin saw as the crucial fact to be explained.
45. Ibid., p. 178.
46. Kölliker, "Anatomisch-systematische Beschreibung," pp. 217–19.
47. Kölliker, "Über die Darwin'sche Schöpfungstheorie," p. 184.
48. Kölliker, "Anatomisch-systematische Beschreibung," p. 224.

In order to cut short all misunderstanding, let me emphasize that for me exactly the same laws underlie organic structures [*Bildungen*] as inorganic nature, and that therefore my fundamental view is the same as that of the great majority of the more recent naturalists. . . . What I call law in organic nature is thus nothing other than what the physicist, the chemist, the astronomer denotes with this name, and I understand under a general law of development of organic nature nothing else than does the mineralogist who speaks of a law of formation of crystals, or the astronomer who speaks of the law of gravitation or the law of development of heavenly bodies.[49]

Just as planets and crystals are formed according to a set of laws that produces the same kind of planet or crystal the same way each time, without any genetic connection to other planets or crystals, Kölliker argued, so, too, could a law of organic development produce a consistent pattern without relying on hereditary transmission or accidental external factors.

It is still not entirely clear whether Kölliker thought the law of organic development was simply similar in kind to other mechanical laws or whether a single law of development covered planets, crystals, *and* organisms, as, for example, Herbert Spencer had posited.[50] It is evident from his insistent tone, however, that he wanted to dissociate himself from vitalism, and late in the discussion he once again explicitly associated himself with a mechanical viewpoint, this time in terms of a chain of necessary causes and effects.[51] But to follow out this argument it is helpful to know more of the specifics of Kölliker's law of development.

Kölliker first presented his "theory of heterogeneous generation" in the 1864 speech, renaming it in the 1872 essay "development through inner causes." He suggested that new species originated by jumps in which one form brought forth a different one that would then reproduce its own new kind. Two processes could bring this about: either a new form would emerge from a fertilized egg under certain changed conditions of its development, or it would appear through parthenogenesis (production of offspring without the contribution of a male). The evidence for the latter possibility came especially from the alternation of generations, which Kölliker had been studying for two decades. He admitted that he lacked facts at present for the first possibility, but thought it likely that the embryo of a

49. Ibid., p. 208.
50. Ibid., pp. 208, 223–24; Herbert Spencer, "Progress: Its Law and Cause" (1891); Spencer's essay was originally published in 1858.
51. Kölliker, "Anatomisch-systematische Beschreibung," p. 224.

higher animal could be jogged from its usual path of development early on so as to head in a different direction and thus produce a new species.[52]

According to Kölliker, the law of organic development worked almost completely through causes operating within the organism. Just as the capacity for all future developmental processes and forms was contained in the fertilized egg of the higher organic individuals, "and just as a mother liquor [*Mutterlauge*] of particular chemical composition will of necessity crystallize into a particular crystal form, so do the original, primitive seeds of all organisms . . . contain within them the potential [*Möglichkeit*] for all later forms and bring these forms lawfully and in a completely determined way to realization."[53] Change is real, but it does not represent the production of true novelty, either in individual organisms or in species. External conditions might alter the direction taken by a particular organism's development, but that course of development had to have been possible already within the organism. And if deterministic laws of nature hold for individual development, so too must they hold for the development of species: "If necessity reigns there, accident cannot rule here."[54]

To Kölliker, this theory of heterogeneous generation held a number of advantages over Darwin's theory of natural selection. Not only was it supported by the facts of polymorphism, but it accounted for those facts in a way that Darwin's assumption of gradual, stepwise changes could not. It further solved the problem that different varieties could always successfully breed, while different species could only do so in some cases. In Kölliker's system, varieties never slipped gradually over into species, but through different modes of generation, two species could arise that might or might not be fertile with one another.[55] And because these developments unfolded entirely through internal causes, Darwin's troubling dependence on accidental, externally caused variations was avoided. Perhaps the aspect of this theory most important to Kölliker was that it did not demand Darwin's premise of monophyletic descent. His theory thus accounted for the

---

52. Kölliker, "Über die Darwin'sche Schöpfungstheorie," p. 183.

53. Kölliker, "Anatomisch-systematische Beschreibung," pp. 223–24. "Mother liquor" is a chemical term that usually refers to the liquid left over from a solution out of which a crystal has crystallized; Kölliker seems to be using it to refer to the solution itself. My thanks to Robert Siegfried for clarifying this point for me.

54. Ibid.

55. Kölliker, "Über die Darwin'sche Schöpfungstheorie," pp. 184–85.

main features of organic development in a lawful way without depending on the unlawful actions of chance or purpose, and incorporated transformation (though not true novelty) without dissolving traditional commitments to the distinctness of species.

In responding to Darwin's theory, Kölliker revealed significant commonalities between his assumptions and those of Bronn and von Baer. Kölliker and von Baer agreed that the "developmental path" of both individuals and species was predetermined, and that the rest of development was the unfolding or working out of that capacity. The terms by which they expressed this determinism were quite different—von Baer emphasized the goal orientedness of both embryos (aimed at yielding mature forms) and life on earth (aimed at producing humanity), while Kölliker emphasized the potentialities existing in the fertilized embryos, waiting to be set into motion—but neither seems to have believed in the possibility of true novelty. External conditions might affect the direction of the developmental path to some extent, even to the point of producing different species, but they could not create new material or new capacities in the embryo. Bronn joined these two scientists in viewing accident as an unacceptable basis for understanding the patterned harmonies manifest in nature. Finally, all three of these men found that Darwin's view of nature was fundamentally flawed because it did not recognize the history of the organic world as a development related in some important way to the history of the individual. According to all three, a law of development necessitated a similarity between individual development and the historical development of animal life.

Despite these similarities, there were also underlying differences. The older men defended the idea of teleology; von Baer was especially vocal about the existence of goal directedness in nature. Kölliker, a child of his teleophobic times, rejected the notion that such unconscious purposiveness acted in nature. At least as significant, he never referred to "forces" or "principles" as underlying his law of development, but chose the intentionally mechanistic language of cause and effect, consistently using the term "*Ursache*" for "cause" rather than the older, more ambiguous "*Grund.*"[56] Although his mysterious "internal causes" may seem to us to serve the same black-boxing function as Bronn's "creative force" or von Baer's "formative power" [*Bildungstrieb*], his self-declaration as a mechanist must be taken seriously. The language implies that these internal causes, whatever they

56. See, e.g., Kölliker, "Anatomisch-systematische Beschreibung," pp. 223–24.

were, should be reducible to physical and chemical causes and not some distinct vital force.

Although most of the scientific zoologists of his generation were less critical of Darwin's theory than Kölliker, some of the same preoccupations appear in their responses as well. Kölliker's Würzburg colleague Franz Leydig, for example, thought Darwin was right "on the main point," and considered it quite possible that natural selection "or a similar factor" played a role in the formation of new species. Nevertheless, he noted that if one followed out the theory, one would come to "a result with which probably nobody can entirely agree. All animal forms then really have originated through accident; . . . an *accidental element* is openly the ruling principle." Although Leydig admitted that it was impossible to disprove such a possibility, he thought it unpalatable, indeed contradictory, to human efforts to understand the world. Here Leydig showed a more mitigated stance than did Kölliker or the older generation: accident was unacceptable not for metaphysical reasons—the world doesn't work that way—but for epistemological ones—"the human mind demands" a more orderly world than that.[57] Nowhere does Leydig make the connection made by the generation of *Naturphilosophen* and many of the next, that the order demanded by the human mind must correspond to the order existing in nature. To Leydig the role of accident made natural selection distasteful, but that could not push it entirely beyond the realm of possibility. Furthermore, despite its unfortunate implications, Darwin's hypothesis was still useful in directing naturalists' attention to hitherto underresearched areas, such as intermediate forms, and it might even yield the true natural system in systematics.[58]

Although Leydig took a less stringent attitude toward the dangers of accident than Kölliker, his stance on this point reflects the same reluctance to take a strong stand on metaphysical issues that Kölliker displayed in his discussion of the law of development. Exactly those silences in Kölliker's writings that create the greatest frustration for the historian—on the precise nature of his law of development, on the causes producing this law, on what he meant by "teleology" and "adaptation"—are themselves indications of the extent to which he, like other "scientific zoologists" of his generation, kept metaphysics at arm's length. As we will see, reactions of this generation to Haeckel also displayed a similar shrinking from strong metaphysical stances.

57. F. Leydig, *Vom Baue des thierischen Körpers* (1864), p. 7.
58. Ibid., pp. 7–8.

This attitude reflects, I think, not only a caution born of the battles of the 1840s but also a certain impatience with those old battles, mirrored in a pragmatic desire to get on with their research. To that end, many of them were willing to accept as working hypotheses a variety of ideas, even perhaps ones with contradictory metaphysical implications, if they might prove fruitful in asking questions or opening up new research areas.

## BELIEVING IN DARWIN: ERNST HAECKEL

At the time of his first major public statement on Darwinism, in 1863, Ernst Haeckel was only twenty-nine years old. Although he had already attained a professorship at this tender age, Haeckel had far fewer long-standing intellectual commitments than his older colleagues. He had much less research experience as well. As a student-assistant of Kölliker, and then later as professor of zoology at Jena, he had conducted a series of studies on marine invertebrates; these formed the bulk of his research to 1863. Over the course of the 1860s and 1870s, Haeckel became the most outspoken supporter of Darwin's theory within the biological community in Germany. But as his *Generelle Morphologie* (1866), his popular *Natürliche Schöpfungsgeschichte* (1868), and his *Anthropogenie* (1874) all attest, his main concern was not to expound Darwin's own theory, but to retell Darwin's theory in terms that were peculiarly Haeckelian.

One of Haeckel's general commitments was already apparent in the speech he delivered at the 1863 meeting of German Naturalists and Physicians at Stettin. The speech was entitled "Über die Entwickelungstheorie Darwins" (On Darwin's theory of development), a choice of words indicating his assumption, common to almost all German writers, that evolution represented some kind of development.[59] A second commitment was also apparent in Haeckel's choice of forum: the general sessions of these national meetings were open to nonscientists, and Haeckel was committed to bringing the ongoing debate over Darwinism into the public arena. In his usual inflammatory manner, Haeckel claimed that scientists were already divided into two parties, that of the Darwinists, whose battle cry was "Development and progress!" and that of the conservatives, who stood beneath the banner

59. He could, e.g., have chosen the term "Descendenztheorie," which was often used by other German biologists (E. Haeckel, "Über die Entwickelungstheorie Darwin's" [1864]).

of "Creation and species!"[60] This explicit association of Darwinism with progressive political and religious views would be a hallmark of Haeckel's Darwinism throughout his career.

Having staked out his political position, Haeckel went on to elaborate the history and theory of descent. Like von Baer, Haeckel pointed out that ideas of the development of all of nature were not new. What Darwin added to the speculations of Lamarck, Geoffroy St. Hilaire, and Oken was evidence and a theoretical explanation for the facts of species development. The key principles, or laws, as Haeckel also called them, were those of variability and heritability; natural selection operated on the concrete manifestations of those laws. The necessary outcome was a "progressive metamorphosis, a progressive reformation and ennoblement of all organisms. . . . The lower, less perfect forms continually become extinct, the higher, more perfect persist."[61] This progressive development, according to Haeckel, was an empirically derived law of nature that could only be explained through Darwin's theory of the struggle for existence and natural selection. The strongest evidence for such a development, in his opinion, came from the threefold parallelism between the embryological, systematic, and paleontological development of organisms.[62]

This speech was only the first salvo in Haeckel's lifelong battle to convert people to evolutionary monism, a philosophy that would do away with mind-body dualism and view all of creation, conscious and unconscious, as a whole unified by evolution. His next major efforts came in the aridly systematic *Generelle Morphologie* (1866) and the much more lively *Natürliche Schöpfungsgeschichte,* published in 1868 from stenographic notes of his popular lecture course at Jena. In these books, Haeckel relentlessly pressed home his message that spirit and matter alike were subject to the mechanical laws of necessary cause and effect. He also retold Darwin's theory of natural selection, at great length. Finally, he developed what was to become the most enduring contribution of his career: the idea that ontogeny (individual development) recapitulated phylogeny (species development), which he would soon dub the "biogenetic law." Each of these points resembled views held by the other morphologists discussed here, but it took Haeckel to combine them into an elaborate theoretical edifice that would shape the morphological thinking of the next two generations.

Kölliker, we will recall, insisted that whatever the laws of develop-

60. Ibid., p. 18.
61. Ibid., p. 26.
62. Ibid., p. 29.

ment might be, they were purely mechanical laws, operating without recourse to a vitalistic *Bildungstrieb* or formative force. Nevertheless, their result was progressive development. Haeckel said just the same thing, only many more times over.[63] For him, development had to be mechanical since all organic and inorganic phenomena were determined by necessary chains of cause and effect. Because of his need to reconcile this determinism with his belief in Darwin's theory, he disputed Kölliker's claim that Darwin's theory gave a prominent place to accident. Instead, he defended the premise that variation itself was determined by necessary chains of events taking place either within the fertilized egg or as the organism responded to its environment.[64]

Despite this specific difference over the legitimacy of natural selection, the assumptions underlying Haeckel's notion of law closely resembled those of Kölliker. Haeckel did not understand the law of development as the manifestation of a force. Instead, seeking to resolve the ambivalence he saw his onetime teacher Johannes Müller having expressed about vitalism and mechanism, he argued that within organisms the "goal-producing *causa finalis* converges with the mechanical *causa efficiens*," with the former becoming subordinated to the latter; "thus the mechanical view of organisms is recognized as the sole correct one."[65] Having asserted (though not fully explained) his commitment to the primacy of mechanical explanations, he presented a hierarchy of laws produced by an endless chain of mechanical causes and their effects. Each law was an empirical fact, a generalization that did not constitute the limit of explanation, but which itself had a causal explanation.[66] Thus the law of development was an undisputed fact visible in the paleontological realm, in individual development, and in the distribution of living forms. According to Haeckel, its cause, in all three realms, was natural selection. And here he elaborated on his ideas in considerably greater detail than his more restrained teacher, Kölliker. Natural selection itself derived from the interaction of adaptation and inheritance. These two were also laws or empirical statements about the workings of nature. They in turn were the expressions of two physiological processes—the

63. In fact, it was probably Haeckel's criticism of Kölliker's 1864 article on Darwin's theory that drove the latter to clarify his stance on this point in 1872 (see Haeckel, *Generelle Morphologie*, vol. 1, p. 101).

64. Ibid., vol. 2, pp. 191–223, esp. p. 192.

65. Ibid., vol. 1, p. 94.

66. For a general discussion of law and cause in biology after Darwin, see M. Adams, "Severtsov and Schmalhausen" (1980), esp. pp. 202–4. On hierarchies of law in British scientific thought in the same period, see T. Broman, "Herbert Spencer."

effects of physiological causes. Adaptation was a consequence of the organism's drive to feed itself, and inheritance a consequence of reproduction. Thus nutrition and reproduction were the causes that yielded the empirical laws of adaptation and inheritance. But nutrition and reproduction "admittedly are founded on the same physical and chemical processes that uniformly rule all of organic and inorganic nature.[67] Therefore the entire chain of laws and causes derived ultimately from physics and chemistry.

Within this chain of events, Haeckel laid particular emphasis on the level of adaptation and inheritance, the interaction of which, he claimed, produced the empirical generalization Darwin called "natural selection."[68] As we have seen, these held an important place in the logic of mechanical cause and effect upon which Haeckel's system was based. But there were also other sources for assigning particular importance to their interaction. In fact, Haeckel traced the idea not to Darwin at all, but to Germany's national hero, Goethe.[69] Goethe had established two *Bildungstriebe* or "driving forces," a conservative one, which Haeckel equated with heritability, and a progressive one, which Haeckel translated as adaptation. Inheritance, the conservative principle, was a process internal to the organism, while adaptation was the process of adjusting to the outside world. He might also have mentioned H. G. Bronn, who, as we have seen, had also honed down the laws of organic development to a progressive and a conservative one. But Haeckel's laws were Bronn's turned upside down: Bronn had seen the internal drive as the progressive one and the necessity for adaptation as a conservative principle. In his interpretation of Darwin's theory, then, Haeckel took existing components in the theoretical vocabulary of earlier German morphologists and reworked them into a new evolutionary language.[70] At the same time, of course, he was changing the terms of Darwinian evolution to accommodate existing beliefs about the laws by which the organic world worked.

The same can be said of the biogenetic law, Haeckel's most important contribution to evolutionary morphology. As we have seen, Bronn, von Baer, and Kölliker all understood evolution as a develop-

67. Haeckel, *Generelle Morphologie*, vol. 2, p. 12.
68. E. Haeckel, *Natürliche Schöpfungsgeschichte* (1874), pp. 172–73; this is the earliest edition to which I have had access.
69. Ibid., p. 270.
70. This is literally true, for Haeckel coined dozens of neologisms, most of which were not picked up by his contemporaries.

ment, governed by the same general laws as individual development. Haeckel agreed. But he also posited something more radical than his older colleagues. The evolution of the animal kingdom was analogous to individual development not just because the laws of development were the same, but also because the animal kingdom itself was a form of individual. In stating this, Haeckel was harking back to a view of nature popular among the German Romantic writers of the early 1800s as well as among some *naturphilosophisch* physiologists.[71] At the same time, he was extrapolating from his own research, justifying the biogenetic law in terms particularly plausible to someone familiar with marine invertebrates. The idea that both the animal kingdom and the organism were in some sense individuals was central to Haeckel's formulation of the biogenetic law. He explained it in the *Generelle Morphologie,* the work in which he first defined evolutionary morphology and elaborated on its tasks.

Haeckel posited three kinds of individuality: morphological, physiological, and genealogical. The usual concept of the individual came from a combination of all three, for the organic individual was at once a morphological unit with characteristic features of form, a physiological unit capable of sustaining its own life, and a genealogical unit constituted in time, from its conception to its death. But there were other levels of individuality as well. Haeckel posited a hierarchy of six "orders" of morphological and physiological individuals ranging from the "plastid" or cell to the "corm" or colonial organism. Each level of individual was made up of an aggregation of individuals of the next lower order. Thus a group of plastids together constituted an organ, or second-order individual, which joined together with other organs to make up an "antimere" or third-order individual, and so on. To these morphological and physiological individuals he added three orders of "genealogical" or temporal individuals: one constituted by the life cycle of a single organism, one constituted by the life span of a single species, which also had a birth, maturity and senescence, and a third consisting of the life history of the phylum or tribe of species sharing a common primitive ancestor. Just as each order of morphological individual was made up of an aggregate of individuals of a lower order, so too were the genealogical individuals organized in a hierarchy such that the phylum was made up of species' life cycles, which in turn were constituted by the sum of individual life

71. Schelling was the most influential Romantic source for the idea of the world as organism; see O. Temkin, "Concepts of Ontogeny and History" (1950).

cycles. Each level of individual at once had an independent existence and constituted an aggregate of individuals of a lower order.[72]

According to Haeckel, ontogeny recapitulated phylogeny. In a system in which the phylum was understood as an individual, it was only natural that the phyletic individual and the physiological individual (what we normally think of as an organic individual) should follow essentially the same course of development. But the biogenetic law said something more. Ontogeny and phylogeny were not simply independent processes subject to the same causal laws, as Haeckel's older colleagues would have had them. In conformity with his hierarchy of laws and causes, Haeckel linked ontogeny and phylogeny in a causal relationship: phylogeny *caused* ontogeny.[73] How was this possible, if the phylum was nothing more than an aggregate of species, which in turn were aggregates of individuals? The critical fact was that these orders of genealogical individuality existed in time. Only by recognizing the essentially historical nature of genealogical individuality could one reconcile phylogeny's causing ontogeny with the idea that the phylum was an aggregate of individuals. One had to start with the beginning of organic history and follow the whole process of development.

According to Haeckel, the form at the origin of all phyla, at the beginning of organic life, was the plastid, which also constituted the initial form present in ontogeny. Everyone agreed that primitive organic forms developed little if at all during their life cycle, but did reproduce. During their lifetimes, the initial forms would differentiate to adapt to different conditions of existence; those adaptations would be passed on through inheritance. In the process of differentiating, the plastids added a stage on to their development. The new species that eventually emerged from this differentiation would have two stages to their development, one corresponding to the plastid and another to the differentiated form. These species constituted the second morphological and physiological order of individuality, in which the individuals possessed tissues and organs. As evolution continued, natural selection of favorable adaptations yielded increasing differentiation and progress. This led to the emergence of successively higher orders of individuality. At the same time, each advantageous new adaptation transmitted through inheritance was added on as an end stage in individual development. Changes and additions in *species*

72. Haeckel, *Generelle Morphologie,* vol. 1, pp. 43–60; he treats the concept of the genealogical individual in more detail in vol. 2, pp. 26–31.

73. Ibid., vol. 2, p. 300; Fritz Müller (*Für Darwin* [1864], pp. 75–77) had previously made a similar argument.

caused by natural selection would thus appear as changes and additions in *individual* development. In this way phylogeny, the development of the phylum, literally caused ontogeny.[74]

There were caveats, of course. Over time, the increased number of stages could produce a disadvantageously long period of development. In the successful higher organisms, therefore, some of the ancestral history was deleted, and ontogeny provided only an incomplete record of the organism's past. Also, in some cases new adaptations were not added on to the end of development, but interpolated into the middle. Careful comparison of developing individuals and adults, Haeckel was sure, would reveal which new forms were out of sequence and which constituted a true ordering of the organism's ancestral past. In any case, neither deletions nor occasional insertions negated the fundamental statements: phylogeny caused ontogeny, and therefore ontogeny recapitulated phylogeny.[75]

In basing the biogenetic law on a hierarchy of individuality, Haeckel was providing a new foundation for the long-standing belief that individual development bore a lawful relationship to the development of the organic world as a whole. But his focus on individuality itself was not unique: in the 1850s, defining the individual was a source of considerable controversy in all realms of biology, and a number of Haeckel's teachers and colleagues had become involved in the discussion.[76] Haeckel had followed his teachers into the fray in 1862 with his first major monograph, *Die Radiolarien*.[77] One peculiar group of these tiny marine invertebrates had a complex life cycle in which units sometimes lived independently but at other stages formed physiologically unified colonies. Haeckel's solution to categorizing these perplexing creatures was to distinguish between a morphological concept of the individual and a physiological one.[78] From a morphological standpoint, he wrote, the radiolaria could only be considered

74. Haeckel, *Generelle Morphologie*, pp. 120, 129.

75. Ibid., p. 300. It is worth noting that this argument is not made in a compact, coherent way in the *Generelle Morphologie* but is scattered throughout the text.

76. Ruth Rinard ("Problem of the Organic Individual," [1981]) has argued that Haeckel's concept of the individual owed much to his teachers Alexander Braun and Johannes Müller. It is worth noting, however, that the discussion went well beyond Braun and Müller, indeed, beyond Germany: T. H. Huxley, for example, was engaged in the same problem; on the broader issue of individuality, see Nyhart, "Problem of the Organic Individual" (1989).

77. E. Haeckel, *Radiolarien* (1862).

78. In the monograph, he does not name names, but he is clearly drawing from Rudolf Leuckart's seminal 1851 work, *Über den Polymorphismus der Individuen* (1851).

as a colony of independent individuals. But from the physiological standpoint, they had to be considered together as organs of a single organism. The two views could not be resolved into one; they had to be accepted as two different aspects of individuality. Thus, Haeckel argued, "the two concepts of the individual and the organ are in nature not nearly so distinct" as one usually assumed: in fact, the decision to call a unit an individual or an organ depended only on the mode of analysis the scientist chose.[79] "Individual" and "organ" were not absolute concepts but relative ones.

Clearly, Haeckel's familiarity with this corner of the organic world led him to accept the distinction between the morphological and physiological individual and to consider the possibility of levels of individuality. The single radiolarian unit was a physiological individual when it split off to found a new colony, but these units, when aggregated, formed a higher, more differentiated order of individual, the colony. This is exactly the argument he generalized in the *Generelle Morphologie*. His decision there to posit six levels of individuals probably arose from his desire for a system general enough to encompass both animals and plants, for it was closely modeled on existing schemes proposed by botanists.[80] By extending that hierarchy to include the upper orders of genealogical individual—the species and the phylum—he was able to accommodate Darwin's theory into his system. It fit in easily, insofar as Darwin viewed species and higher taxonomic categories as relative rather than absolute entities.[81] Understood in the context of his system, Haeckel's formulation of the biogenetic law falls out as a natural consequence of his meshing Darwin's theory with his own views on individuality as they had been developed through his experience studying marine invertebrates.

As this account suggests, the *Generelle Morphologie* was far more than simply a response to Darwin's theory; it was nothing less than an attempt to build a logically coherent system of morphology, incorporating Darwin's theory, that would serve as the theoretical basis and guide for all future morphological work. Haeckel presented a tightly coherent system that bound up his ideas of law and cause, his interpretation of Darwin's theory, and his biogenetic law into one package. This formed the theoretical basis of his life's work, both for his own research monographs on siphonophores and sponges and for

79. Haeckel, *Radiolarien*, p. 122.

80. Haeckel, *Generelle Morphologie*, vol. 1, pp. 250–51.

81. In the *Radiolarien*, pp. 232–33, Haeckel noted approvingly that Darwin had shown that the term "species" was "a no less arbitrary abstraction . . . than the concept of the individual, the concept of the genus, family, order, and so on."

his more popular works.[82] Like Bronn and Kölliker, Haeckel found in Darwin's theory a focus for his own thoughts about the nature of evolution, the laws of form, and the more general problem of how the laws of organic nature operated. Unlike them, he also found in it—or in his revision of it—the inspiration for a new kind of morphology, whose goal would be to search for the phylogenetic connections between current and past forms, aided by the biogenetic law.

As we will see in chapter 6, Haeckel's contemporaries found various other ways of integrating Darwin's theory into their thinking, and here Haeckel was characteristic of his generation. Whereas men of earlier generations tended to talk about his theory almost entirely in terms of accepted evidence and theory, this younger group actively incorporated language and ideas labeled "Darwinian" into their ongoing research. But after 1866, it was almost impossible for German zoologists to respond to Darwinism without responding to Haeckel, and indeed, one might argue that it was Haeckel's unremitting effort to create a Darwinian program—even a worldview based on evolution—that focused much more attention onto Darwinism. Although von Baer directed his 1876 arguments at Darwin's theory, for example, it seems likely that some of his fervor was aroused by claims pressed harder by Haeckel than by Darwin himself. Not only did Haeckel work to make evolution a key part of a broad attack on organized religion, something that must have seemed to von Baer outside the bounds of proper scientific behavior, but his biogenetic law too closely resembled the old "law of parallelism" associated with the Naturphilosophen and opposed by von Baer. In addition, Haeckel considered both descent and individual development the products of an aggregation of individual parts rather than the differentiation of an already organized whole. In this way, he took what Reichert had called the "atomistic" view rather than the "systematic" one (see chap. 3), which seemed to place him in the camp of the physicalist physiologists. For all these reasons one might reasonably expect hostility from the older camp to Haeckel's ideas, even if it was not aimed directly at him.

Other German zoologists who had kept quiet (at least in print) about Darwin's theory before 1866 more often than not first hinted at their position on it in reviewing Haeckel's work. An especially instructive case in point is Rudolf Leuckart, a contemporary of Kölliker who did not discuss Darwin's theory in his annual reviews of

82. A bibliography of Haeckel's most important works is printed in Uschmann, Geschichte der Zoologie, pp. 239–40.

literature on the lower invertebrates until he reviewed the *Generelle Morphologie*. There he said for the first time in print that, "on the main issue, on the question of descent," he was "decidedly on [Haeckel's] side."[83] He also offered a defense against Haeckel's attack on teleology that illustrates well the stance of Leuckart's (and Kölliker's) own generation. Haeckel had denigrated teleology as expressing a conservative dualism of spirit and matter; this was what he wanted to replace with his monistic philosophy. Leuckart expressed annoyance with Haeckel's trumpeting of monism, noting that no one would disagree with Haeckel's call for mechanical explanations in the organic world—in fact, such explanations were widely assumed to be the goal. But Haeckel had seriously misconstrued the role of teleology in scientific thinking:

> Teleological reasons [*Gründe*] are not mechanical causes, and can only serve to justify a phenomenon, that is, to indicate the connections that they present to other phenomena. . . . One needs only to change certain key words of the present-day teleological viewpoint—which, we repeat again, has long since exchanged its earlier principal meaning with a purely formal one—instead of saying "purposeful" [*zweckmässig*], merely to say "useful," to assure them a place in the so-called monistic system. What modern teleology particularly pursues in our science is substantially the demonstration of the same harmonious relationship between structure and function that the author, with Darwin, refers to under the name of adaptation [*Anpassung*].

Teleology, Leuckart maintained, had "great heuristic value," even if it possessed "absolutely no justification in principle."[84] Like his contemporary Franz Leydig, Leuckart was willing to employ an idea for its heuristic uses without committing himself to its underlying metaphysics. As we will see in chapter 6, the pragmatic, flexible attitude of his generation contrasted sharply with the more strident demands for allegiance to a program made by Haeckel.

## CONCLUSION

The analysis developed in this chapter suggests that it may be misguided to look for a unified "German response" to Darwin's theory. Even those who shared a deep interest in the study of form differed in certain important assumptions; significantly, these differences appear to fall along generational lines. To be sure, one common underly-

---

83. Leuckart, "Bericht 1866–1867" (1867), p. 167.
84. Ibid., pp. 164–65.

ing theme does emerge: the leading figures discussed here—Bronn, von Baer, Kölliker, and Haeckel—all argued for the existence of a "law of development" in organic nature, a commitment which, at one level, united them. This law, they agreed, accounted for the similarities they saw between individual development and the more general history of life on earth. Thus for these men, the term *Entwicklung* did not have two different meanings, evolution and individual development. Instead, it had one meaning, development, which could be manifested in two ways, by species and by individuals. While in English the ideas of evolution and individual development came to mean quite separate things, the Germans never lost the connection between the two. At least until the end of the century, descent was considered one facet of the more general phenomenon of development.[85] This German understanding of evolution also explains why Haeckel's neologisms "ontogeny" and "phylogeny" were promptly accepted by German scientists: these terms permitted a clear and concise distinction between the two kinds of development while preserving the long-held belief in their essential similarity by defining both as subsets of development.[86]

Nevertheless, this broad similarity in their views is undermined when one realizes that the scientists under discussion here did not agree on much about the nature of this law of development. To take the most straightforward difference, the specific analogies they drew between individual development and the history of life depended on their special familiarity with different realms of zoological inquiry. Bronn's criticisms of Darwin reflected his orientation toward paleontology, in which the crucial phenomenon to be explained was the succession of progressively higher and more differentiated forms over geological time. His solution depended partly on the idea that species were analogous to developing individuals, particularly in their need to adapt to ever narrower and more specific conditions of existence. This led both species and individuals, over time, to differentiate and to progress.[87] Kölliker's alternative to Darwin's theory, "heterogenous generation," similarly looked to individual development for the key to understanding the transformation of species. But it was his particular

85. For a detailed historical study of the word "evolution" and its relationship to individual development, see P. Bowler, "Changing Meaning of 'Evolution'" (1975). For an opposing view, see Richards, *Meaning of Evolution*.

86. At least one recent German dictionary still defines evolution as a "gradual progressive development" (*allmählich fortschreitende Entwicklung*) and *Evolutionstheorie* as "the biological theory of the development (*Entwicklung*) of the organic world from lower to higher forms" (Klappenbach and Steinitz, *Wörterbuch* [1967], vol. 2, p. 1168).

87. Bronn, *Morphologische Studien*, pp. 144–46.

experience with alternation of generations in marine invertebrates that provided the analogy. Finally, Haeckel's familiarity with radiolarians and other marine invertebrates led him to a peculiar concept of individuality that provided the causal basis for his biogenetic law.

At least as serious were the differences among the meanings these men applied to the terms "law" and "development." The oldest generation remained committed to a view of law as a manifestation or representation of an underlying force, a force that had the unique property of moving the organic world to ever higher levels of complexity and perfection. Standing just behind this force was a rational Creator, the discussion of which was beyond the bounds of science, but a faith in which was necessary for the entire framework. And development was by definition a guided process with an outcome known, if not to the scientist, at least to God.

The middle generation displayed a much greater metaphysical modesty, expressed by Kölliker in his focus on laws rather than on trying very hard to explore the forces or causes underlying them; by Leydig in his willingness to consider natural selection as a fruitful hypothesis despite the discomfort one might feel at its philosophical consequences; and by Leuckart in his discussion of the heuristic value of teleology. This generation—still operating in a style probably called forth by the intellectual turbulence of the 1840s, their professionally formative years—was in retreat from anything that threatened to be called metaphysical. They consistently talked in more measured terms about what might be "useful" or "fruitful" in guiding research and consistently focused on the inductive side of generating laws, rather than commenting very much on what might lie behind them. Although they seem to have shared the view that laws were the expressions of mechanical causes rather than immaterial forces, they did not spend much time looking for such causes.

Haeckel's interpretation of Darwin's theory moved attention from the laws to their mechanical causes, highlighting adaptation and heredity as the two important causes of evolution, in addition to being empirically demonstrable "laws." These were not the deepest underlying causes—they themselves could be further unpacked—but they were "mechanical" in three senses: they were emphatically not the result of immaterial forces; as causes, they produced their effects in a deterministic fashion; and as empirical generalizations or laws themselves, they were derived in a determined way from a previous chain of mechanical causes that could be followed all the way back to fundamental physical and chemical processes. Other zoologists of Haeckel's generation were more reticent than he about the reasoning underlying the association of Darwinian adaptation and heredity with mechanical causation, but they generally seem to have accepted it.[88]

Despite all this determinism, Haeckel did allow for true novelty in both individual development and the history of life. This placed his ideas in opposition to Kölliker, for whom all the possibilities for individual development were there *in potentia* within the fertilized cell, and who extrapolated from the individual to argue against the production of any true novelty in nature. For Haeckel, the interplay of adaptation and inheritance, in part through the inheritance of acquired adaptations, allowed for genuinely new features to arise. Development understood historically, at least, could be an open-ended process. In this way, Haeckel's view of development—at both individual and species levels—was strikingly different from those of his older colleagues.

The various German interpretations of the *Origin* discussed here offer a revealing glimpse into the community of scientists concerned with animal form, for Darwin's book provided a rare opportunity for morphologists to reassert their normally implicit scientific values. But morphologists were not at one in their values; rather, the community divided along generational lines. And thus the significance of the *Origin* was different for the different generations. For the oldest generation, who believed in a universe guided by a rational God, Darwin's theory represented a profound challenge—so profound that von Baer thundered against it, while Bronn, though more favorably disposed toward it, still could not overcome his discomfort about the lack of order inherent in a Darwinian world. For the generation who had founded scientific zoology, by contrast, Darwin's book held a different meaning. Darwin's assertions that morphology was "the most interesting department of natural history, and may be said to be its very soul," and that embryology was "second in importance to none in natural history,[89] lent weighty authority to this group's claims that their endeavors were truly scientific. Darwin's theory seemed to demonstrate clearly that the study of form and development did not have to be teleological in the sense of spilling over into metaphysics; rather, the search to understand how an organism's form and function fit with its environment, and the way these relationships changed over time, could be undertaken very much within the boundaries of science. Furthermore, the book demonstrated how fruitful a good heuristic could be, a point that fit well with this generation's more pragmatic

88. See chap. 6, esp. pp. 190–92.
89. Darwin, *Origin*, pp. 434, 450.

and antimetaphysical style. Finally, as is suggested by Haeckel's case and as is explored more fully in subsequent chapters, for members of the generation just embarking on their professional careers in 1860, Darwin's theory offered a new takeoff point for developing a personal program of research, for it provided a way of justifying the long-held belief that embryology, classification, and the study of the history of the organic world were all legitimate and mutually reinforcing realms of morphological inquiry. As we shall see in the following chapters, these different meanings attached to Darwin's theory, and the concerns they reflected, would come into play as morphologists worked to gain and maintain institutional and intellectual ground in the rapidly shifting disciplinary environment of the German university.

## CHAPTER FIVE

# Evolutionary Morphology at Jena

T he appearance of Charles Darwin's *Origin of Species* in 1859 turned out to be a pivotal event in the history of German morphology, for as scientists integrated it into their research over the next two decades the meaning attached to studying animal form shifted considerably. The process by which this change took place was simultaneously intellectual—the work of individuals trying to find acceptable accommodations between Darwin's theory and their own existing research commitments—and social—the development of groups devoted to working out the new research programs and the shaping of institutional forms and priorities to support them.[1] Whereas in the previous chapter I examined the variety of intellectual preoccupations of the three generations of researchers active during the 1860s, in this chapter and the following two I will focus on those who carried evolutionary morphology forward within the disciplines of anatomy and zoology up to the early 1880s. By the time the *Origin* appeared, these two disciplines had become the primary institutional homes for researchers interested in animal form and development.

---

1. I do not mean to suggest here that intellectual activity takes place entirely in the absence of social interaction—I find social constructionist accounts of scientific change through processes of negotiation and exchange to be extremely persuasive. What I want to emphasize, rather, is that however much a scientist's ideas are shaped in response to the ideas of others, they are nevertheless the unique product of that person's processing of his experience, and may be analyzed at an individual level. The development of research groups, disciplines, and the institutional structures that support them, by contrast, are group categories that demand analysis at a social level.

Within both disciplines, advocates of evolutionary morphology sought to claim the high ground of *Wissenschaftlichkeit,* but because of differences in the functions of the disciplines within the universities, in the internal structure of the disciplines, in the personalities of the leaders, and in the characters of the research groups that formed around them, evolutionary morphology took on a somewhat different shape and had different success in the two disciplines.

Although most researchers on animal form in the 1860s and 1870s agreed that comparative anatomy and embryology would provide the main clues to fundamental morphological laws, two important differences were already beginning to separate the zoologists from the anatomists as they began to absorb Darwin's theory. First, the zoologists generally concentrated their attentions on the invertebrates, while the anatomists attended to the vertebrates. There was nothing inherently necessary about this division of labor, but it followed a larger pattern that was emerging by mid-century, in which professors in neighboring disciplines avoided duplication of teaching and research efforts by carving out distinct territory. Because of basic differences in form, function, and development among these animals, the specific problems facing researchers differed quite substantially between the two groups, and points of controversy in one group came to have little to do with those in the other. Although a number of issues were of continuing interest to both groups—the nature of cell division, the classification of forms that seemed to lie between the vertebrates and invertebrates, and the development of the germ-layer theory, which also appeared to unite vertebrates and invertebrates—even those problems came to separate out along disciplinary lines that also corresponded to the material with which the scientists were working.

There were exceptions, of course. Kölliker, as we have seen, ventured into the realm of the invertebrates, as did his near contemporary Alexander Ecker and his student Carl Gegenbaur. And a number of zoologists conducted research on vertebrates in the 1850s and early 1860s. But this period marked the end of easy crossover between the disciplines, and by the 1870s it was commonly accepted that anatomists worked only on the part of the animal kingdom directly relevant to the study of the human form. Zoologists were obliged to cover the entire spectrum, since they still had to teach prospective *Gymnasium* and *Realschule* teachers about local fauna, vertebrates and invertebrates alike. But within the animal kingdom as a whole, the vertebrates took up just one corner, and most zoologists chose to focus on the much larger realm of invertebrates instead. In addition to allowing zoologists to avoid competition with anatomists, the leading appeal of the invertebrates was that many invertebrate groups were less well studied, both taxonomically and morphologically, than the vertebrates and therefore offered promise for publishable results. As a result of these diverging interests and pressures, the bread-and-butter research of anatomists and zoologists branched apart into separate realms.

There was a second difference as well. As we have seen, by the 1850s morphologists were viewed by many of their medical colleagues as holding an outmoded, even reactionary view of life. Rudolph Wagner came to represent the worst, when he made explicit the links between his morphological views and his religious ones in 1854. But the repeated defenses of teleology by scientists such as Reichert and von Baer also marked morphology as the old-fashioned approach of old men.[2] Only a few contemporaries of the 1847 group, born in the late teens, perpetuated the morphological approach as anatomists. Kölliker, for example, adjusted to the times by concentrating on producing a wealth of empirical data and confining his theoretical statements as much as possible to specific problem areas. Where necessary, as in his response to Darwin's theory, he asserted his allegiance to a mechanical worldview and his rejection of teleology. Like Kölliker, most anatomists chose to avoid metaphysical controversy, some by retreating to the level of description—a field that was still wide open at the level of cells and tissues—others by seeking to trace the changes undergone by cells, tissues, and organs during development, and still others by pursuing the more explicitly "mechanical" anatomy developed by Hermann von Meyer in the 1850s.[3] All of these research paths avoided the teleological, "unscientific" associations of the older morphology.

Those morphologists who crossed the institutional line into zoology, however, found themselves in a different world. There, morphology became the watchword of the self-appointed vanguard, those people who were going to bring *Wissenschaftlichkeit* to zoology. The leaders of this group, mainly belonging to Kölliker's intellectual generation, identified morphology as zoology's path out of the swamp of arbitrary classification and anecdotal natural history to the high ground of scientific law. In this setting, zoologists could defend a teleological perspective as an aid to understanding adaptation, even if they conceded that it held merely heuristic value.

2. On Reichert, see chap. 3, esp. pp. 72–73; on von Baer, see chap. 4, esp. pp. 117–21.

3. For more discussion of these orientations, see chap. 7.

As we will see, Darwinian morphology would play different roles in these two settings. Although in both fields Darwin's theory was incorporated into efforts to make morphological inquiry more scientific and synthetically unified, such efforts confronted different alternative approaches and values in zoology and anatomy. Evolutionary morphology consequently had a slightly different history in each discipline.

The beginnings of that bifurcated history can be traced to a common root, however, in a university where the professors of zoology and anatomy treated disciplinary boundaries lightly. At the tiny University of Jena, nestled in the Saale River valley near the center of the German-speaking states, Ernst Haeckel and Carl Gegenbaur founded a new program of Darwinian evolutionary morphology in the 1860s. Although, as we will see in later chapters, some of their contemporaries practiced a science that was both evolutionary and morphological in the terms of the time, Haeckel and Gegenbaur nevertheless set the research agenda that became identified with "evolutionary morphology" by 1870. In the later 1870s and 1880s, after Gegenbaur had left Jena, differences in their approaches would emerge more clearly, as would the different disciplinary pressures affecting morphologists in zoology and anatomy more generally. But to understand the early development of the post-*Origin* evolutionary program of morphology, we must start with Haeckel and Gegenbaur at Jena.

## CARL GEGENBAUR AND ERNST HAECKEL: BIOGRAPHICAL BACKGROUND

The story of the new program is inextricable from Haeckel and Gegenbaur's many-leveled relationship. As researchers and teachers of research, the two cooperated closely during the twelve years they were together at the University of Jena. Although Gegenbaur (1826–1903) was older than Haeckel by eight years (falling between the cohorts of Kölliker and Haeckel), the two became intimate friends who shared their hopes and ambitions, and who consoled each other when both suddenly lost their wives in the same year.[4]

Even before Gegenbaur brought Haeckel to Jena in 1861, the two men's lives shared numerous common threads. Both were the eldest sons of civil servants and followed paths typical of academically in-

4. M. Fürbringer, "Gegenbaur" (1903), pp. 407–10; on Gegenbaur, see also W. Coleman, "Gegenbaur, Carl"; and C. Gegenbaur, *Erlebtes* (1901).

clined young men of the educated middle class or *Bildungsbürgertum*. Gegenbaur was born in the Bavarian town of Würzburg, where he later attended a local *Gymnasium*. Haeckel was born in Merseburg, Prussia, and attended the *Gymnasium* there. During their school years, each became interested in natural history, an interest they expressed by building up herbaria. Both decided to study medicine because they were interested in natural history but needed a more secure occupation to earn them a living while they pursued it. Gegenbaur matriculated at the University of Würzburg in 1845, Haeckel in 1852.[5]

Würzburg was an exciting place for a scientifically minded young man to study medicine in the late 1840s and early 1850s. Albert von Kölliker arrived in 1847 to teach anatomy, physiology, comparative anatomy, histology, and embryology, and his slightly younger contemporary, the radical Rudolf Virchow (1821–1902), was appointed in 1849 to a new chair in pathological anatomy. Gegenbaur studied with both of them and was increasingly drawn to the theoretical and scientific side of medicine, just as Haeckel would be a few years later.

Gegenbaur hoped to follow an academic career, but his finances made that possibility doubtful. To make himself financially independent of his parents, in 1850 he took the post of assistant at the *Juliusspital,* the hospital associated with the university. For over a year, as he completed his doctoral work, he struggled to learn to like practical medicine. The prospects for a practice were uninspiring—a friend of his father had said he could help Gegenbaur's career if he would work in the typhus-ridden Rhön Mountains north of Würzburg. Unenthusiastic about leading the life of a healer, the young doctor decided in 1851 it was time to take a break from medicine and see the world.[6]

Two experiences were especially important to Gegenbaur in making the shift from practical medicine to academic morphology. In Berlin in 1851, he met the famed Johannes Müller, whose intellectual shift from physiology to comparative anatomy, especially of marine invertebrates, was taken by Gegenbaur as a model for his own career. In 1852 he traveled to Messina, Italy, as member of a project led by his former teacher, Kölliker, to study and collect marine fauna. Both events confirmed his interest in natural history and reinforced his

5. Gegenbaur, *Erlebtes,* pp. 13–44; E. Haeckel, *Story of the Development of a Youth* (1923); see also G. Uschmann, "Haeckel, Ernst Heinrich Philipp August."

6. Gegenbaur, *Erlebtes,* pp. 40–55; for a clearer chronology, see Uschmann, *Geschichte der Zoologie,* pp. 28–29.

intention to leave the world of practical medicine. His work on Kölliker's project resulted in fourteen publications on marine animals in 1853, including a comprehensive essay on the alternation of generations and reproduction in medusae and polyps, which served as his *Habilitationsschrift*. At the end of the winter semester of 1853–54, Gegenbaur was named *Privatdozent* for anatomy and physiology at Würzburg.[7]

These experiences, too, had their parallels in Haeckel's life. Already filled with enthusiasm for marine research by Kölliker's reports of the 1852 trip to Messina (and probably urged on by Gegenbaur, with whom he became acquainted in the summer of 1853), Haeckel went to Berlin in 1854 to study under Müller. The next summer he followed Müller to Helgoland, where the two fished for specimens together and Haeckel absorbed as much as he could of Müller's knowledge about marine invertebrates. After another semester at Berlin, Haeckel returned to Würzburg in the spring of 1855. Like Gegenbaur, he later traveled with Kölliker to continue his seaside studies of marine invertebrates.[8]

Meanwhile, Gegenbaur's career was developing along academic lines, if not exactly in the way he had expected. As stimulating as it may have been to be a student at Würzburg, it was not a promising location for a young medical *Privatdozent* in the mid-fifties. Kölliker and his two colleagues Franz Leydig and Heinrich Müller (1820–64) between them covered anatomy and physiology so thoroughly that a beginning *Privatdozent,* bound by courtesy and custom not to tread on occupied territory, could scarcely find a topic to lecture on. So Gegenbaur spent three semesters lecturing on zoology. (Apparently his scruples did not extend to the zoology professor Valentin Leiblein, whom Gegenbaur later described as "an exceedingly boring teacher and also not situated at a very high scientific level, hardly to be considered an opponent.")[9] Following the advice of older colleagues, he also gave a popular course in anatomy and physiology for law students, but after a single semester, Kölliker informed him he should not announce such a course again because it fell within Kölliker's official jurisdiction. A release from this uncomfortable situation finally came in the summer of 1855, when he was called to Jena as *ausserordentliche* professor of zoology. Although his parents still lived in Würzburg and a number of colleagues also urged him to stay, Gegenbaur

7. Fürbringer, "Gegenbaur," pp. 401–3.

8. Uschmann, *Geschichte der Zoologie,* p. 34; G. Uschmann, "Über die Beziehungen" (1976), p. 126.

9. Gegenbaur, *Erlebtes,* p. 83.

knew that he would find neither intellectual nor institutional breathing room so long as he remained. He left with few regrets.[10]

Jena contrasted sharply to Würzburg. Würzburg had a population of roughly twenty-two thousand, and was situated on an important transportation route, the river Main; after 1854 it was linked to other towns by a railroad. Jena, buried in central Germany's Thuringian States, was an "academic village" of some sixty-five hundred residents in the mid-1850s, three hours by foot from the nearest railway station. Würzburg boasted Bavaria's most distinguished university, and in the 1850s, the reputation of its medical faculty was reaching new heights. Jena was known, at least to students, less for its intellectual life than for its bad beer, rowdy fraternities, and romantic past.[11] Nevertheless, Gegenbaur found the intellectual freedom and collegiality at Jena a welcome change from Würzburg. For the first time, he also enjoyed autonomous control over his subject.

During his three years as *ausserordentliche* professor at Jena, Gegenbaur alternated semesters lecturing on zoology and comparative anatomy and conducted laboratory courses in histology and zootomy (animal dissection). At the same time, he continued his research on invertebrates and undertook a major reorganization of the zoological museum.[12] This cluster of responsibilities proved only temporary, however. In 1858 Jena's professor of anatomy and physiology, Emil Huschke, died. Gegenbaur accepted his chair on the condition that someone else take over physiology, which he did not feel competent to teach. That request was granted, and Gegenbaur became the first professor at Jena in recent institutional memory to lecture on anatomy and comparative anatomy without sharing responsibility for physiology as well.[13]

At first, Gegenbaur continued to teach zoology, but that soon changed.[14] Already in the summer of 1858, he had brought Haeckel

10. Ibid., pp. 83–88.
11. On Jena, see H. Koch, *Geschichte der Stadt Jena* (1966), esp. pp. 252–53. On the University of Jena's reputation, see Fürbringer, "Lebenserinnerungen" vol. 1, pp. 159–60, UB Heidelberg, Fürbringer Nachlass; see also R. Hertwig, "Wie ich Ernst Haeckels Schüler wurde" (1914); on Würzburg, see R. A. Pistner, "Würzburg als Handelsstadt" (1968), pp. 69–70.
12. Uschmann, *Geschichte der Zoologie*, pp. 29–31.
13. Although Huschke lectured on physiology, the physiological institute that was set up in 1845 was led by the botanist Mathias Schleiden and the mineralogist E. E. Schmid; the two led a physiological practicum beginning in 1843 that appears to have lasted, along with the institute, until 1856 (see I. Jahn, "Geschichte der Botanik" [1963], pp. 423–36). I am grateful to Dr. Jahn for making these pages of her Ph.D. dissertation available to me.
14. Ibid., pp. 31–32.

to Jena to visit, and he was apparently considering Haeckel's potential as an assistant even then. It was only in early 1861, however, after lengthy official negotiations and private assurances that a professorship would soon be in the offing, that Haeckel came to Jena as *Privatdozent* for comparative anatomy and zoology.[15] A year later he received the promised promotion to *ausserordentliche* professor in the medical faculty. And in 1865, when he was being considered for a professorship back at Würzburg, the university administration at Jena capitulated to his demand for a full professorship of zoology in the philosophical faculty.[16] For the next eight years, Haeckel and Gegenbaur worked closely together as professors and friends, developing their approaches to evolutionary morphology.

## DARWINIAN MORPHOLOGY UNDER HAECKEL AND GEGENBAUR, 1865–1873

Haeckel joined Gegenbaur at Jena just when both men were absorbing the impact of Darwin's theory, and the two quickly molded their research into a new, evolutionary program. As morphologists, their goals differed from those of Darwin. Whereas the *Origin* sought to explain speciation in terms of variation, adaptation, and divergence, for Haeckel and Gegenbaur the compelling mysteries were the connections between and within the largest taxonomic groups in each kingdom, which Haeckel dubbed "phyla" (from the Greek word for root or stock). Their reading of Darwin's theory led them to believe that by determining phylogenetic linkages, they would also be establishing the evolutionary laws of form.

To do this, Haeckel and Gegenbaur relied on a method already available to them: comparison. As Gegenbaur put it in the revised edition (1870) of his *Grundzüge der vergleichenden Anatomie*, "The task of comparative anatomy lies in explaining the manifestations of form in the organisation of the animal body. . . . In this way comparative anatomy offers evidence for the continuity of entire series of organs."[17]

15. Haeckel actually habilitated for comparative anatomy only but taught courses in zoology from the beginning of his career at Jena (Uschmann, *Geschichte der Zoologie*, pp. 34–43).

16. Ibid., pp. 46–51.

17. C. Gegenbaur, *Grundzüge* (1870), p. 6; these sentences are not in the first edition of 1859; indeed, whereas in the first edition Gegenbaur places comparative anatomy in the larger context of a morphology with its own contributions to make that are distinct from physiology, in the second edition he devotes significantly more space to comparative anatomy as a method distinct from other aspects of anatomy.

Haeckel added a second means of pursuing this aim with his biogenetic law, "ontogeny recapitulates phylogeny." The assumption of recapitulation opened a grab bag of comparative approaches that he hoped would bring to light the true paths of evolution. Implicit in the law itself was the legitimacy of comparing an individual's various stages of development with its ancestral line. Since paleontology provided only very limited fossil material for direct comparison with present-day embryos, however, it was fortunate that there were other avenues one could take. Zoologists generally agreed that some present-day forms were more primitive than others, meaning not only that they were less complex, but also that they represented evolutionarily more ancient forms. This assumption appeared to be borne out by the increasing complexity of the organisms in the fossil record as it approached the present. So, for example, sharks were considered among the most primitive vertebrates on the basis of both their apparently simple form and the ancient location of sharks in the fossil record. Similarly, present-day amoebas and other single-celled animals could be used as proxies for the ancient animals that Haeckel presumed to be the Ur-types of all animal life. On the basis of such assumptions, one could compare present-day adults of "primitive" forms with present-day embryonic stages of such "highly evolved" creatures as man and expect to find true homologies.[18]

The assumption of recapitulation also allowed one to compare the developmental stages of two organisms to each other in order to answer phylogenetic questions. If one assumed that the stages manifested in each embryo reflected adult ancestors, then one could dispense with adults altogether. By this reasoning, the longer two ontogenies paralleled each other, the more recently the types they represented branched off from a common ancestor.[19]

Haeckel stressed comparative embryology while Gegenbaur emphasized the comparison of adult structures, but both believed the two methods would work together to achieve the common goal of evolutionary morphology. As Gegenbaur wrote: "The results of morphology join together into a theory of consanguinity (genealogy) of

---

This shift reflects nicely the difference in what was at stake for anatomists in 1859 and in 1870 (see chap. 7); for a comparison of the two editions focusing on different issues, see Coleman, "Morphology."

18. See E. Haeckel, *Anthropogenie* (1874).

19. For a thorough analysis of various theories of recapitulation in the nineteenth century, see Gould, *Ontogeny and Phylogeny;* for a more recent treatment of the same subject from a quite different and challenging perspective, see Richards, *Meaning of Evolution.*

the organisms, and this finds its expression through systematics."[20]
Both men used evidence from comparative anatomy and embryology
in their large, synthetic works of the 1870s, just as they often made
laudatory references to each other's writings.[21]

In addition to developing complementary research methods, the
two also chose to focus their empirical investigations on different
subject areas. Although Gegenbaur had originally concentrated on
marine invertebrates, once he was named professor of anatomy, he
soon shifted his attention to the vertebrates. And although Haeckel
was to increase his fame with his synthetic *Anthropogenie,* which
deployed the biogenetic law to trace the history of humankind from its
earliest single-celled origins, in his own research he confined himself to
the invertebrates. This division of labor was becoming customary for
anatomists and zoologists in the 1860s and did not hold special importance for Jena. It would, however, make a difference later on,
when Gebenbaur developed a closely knit school of vertebrate morphologists and Haeckel produced no real equivalent for invertebrate
morphology.

More significant for the immediate development of their program
was the peculiar combination of Haeckel's and Gegenbaur's personalities and teaching talents. Both men impressed their students power-
fully, but for different reasons. Gegenbaur, with his dark hair, stocky
body, and intense brown eyes, was an imposing character; his student
Max Fürbringer described him privately as "the most awe-inspiring
man, after Bismarck, that I have met in my life. . . . His lectures
were not brilliant, but always deep and testifying to concentrated
reflection . . . ; he belonged to those few men, who never have said
anything stupid or ill-considered."[22] Haeckel had a more charismatic
and buoyant manner. In unforgettable prose, Fürbringer described
him as he appeared to his students in the mid-1860s, at about the age
of thirty:

> And now he stepped into the auditorium, not with the measured step of
> the professor, but with the triumphant charging-along of an Apollonian
> youth, hurrying toward the cathedra, a tall, slender, impressive form;
> a rosy countenance, certainly telling of much reflection and work, but

20. Gegenbaur, *Grundzüge,* p. 6.
21. Haeckel dedicated the first volume of his *Generelle Morphologie* to Gegenbaur;
Gegenbaur dedicated vol. 3 of *Untersuchungen zur vergleichenden Anatomie der Wirbelthiere* to Haeckel; compare also Haeckel, *Anthropogenie* (1874) and Gegenbaur,
*Grundzüge.*
22. Fürbringer, "Lebenserinnerungen," vol. 1, pp. 144–45, UB Heidelberg, Fürbringer Nachlass.

not made sickly by it; an immense brow attesting to a great forebrain; golden, flying locks, large, blue, flashing eyes—probably the most beautiful man that I had ever seen, and it seemed to me as if the room, which had already been bright, became noticeably lighter. And then the lecture began, not with the finishing touches already in place and fully elaborated, but a direct effusion, a showering and sparkling of new revelations. The appearance, the brilliance of thought, and the particular manner of his lecture at first worked exclusively on me; only later did the content grasp me.[23]

In his lectures, Haeckel conveyed to beginning students the excitement of doing science, of asking big questions and framing a new way of looking at the natural world, while Gegenbaur laid greater emphasis on the careful observation and comparative method that was necessary to undergird and flesh out the theory.

Haeckel's enthusiam contributed to his reputation outside of academia, as did his popular works and his outspoken denunciation of organized religion. According to Haeckel, Darwin's doctrine of descent with modification through natural selection would fully explain the harmony manifested in nature without recourse to God or some other immaterial force. His explicit linkage of Darwinism with materialism, coupled with his attacks on church authority and "religious superstition," quickly gained him notoriety, and over his long career hundreds of students were drawn to Jena to hear his provocative lectures.[24]

If Haeckel was evolutionary morphology's most voluble spokesman, Gegenbaur was its greatest practitioner. This was amply demonstrated by the papers and monographs he produced during his tenure at Jena, especially from the mid-1860s, when his evolutionary orientation came to the fore. Indeed, if we want to know what detailed empirical research in evolutionary morphology looked like, it is easier to recover it from Gegenbaur's writings than from those of Haeckel. The period from 1860 to 1873 encompassed nearly all of Gegenbaur's

23. M. Fürbringer, "Wie ich Ernst Haeckel kennen lernte" (1914), pp. 336–37.

24. In H. Schmidt, ed., Was Wir Ernst Haeckel Verdanken, a collection of testimonials honoring Haeckel in 1914, an astonishing number of people recalled that they were introduced to his ideas through his popular Natürliche Schöpfungsgeschichte (1874). The experimental embryologist Hans Driesch was also originally inspired by Haeckel to study natural history (see Driesch, Lebenserinnerungen [1951], p. 32). Haeckel's anticlericalism fit easily into Jena's existing reputation for religious free thought, which extended back at least as far as Goethe's time and in the 1860s made it especially attractive to students of science. For Fürbringer, Jena's attraction lay as much in its ideological and academic openness as in its religious freedom (see Fürbringer, "Lebenserinnerungen," vol. 1, pp. 79, 171, UB Heidelberg, Fürbringer Nachlass).

publications on vertebrate embryology and histology as well as the bulk of his work on the comparative anatomy of the skeleton.[25] His research in these two areas revolved around the appearance of characteristics that define vertebrate form: the development of bony tissue, which marks off higher vertebrates from their cartilaginous forebears, and the emergence of the head and limbs, structures that are first morphologically differentiated from the trunk in the vertebrates. Typically, Gegenbaur combined embryology and comparative anatomy to present a smooth and plausible transition from an Ur-form or primitive feature to be a "higher" one. In doing so, he continually sought to show evolutionary unity and connectedness where others had seen separate and diverse structures.

A good example is his first major publication on the comparative anatomy of the skeleton, published in 1864. Gegenbaur began by setting out the general and specific aims of his work. The general task was that outlined in his textbook, to determine "whether and how the conditions evidenced by higher vertebrates were derivable from the lower forms, whether to some extent common conditions lay at the base of the organizations of all classes, and in what manner the modifications occurred that determined the characteristic features" of the different vertebrate taxa. The specific task was to compare the size, shape, and number of the major hand and foot bones to discover what unified understanding of hand and foot form could be derived that would embrace both higher and lower forms. In seeking to develop a continuous succession of forms, Gegenbaur sometimes drew on embryonic series and sometimes compared adult forms of different classes.[26] Underlying the entire project was a conviction that a particular form represented one moment in a continuous series, and that an understanding of that form was to be achieved by uncovering the succession of forms in the series.

This principle of succession played an even greater role in his 1866 essay on the development of primary and secondary bone tissue. Gegenbaur challenged the prevailing notion that these were fundamentally different types of bone, arguing instead for their identity. The two types of tissue were indeed identical in their mature state; the earlier distinction between primary and secondary bone rested on an apparent difference in their mode of formation. Primary bone was believed

25. Gegenbaur's three-volume *Gesammelte Abhandlungen* (1912) contains two helpfully organized lists of his works at the back of vol. 3, one in chronological order, pp. 575–88, one according to subject matter, pp. 588–97.

26. Gegenbaur, *Untersuchungen*, vol. 1, pp. iv–v.

to form through a transformation of cartilage, while secondary bone grew from a buildup of osseous layers unconnected to the cartilage.

But this understanding of bone growth presented a morphological and developmental puzzle: why should bony tissue that is histologically identical in the adult be produced in two quite different manners, sometimes both taking place in the same structure? In Gegenbaur's view, morphological identity could be produced only through fundamentally identical processes of development. Thus the only solution acceptable to him was that previous researchers were mistaken in their understanding of the tissue's genesis. Through close observation, Gegenbaur discovered that cartilage cells did not transform themselves by development into bone cells (the presumed genesis of primary bone tissue), but rather that secondary bone tissue invaded the cartilaginous areas, destroying the cartilage and replacing it, thereby becoming primary bone. Thus the difference between primary and secondary bone tissue was a matter of degree of ossification—secondary bone tissue merely layered over part of the surface of the cartilage, whereas primary bone completely surrounded and penetrated it. This remarkable discovery meant that primary and secondary bone tissue indeed shared a single mode of development.

Gegenbaur might well have rested content with showing the common developmental history of the two kinds of bone tissue, but he did not. Instead, as further support for his view, he added an evolutionary explanation based on the comparative anatomy of adult vertebrates. By comparing the development of particular bones across the major vertebrate orders, he was able to show that as one progressed from forms commonly accepted as more primitive to ones viewed as higher, both within the individual and in the vertebrate phylum, one could trace intermediate stages connecting primary and secondary bone formation into a single process. The evidence from adult comparative anatomy reinforced his embryological case and brought his discovery into the broader context of evolution, at once lending theoretical weight to his new view of bone formation and reinforcing the value of the evolutionary framework.[27]

This study exemplifies the sort of approach that Gegenbaur and Haeckel sought to teach at Jena, as they worked closely together to nurture evolutionary and morphological interests among their students. Although their lectures provided the initial hook into evolutionary thinking, both men also developed closer contacts with students

27. C. Gegenbaur, "Über primäre and secundäre Knochenbildung" (1866).

through their smaller laboratory courses. Beginning in the winter of 1862–63, Haeckel offered a laboratory class on human histology; starting in 1865–66, his first year as a full professor, he offered "zoological exercises" as well, in the newly founded zoological institute. According to Haeckel's own description, he would give each student "a characteristic animal species from a typical major taxonomic group (class, order) of the animal kingdom, and direct them in exact anatomical (both organological and histological) examination and drawing under my constant supervision."[28] To this end, the zoological laboratory had already acquired at least a dozen new microscopes by 1868, including three compound ones. Student numbers in these courses alone do not seem to have demanded them—Haeckel's formal laboratory enrollments in the late 1860s typically ranged from three to six students—but apparently other students and local schoolteachers made use of the lab as well.[29] During the time Gegenbaur and Haeckel both worked at Jena, Haeckel had no money for a paid assistant, though five advanced students—Anton Dohrn, Nicolai Mikloucho-Maclay, Nicolai Kleinenberg, Benjamin Vetter, and Gottlieb von Koch—acted consecutively as unpaid assistants between 1865 and 1873.[30]

Gegenbaur also ran practical laboratory classes in human anatomical dissection and histology, as would have been expected of an anatomy professor, though as was also typical, he did not run it alone. Instead, he hired a long-term salaried assistant and two other short-term assistants from among his former students, who worked together to supervise the laboratory courses (which typically had enrollments of thirty to forty) and prepare specimens for Gegenbaur's lectures.[31] Gegenbaur also ran a separate practice course for advanced students[32] and had a series of more senior students conducting more or less independent research in his lab; he drew many of his assistants from this pool.[33]

Although each professor had his own laboratory, and most advanced students conducted their research primarily under one professor or the other, the influences of both Gegenbaur and Haeckel can

28. Haeckel, "Jahresbericht für 1868," dated 25 March 1869, quoted in Uschmann, *Geschichte der Zoologie*, p. 62.

29. Uschmann, *Geschichte der Zoologie*, pp. 46, 57, 59, 63.

30. Ibid., pp. 65–66, 70, 126, 87.

31. Fürbringer, "Lebenserinnerungen," vol. 1, pp. 250–53, UB Heidelberg, Fürbringer Nachlass.

32. Fürbringer, "Gegenbaur," p. 412.

33. Fürbringer, "Lebenserinnerungen," vol. 1, p. 276, UB Heidelberg, Fürbringer Nachlass.

be detected in the work of their advanced students. Max Fürbringer, for example, was inspired initially by Haeckel but eventually considered himself a Gegenbaur student. Another Jena student, V. O. Kovalevskii, conducted comparative anatomical research under Gegenbaur but later received his doctorate in zoology under Haeckel. Benjamin Vetter, who arrived with his doctorate already in hand, worked as Haeckel's assistant but conducted his research under Gegenbaur's supervision. Other students wrote dissertations under Haeckel on topics that clearly derived from Gegenbaur's theories. Such stories could be told of at least a dozen advanced students who worked at Jena between the mid-1860s and 1873.[34]

The intellectual closeness of anatomy and zoology at Jena were reinforced in various ways outside the lab and lecture hall. The institutional links between the medical and philosophical faculties were already stronger at Jena than elsewhere: whereas in Prussia by 1861, zoology had been relegated along with the other natural history subjects to an extremely minor part of the basic science exam for medical students (*tentamen physicum*), at Jena, the university for the principalities of Weimar-Eisenach, Altenburg, and Gotha, the basic exam still included zoology, mineralogy, and botany as separate examination subjects. Haeckel was the examiner in zoology (and at least on one occasion, in botany, too). In addition, medical students at Jena were required to take courses in both human anatomy and zoology.[35] The fluidity between the fields that could result is best exemplified by the brothers Oscar and Richard Hertwig, who collaborated on nearly all of their early works and spent so much time together that they were known during their student days as "the Siamese twins." With virtually identical qualifications, Oscar gained his lecturing credentials for anatomy in Jena's medical faculty, while Richard did the same for zoology in the philosophical faculty.[36]

Several of the men who worked closely with Haeckel and Gegenbaur in this period, most of whom were born between 1845 and 1852,

34. On Haeckel and Gegenbaur's joint influence on students at Jena, see Uschmann, *Geschichte der Zoologie*, pp. 102–3; on Vetter, see ibid., p. 127; on Fürbringer, see "Lebenserinnerungen," vol. 1, p. 276, UB Heidelberg, Fürbringer Nachlass.

35. Uschmann, *Geschichte der Zoologie*, pp. 63–65. Although the German states had agreed to honor each other's medical degrees and exams in 1869, the subject matter of the preliminary exams was not regularized across the different states until 1883 (see "Bekanntmachung des Reichskanzlers, betr. die ärztliche Prüfung, 2 June 1883," reprinted in G. Liebau, *Das Medizinal-Prüfungswesen* [1890]), pp. 22–94).

36. Fürbringer recalled the Hertwigs' nickname in "Lebenserinnerungen," vol. 1, pp. 320–21, UB Heidelberg, Fürbringer Nachlass; on the Hertwigs' habilitation, see Uschmann, *Geschichte der Zoologie*, pp. 89–96.

went on to attain professorships at German universities. The most prominent were Richard Hertwig (1850–1937), who became one of the leading zoologists of his generation and professor of zoology at Munich; Oscar Hertwig (1849–1922), who distinguished himself as one of Germany's most prominent researchers and eventually became professor of comparative anatomy and embryology at the University of Berlin; Theodor Wilhelm Engelmann (1843–1909), professor of physiology first at the University of Utrecht, then at Amsterdam, and finally at Berlin; and Max Fürbringer (1846–1920), who taught anatomy at Jena and Heidelberg. A few others also maintained academic careers: Georg Ruge (1852–1919) taught anatomy, first in Amsterdam and then at the Swiss university of Zurich; Gottlieb von Koch (1849–1914) became professor of zoology at the technical school at Darmstadt; and Benjamin Vetter (1848–93) took a similar professorship at the technical school in Dresden.[37]

Still other morphological researchers from this period at Jena remained independent of university positions, though not always by choice. V. O. Kovalevskii (1842–83) was unable to find an academic position in his homeland of Russia until 1881; two years later he committed suicide. The wealthy friends Nikolai Mikloucho-Maclay (1846–88) and Anton Dohrn (1840–1909) did not even try for university positions but conducted research on their own and devoted much of their energies to developing independent zoological research stations. Dohrn founded and directed the Zoological Station at Naples, Italy, while the Russian Mikloucho-Maclay founded a similar station in Sydney, Australia, where he spent much of his short career.[38]

These students absorbed the evolutionary outlook of their teachers but did not necessarily stay tied to Haeckel and Gegenbaur's particular program of determining phylogenies by means of comparative anatomical and ontogenetic studies. Dohrn remained true to the aim but, like many others, called for increasing the attention given to physiology and adaptation. Mikloucho-Maclay, traveling in the southern Pacific, took up anthropometric studies of aboriginal peoples

37. All of these men are mentioned by Fürbringer in "Gegenbaur," p. 413; further information and references on O. Hertwig may be found in P. J. Weindling, *Darwinism and Social Darwinism* (1991); for a list of articles on R. Hertwig, see R. Weissenberg, *Oscar Hertwig* (1959), pp. 56–57. I have been unable to find any extensive biographical information on Gottlieb von Koch or Benjamin Vetter.

38. On Kovalevskii, see D. Todes, "V. O. Kovalevskii" (1978); on Mikloucho-Maclay and Dohrn, see I. Müller, ed., *Mikloucho-Maclay* (1980).

there. Richard and Oscar Hertwig turned their research to issues of fertilization and transmission.[39] To some extent, these diverse interests are a sign of the multiplicity of questions raised by evolutionary theory. These men were responding to problems raised by evolution that Haeckel and Gegenbaur's program did not address.

At the same time, their range of interests also reflects the tone of their training. As advanced students, most of these men worked less closely with Gegenbaur than with Haeckel, who by all accounts ran a highly unstructured laboratory. He often let his students choose their own research projects (practically unheard-of in a German scientific institute) and gave them little supervision.[40] By contrast, those who worked primarily under Gegenbaur, namely, Max Fürbringer and Georg Ruge, remained much more tightly allied with the research program of their teacher (see next section below).

This divergence between the styles of Haeckel and Gegenbaur became more apparent beginning in the early 1870s, as the two men began to turn their own efforts in new directions. Each came out with a new, substantive theory in those years, the defense and elaboration of which would occupy much of their individual efforts over the next twenty years. In a research monograph of 1872 and then in a series of articles published between 1873 and 1875, Haeckel introduced the world to his *Gastraea* theory. In his eyes the concrete demonstration of the biogenetic law, the *Gastraea* theory was his major contribution to biological theory. Arguing that the early cup-shaped gastrula stage of development was a universal feature of multicelled animals, he posited the existence of an ancestral form, the gastraea, that corresponded to the gastrula and that was a common ancestor to all subsequent animals.[41] At just about the same time, Gegenbaur brought out two related theories of his own to explain the evolution of the vertebrate skeleton. In 1870, he presented the first version of his "archipterygium" theory of the extremities, which derived both the side-fins of fishes and the fore- and hind limbs of the higher vertebrates from a common primitive form.[42] And in 1872, he published the first statement of his theory of the vertebrate skull, in which he derived its

39. I. Müller, *Mikloucho-Maclay;* Weindling, *Darwinism and Social Darwinism.* On Dohrn's views, see Kühn, *Anton Dohrn* (1950).

40. Uschmann, *Geschichte der Zoologie,* p. 106.

41. For a more detailed analysis of the gastraea theory and its reception among German zoologists, see pp. 181–200.

42. Gegenbaur, "Über das Skelet der Gliedmassen" (1870); On further developments of this theory, see chaps. 7 and 8 below.

bony parts from modified vertebrae and derived the cartilaginous parts separately.[43]

Perhaps had Gegenbaur remained at Jena, their paths would not have diverged as much as they did, and there would have remained a single, well-integrated program of evolutionary morphology based at Jena. As it happened, however, Gegenbaur left Jena just at the time that he was developing a theoretical framework that warranted a narrower focus on the vertebrates; it was at Heidelberg that Gegenbaur's distinct school of evolutionary morphology flourished. Focusing on the evolution of the basic features of vertebrate form—the head, the limbs, and the vertebral column—Gegenbaur's school was closely tied to his theory of vertebrate evolution. Although Gegenbaur continued to exercise an influence in academic politics at Jena, the subsequent intellectual character of evolutionary morphology there was shaped more clearly in Haeckel's mold.

## AT JENA AFTER GEGENBAUR

After Gegenbaur's departure in 1873, Haeckel, too, harbored some hope of leaving Jena for a larger university, but the right call never came.[44] Instead of allowing himself to feel isolated at Jena, Haeckel worked to promote his program within the university. Not only did he continue to teach his own version of evolutionary morphology in his lectures and practica, in addition, whenever new professorships in related areas opened up within the university, he wielded his influence to tilt the hiring decisions toward candidates actively sympathetic to his own approach and values.

The easy cooperation between anatomy and zoology encouraged by Haeckel and Gegenbaur threatened to be disrupted by the latter's departure in 1873. Between then and Haeckel's retirement in 1909 the professorship of anatomy turned over four times, and the medical faculty did not always want someone quite in Gegenbaur's mold. At least once, Haeckel's influence was critical in shifting the balance to favor a Gegenbaur student, who could be counted on to be amenable to Haeckel's own approach. These episodes and those surrounding the appointment of other professors in the medical and philosophical faculties illustrate well Haeckel's power at Jena, as well as the abiding influence that Gegenbaur wielded from Heidelberg. Both help us un-

43. Gegenbaur, *Untersuchungen,* vol. 3.
44. Haeckel did enter into negotiations for a position at Bonn in 1874, but the Weimar government's counteroffer was enough to keep him in Jena (Uschmann, *Geschichte der Zoologie,* pp. 80–82).

derstand how Jena remained such a strong center for evolutionary morphology.

Little information is available on the appointment of Gegenbaur's immediate successor, Gustav Schwalbe, but it seems clear that the medical faculty was willing to try for a change. Schwalbe's orientation was decidedly histological and physiological: not only had he studied with the great technical microscopist Max Schultze in Bonn, but he had also worked as an assistant in the institute of the physiologist Willy Kühne in Amsterdam and as extraordinary professor for histology in Carl Ludwig's physiology institute in Leipzig. Although Schwalbe's student and memorialist Franz Keibel stressed Schwalbe's physiological turn, he was also recalled by another student as a member of a group of professors in the biological sciences (including Haeckel) whose "outlooks and grasp of principles harmonized with one another."[45] Indeed, Schwalbe gave his strongest support to the advancement of Haeckel and Gegenbaur's former student Oscar Hertwig to *Privatdozent* for comparative anatomy, quite possibly because he was happy to turn over a subject that wasn't his specialty to someone else.[46]

By the time Schwalbe decided to leave Jena to accept the anatomy professorship at Königsberg in late 1880, the importance to Haeckel of having an anatomist sympathetic to his own views was clear. Haeckel hurried back from a vacation in Italy to lobby the medical faculty to alter their list of candidates, which favored anatomists uninterested in Darwinian morphology. He pressed instead for Oscar Hertwig, who by then was working as an *ausserordentliche* professor at Jena. With the help of Gegenbaur, who fired off appropriately strongly worded letters from Heidelberg, Haeckel successfully pressured the physiologist Preyer to shift his swing vote, while also waging a campaign to persuade the university's highest official (the *Curator*) and the ministry in Weimar to hire Hertwig. Although the medical faculty could not come to unanimity over the candidates, Haeckel's higher-level efforts succeeded, and in early 1881 Hertwig was appointed to the chair of anatomy.[47]

When Hertwig left for Berlin in 1888, the composition of the medical faculty and its attitude toward evolutionary morphology had shifted sufficiently that another Gegenbaur disciple, Max Fürbringer, was nominated without much controversy. Although Fürbringer re-

45. Hans Gadow, quoted in ibid., p. 105; on Schwalbe, see F. Keibel, "Schwalbe" (1916–17).

46. Uschmann, *Geschichte der Zoologie*, pp. 91–92.

47. On Hertwig's appointment as *ordentliche* professor, see ibid., pp. 96–100.

called that there had been a faction favoring another candidate, Gegenbaur's recommendation, combined with the local support of Hertwig, Haeckel, the pathologist, the professor of surgery, the physiologist, and the professor of ophthalmology (who was also the son-in-law of the minister of state), prevailed.[48] And in 1900, when Fürbringer departed for Heidelberg to take up Gegenbaur's mantle there, he successfully managed, with Haeckel's help, to install yet another Gegenbaur student, Friedrich Maurer, as professor of anatomy.[49]

Although the closest bonds supporting evolutionary morphology as an interdisciplinary enterprise were initially those between Gegenbaur and Haeckel, the history of the program at Jena after Gegenbaur's departure in 1873 shows that it was not wholly dependent on this particular relationship, or indeed on the close relationship between anatomy and zoology. Haeckel was enormously successful in using his considerable influence to gather around him in the medical and philosophical faculties a collection of professors sympathetic to evolution, animal morphology, and himself.

In 1888, for example (the year Hertwig left), the physiologist Preyer gave his notice as well, providing another opportunity for Haeckel to exercise his clout. He had his eye on Theodor Wilhelm Engelmann, who had studied at Jena in the early 1860s and had heard Haeckel give his first course of lectures on Darwin's theory in 1862–63.[50] Engelmann had since become professor of physiology at the University of Utrecht in the Netherlands but had expressed interest in returning to Germany if the conditions were right. For his part, Haeckel viewed Engelmann as one of the very few active physiologists with any sympathy for morphology and exerted himself to persuade both his Jena colleagues and Engelmann himself that Jena and Engelmann were a perfect match. Although Engelmann did visit Jena that summer, his financial demands could not be met, and Haeckel regretfully told him that the medical faculty had replaced him with a new candidate.[51]

The physiologist who accepted the Jena chair was Wilhelm Biedermann (1852–1929), a man upon whom Haeckel also had reason to

48. Fürbringer, "Lebenserinnerungen," vol. 3, pp. 1a–2a, UB Heidelberg, Fürbringer Nachlass.

49. See pp. 213–14.

50. Uschmann, *Geschichte der Zoologie*, p. 65, n. 306.

51. Haeckel to W. Engelmann, 2 July 1888 (no. 66), 15 July 1888 (no. 68), and 27 July 1888 (no. 71), SBPK Lc 1875 (13) E. Haeckel.

look with favor. Biedermann had trained under the charismatic Ewald Hering, who viewed himself as an outsider to the physiological establishment, and who introduced Biedermann to the possibilities of comparative physiology, a less well developed side of the subject that could not help but appeal to Haeckel.[52] Once at Jena, Biedermann quickly joined the social and intellectual circle around Haeckel, which now included Fürbringer, the professor of pathology Wilhelm Müller (Fürbringer's cousin by marriage), the botanist Ernst Stahl, and their families.[53]

Finally, the professorship in botany, the subject in the philosophical faculty most closely related to Haeckel's own, was also occupied by men who could be counted on to support his general evolutionary direction. In 1869, the first time the botany chair opened up after Haeckel was a professor, the position went to Eduard Strasburger (1844–1912). As a student at Jena in the mid-1860s Strasburger had studied under Haeckel and taken a secondary field in zoology for his doctoral exam. By the time he was promoted from *ausserordentliche* professor to full professor in 1873, he had become a complete convert, not just to evolution but to Haeckel's phylogenetic program. Certainly his inaugural speech, "On the Meaning of Phylogenetic Methods for Research into Living Beings," which called for using individual developmental stages to draw evolutionary inferences, could have been dictated by Haeckel—only the examples were from botany rather than zoology.[54] One might suspect that the occasion dictated the thrust of his remarks, and it is certainly true that even in this period, Strasburger was best known for his work on cellular morphology and division in plants, but his sympathy with Haeckelian evolution and with Haeckel himself persisted through his career.[55] When Strasburger moved on to Bonn in 1881 he was replaced by Ernst Stahl (1848–1919), a botanist whose career was occupied with the Darwinian problem of plant adaptation. Again, his research was not in the Haeckelian mold—he had little interest in constructing plant phylogenies—but his more physiologically oriented work on the relationship

52. On Hering as leader of a school, see R. S. Turner, "Vision Studies" (1993); on Hering's suggesting comparative physiology as a fruitful area, see Wilhelm Biedermann, quoted in E. Giese and B. von Hagen, *Geschichte der medizinischen Fakultät* (1958), p. 495.

53. Uschmann, *Geschichte der Zoologie*, p. 182 n. 916; Giese and von Hagen, *Geschichte der medizinischen Fakultät*, pp. 494–95.

54. Strasburger, "Über die Bedeutung phylogenetischer Methoden" (1874).

55. Uschmann, *Geschichte der Zoologie*, pp. 67–68; on Strasburger's work in plant cytology, see W. Coleman, "Cell, Nucleus, and Inheritance" (1965).

of plants to their environments was at the Darwinian end of the range
of contemporary botanical research.[56]

Haeckel's ability to attract a nearly continuous supply of supporting professorial characters in complementary fields during the nearly
fifty years he taught at Jena testifies to his great personal power there.
This was apparent not only in the large numbers of students who
faithfully attended his charismatically delivered lectures but also in
the ways that he and his allies were able to assure continued support
for professors sympathetic to evolution in general, and Haeckelian
evolutionary morphology in particular. Yet despite his great success
in creating and sustaining an atmosphere congenial to his program,
one of the most notable aspects of Haeckel's tenure at Jena after
Gegenbaur's 1873 departure is the dearth of Haeckelian students who
went on to work as professors of zoology in Germany. In his detailed
study of zoology at Jena, Georg Uschmann discussed the students
who wrote dissertations under Haeckel. Of the thirty who completed
zoology dissertations in the decade following Gegenbaur's departure,
not a single one obtained a full professorship of zoology in Germany.
Those who gained reputations in zoology found their long-term niches
elsewhere: Paul Mayer (1848–1923), who completed his dissertation
in 1874, became a permanent employee at the Naples Zoological
Station and editor of the station's publications as well as the author
of important textbooks on microscopic technique; the Swiss student
Arnold Lang (1855–1914), after being named the first "Ritter Professor of Phylogeny" at Jena in 1886 (a position carrying the rank of
*Extraordinarius* only), moved on to an *Ordinarius* in Zurich in 1889;
Hans Gadow (1855–1928) ultimately found employment at the zoological museum at the University of Cambridge; Wilhelm Haacke
(1855–1912) eventually (and relatively briefly) became director of
the Frankfurt Zoo; and Richard Semon (1859–1918), who would
later coin the term "Mneme," received doctorates at Jena in both
zoology and medicine and gained an *ausserordentliche* professorship
in the medical faculty there in 1891, only to retire to private life
and study in 1897, after running off with the wife of another Jena
professor.[57]

Nor did the situation improve substantially in subsequent years.
Between the winter of 1883-84, when Haeckel's new institute building
was completed, and his retirement in 1909, Haeckel increasingly removed himself from his students, concentrating instead on his intro-

56. On Stahl, see E. Cittadino, *Nature as the Laboratory* (1990), pp. 83–96.
57. Uschmann, *Geschichte der Zoologie*, 105–24.

ductory lectures, his own research, and his ever more philosophical and popular writing. The zoological practica continued, but after the mid-1880s Haeckel continued only with the "small practicum," leaving the larger introductory laboratory class and technical training of more advanced students to the Ritter professor.[58] During this last quarter-century of his teaching, even fewer German zoologists of note worked with him, either at the doctoral or at the advanced level.[59]

Thus one must agree with Uschmann, who noted that Haeckel never produced a "school" in the usual sense of a group of students dedicated to pursuing a particular research program[60]—this despite the fact that his gastraea theory, which will be discussed in the following chapter, could have served as the foundation for just such a school. Rather, his influence was diffused in other ways: through his contact with students in different disciplines in the small and interdisciplinary atmosphere of Jena, through the advanced students who went on to work in different countries, and most important, through his theoretical writings and those of his popularizers.[61] This is not to say he was without impact on the discipline of zoology—as we will see in the next chapter, his ideas were certainly noticed in his own field! But to the extent that one might look for a distinctive Haeckelian school of professional zoologists trained under his tutelage, one looks in vain. One can attribute this partly to Haeckel's own ineptness in newer microscopic techniques involving staining and slicing of preserved tissue, which meant that students had to rely on the zoology assistants and later, Ritter professors, or on teachers in other fields (such as Strasburger in botany) to teach them the microscopic techniques that were viewed increasingly as supplying a necessary credential for a future zoologist. But also, part of Haeckel's style was not to dominate in the laboratory anyway—he did not seem especially interested in thinking up small-scale empirical topics for dissertators to pursue. He seemed far readier to stand as an intellectual catalyst and let people pursue their interests from there on.[62] It is one of the contradictions

58. Ibid., pp. 153, 155, 160, and esp. p. 175.

59. Among the more notable zoologists who took their doctorates under Haeckel during this period were Ludwig Plate (1862–1937), Haeckel's successor at Jena; Hans Driesch (1867–1941); and Curt Herbst (1866–1946), ibid., pp. 176–96.

60. Ibid., pp. 123, 175.

61. Alfred Kelly has argued persuasively that Haeckel's greatest popular influence came not from his own writings but from those of such popularizers as Wilhelm Bölsche (see Kelly, *Descent of Darwin*, esp. chaps. 2 and 3).

62. On the role of intellectual style in the sort of influence a teacher has on a student, see J. Fruton, *Contrasts in Scientific Style* (1990); and D. Krantz and L. Wiggins, "Personal and Impersonal Channels of Recruitment" (1973).

of his character that despite the lack of supervision he gave advanced students in the laboratory, he was so often disappointed in those who did go in different directions, viewing them as having betrayed him personally when they came to disagree with him.

## CONCLUSION

As a teacher of future zoologists, the most fruitful period of Haeckel's career was near its beginning, that first decade or so at Jena when he was young and most actively engaged in new empirical research, when Darwinism was a novel and possibly dangerous idea, when his theorizing was more exploratory than it later would become, and when he was cooperating closely with Gegenbaur. None of these elements alone seems to be able to account for the relatively greater number of distinguished zoological researchers to come out of Jena in that decade as compared to later. In addition, as will be made clearer in later chapters, Haeckel's apparent inability to produce students suitable for academic positions after 1873 may be viewed as part of a larger phenomenon in German zoology between the mid-1870s and early twentieth century, namely, the lack of suitable zoological positions to be had. Already in the mid-1870s, zoologists were beginning to worry about the lack of professorships to meet the supply of advanced students.[63] And the situation only grew worse over time, continuing well into the second decade of the twentieth century. Thus any discussion of the relative success (or lack of it) of the morphological program begun at Jena must take into account not only its intellectual and social character but also the larger disciplinary contexts into which the program fit in the late nineteenth century.

When Gegenbaur left Jena, the remarkable collaboration that produced the Ur-program of evolutionary morphology diverged into two intellectual and social streams that continued in the same general direction but came to be of quite distinct character. The group that coalesced around Gegenbaur at Heidelberg after 1873 conforms to the model of a typical research school, with a well-defined theoretical problem and set of disciples; the group around Haeckel never had that character. This means that if we want to place these two men and their ideas in their disciplinary contexts, they require somewhat different treatment. Among anatomists, Gegenbaur and his school represented evolutionary morphology within their discipline; it is relatively easy to treat evolutionary morphology within anatomy as a

63. See pp. 193–95.

mostly self-contained package represented by his school.[64] Because Haeckel did not really found a school, understanding the place of his program in the discipline of zoology is not so much a matter of examining how he and his followers stood up against proponents of alternative programs. It is, rather, a question of seeing how his program was received, reshaped, and appropriated by other zoologists. Furthermore, even when the program was first being developed in the 1860s, zoology had a different character than anatomy. Whereas anatomy was organized around fairly distinct problem areas, zoology was characterized by a certain diffuseness in its problems. The next two chapters will turn on the theoretical programs, respectively, of Haeckel and Gegenbaur from the mid-1860s to the early 1880s and their fit into the disciplines of zoology and anatomy.

64. The further development of Gegenbaur's school and its relationship to the discipline of anatomy is the subject of chap. 7.

## CHAPTER SIX

## Evolution and Morphology among the Zoologists, 1860–1880

The reception of Haeckel's program among German zoologists in the 1870s and 1880s is best framed by beginning in the 1860s, the decade when the entire zoological community was beginning to grapple with Darwin's theory and its consequences for the study of animal form. In 1860, it will be recalled, morphology was a core element in the research program of scientific zoology, whose proponents set the search for laws of form, adaptation, and development over against the older project of classification. Even as scientific zoology persisted, by the early 1880s a different approach to form had emerged as well, identified with Ernst Haeckel's program of creating evolutionary taxonomies guided by the study of individual development and separated from the study of the organism's function and its relation to its external environment. Although Haeckel's program may be viewed as having evolved out of the broader orientation of scientific zoology, zoologists seeking to take the older orientation in other directions found its peculiarities of content and tone troublesome.

Haeckel was not the only researcher to propose a way of integrating evolutionary theory into zoology; as they absorbed the initial impact of Darwin's theory in the 1860s, zoologists came up with a variety of connections between evolution and their ongoing research problems. The reception of Haeckel's program in the zoology community must therefore be seen partly in relation to these other efforts of his contemporaries and his older colleagues, who all tended to hew more closely to the established empiricist tenets of scientific zoology.

Most of these zoologists found Haeckel's speculative, dogmatic approach offensive; it exceeded the bounds of intellectual modesty that they viewed as essential to good science. Furthermore, Haeckel's theory steered attention away from the questions about adaptation that his contemporaries thought must be taken into account in a truly scientific explanation of form.

If zoologists of Haeckel's generation and older tended to view his program with deep mistrust, the cohort of zoologists coming into the field in the late 1860s and after—those young enough to be students of Haeckel and his contemporaries—saw his theory from a different perspective. The overwhelming fact of their professional life was that they were entering a newly crowded discipline, with many more job aspirants than positions. Under these circumstances the pressure was great to publish rapidly work of both theoretical and empirical significance, and many younger zoologists found in Haeckel's framework a guide for doing just that. In the process, they changed the cluster of problems and concepts with which morphology was associated. It was not Haeckel's theory alone that wrought this new meaning of morphology; rather, the younger generation adapted his program to their professional needs, taking his framework in a direction with which even Haeckel himself was not entirely comfortable.

## THE CHARACTER OF ZOOLOGY IN THE 1860s

For the self-declared scientific zoologists who had left the medical faculties for philosophical faculties beginning in the 1850s, the great programmatic division in the discipline was that between themselves and the systematists, between morphology and old-fashioned classification.[1] This split continued to be invoked in hiring decisions as late as 1868, when a search committee at Leipzig argued that zoology was no longer "a 'descriptive science': the structure, development and physiology of animals are the modern tasks of zoology . . . ; without a thorough knowledge of these parts of zoology a scientifically trained zoologist is unthinkable."[2] Even in 1893 a zoologist could describe the field in these terms, as Richard Hertwig did when he wrote that

1. The leading scientific zoologists of the 1850s and early 1860s were Albert von Kölliker and Carl Theodor von Siebold, coeditors of the *Zeitschrift für wissenschaftliche Zoologie*; Franz Leydig; Julius Victor Carus; and Rudolf Leuckart. Johannes Müller followed much the same intellectual path but does not appear to have been very close socially to this group (for more on scientific zoology, see chap. 3).

2. Philosophische Fakultät Universität Leipzig to Ministerium des Cultus (Dresden), 5 December 1868, UA Leipzig, Personalakte R. Leuckart.

"at the present time, nearly all chairs in zoology are taken by 'morphologists'" rather than by practitioners of systematic zoology.[3]

In this division, a morphologist was anyone who did not pursue traditional problems organized around systematics but concentrated instead on problems concerning the "structure, development and physiology of animals." This focus did not preclude making statements about the taxonomic organization of the animals under investigation, of course, but such classificatory results were not the primary aim of the scientific zoologists, who sought instead the laws governing the internal organization, development, and generation of animals. Most important, as the statement by the Leipzig search committee indicates, this configuration of zoological problems linked the study of an organism's structure and development to the study of its function rather than separating the two.[4] As we will see, this feature of morphology was one that Haeckel's program would dissolve.

The oldest active generation of zoologists in this period, comprising the systematists and morphologists who had been at odds with one another in the 1850s, continued to perceive the field in terms of this division. Referring to several systematists, Carl Theodor Ernst von Siebold wrote in 1867 of the "zoologists of the old school, such as Troschel, Peters, Grube etc." who haven't kept up with the "present-day standpoint" of zoology, and again in 1876 complained, "When will Berlin realize that a zoological cabinet is not the means by which the science of zoology can be supported?"[5] From the other side, the systematist Gustav Zaddach eulogized his recently deceased colleague A. E. Grube in 1881 as follows: "It is much easier to swim with the dominant stream in science than to turn onto a lonely path; it is much more enticing to track down general laws of transmission and development in animals . . . than painstakingly to weigh the common and divergent characteristics of various species, to compare each specimen with many others and to alter and polish the diagnosis for each species until the determining characteristics are razor sharp."[6]

For those who had not been among the first generation of scientific zoologists, however, the division was less significant. Indeed, the cohort that was establishing itself in the 1860s and that would dominate

---

3. R. Hertwig, "Zoologie und vergleichende Anatomie" (1893).

4. It will be recalled that such a separation between the studies of form and function was what the same scientific zoologists, when still in the medical faculties, had found so objectionable in the commitments of the physicalist physiologists (see Chap. 3).

5. Siebold to Claus, 10 November 1867 (Bl. 57), SBPK, 3k 1855(5): C. T. E. von Siebold. Siebold to Ehlers, 26 October 1876, UB Göttingen, Cod. MS E. Ehlers 1815.

6. G. Zaddach, "Adolph Eduard Grube" (1880), p. 121.

German zoology in the 1870s and 1880s did not view classification and morphology as ideologically opposed. Nearly everyone in this group, which, besides Ernst Haeckel (1834–1919), included Anton Schneider (1831–90), Carl Semper (1832–93), August Weismann (1834–1914), Carl Claus (1835–99), and Ernst Ehlers (1835–1925),[7] published research in both areas in the 1860s. One does not have to look far to find direct influences on the morphological side of their research: Haeckel was trained by Johannes Müller and Albert von Kölliker, two of the great morphologists who remained in medical faculties; he also became close to Siebold. Schneider studied with Müller. Semper spent his career at Würzburg, first as a student of Kölliker, Franz Leydig, and Carl Gegenbaur, and then as a professor. Weismann and Claus both claimed Leuckart as their most important teacher. Ehlers felt the abiding influence of Siebold and the morphologically oriented physiologist Rudolph Wagner. One might almost think that in the middle and late 1850s it was hard to avoid being trained in the functional morphology of the scientific zoologists.

Their rapprochement to classification seems to require more explanation, and here one can quite reasonably look to the influence of Darwinism. The theory of descent invited people to see one set of laws guiding development, generation, and taxonomic divisions; classification suddenly appeared to have just the lawful, scientific basis that the tradition of scientific zoology would have required. But other factors counted as well: first, the practices of classification and morphology were not nearly as separate as the rhetoric would have them. As Zaddach's statement suggests, the general technique of comparing specimens was just as much the property of systematists as of morphologists. For those not personally engaged in old programmatic battles, holding them apart may have seemed unnecessary. Second, where zoology had been institutionally grouped with comparative anatomy, anatomy and/or physiology as part of the medical curriculum, its classificatory side could remain subordinated. But when it gained institutional independence, it almost necessarily took on classification as one of the major topics under its jurisdiction. Classification had been the central task of zoology for too long for it to be dropped out of hand. Finally, as we will see, the zoological community placed a high value on the breadth of a scholar's research, and this may also

---

7. Wilhelm Keferstein (1833–70) was also a member of this cohort but died before he could develop his anti-Darwinian arguments into a full-blown program. Another member of the cohort active in working out Darwinism was Gustav Jäger (1832–1917), who became teacher of zoology at the Polytechnic in Stuttgart and the agricultural school at Hohenheim.

have invited individuals educated in scientific zoology to try their hand at classification as well.

In any event, the publications of the cohort establishing itself in the 1860s suggest that its members felt it necessary to demonstrate their facility at classifying in addition to their mastery of morphological problems and techniques.[8] So many published an early major work containing such demonstrations of multiple mastery that one might suspect that the older scientific zoologists, despite their rhetorical claims, thought it a good idea for their students to establish their credentials by means of such a work. Indeed, this sort of study can be traced back well before 1859 and would continue to form the bread-and-butter research of zoologists for decades to come. The enduring importance of such basic empirical embryological, anatomical, and classificatory research is evident not only from the larger monographs but also from a glance at the two leading zoological journals, the *Archiv für Naturgeschichte* and the *Zeitschrift für wissenschaftliche Zoologie,* the pages of which are filled with contributions along these lines.

To suggest that zoology in the 1860s consisted only of descriptive anatomy, embryology, and classification, however, would be to ignore three other important clusters of problems that gave the field a much greater richness. The most important of these was the study of generation, which had been one of the liveliest areas of research in the 1850s and remained the chief interest of Carl Theodor von Siebold and Rudolf Leuckart, the two leading scientific zoologists. Often integrated into developmental and classificatory studies, generation was also frequently treated separately. In fact it was not just one topic but a cluster of problems encompassing the various different modes of generation expressed by sexual, hermaphroditic, budding, and parthenogenetic forms; metamorphosis; alternation of generations and the physiological division of labor, and the tracing of sex cells.[9] A second problem area was histology, which, although dominated by anatomists, was also, under the leadership of Leuckart's contemporary Franz Leydig, gaining interest among zoologists. It, too, could be closely joined with studies of development, generation, and even clas-

---

8. A. Schneider, *Monographie der Nematoden* (1866); E. Ehlers, *Borstenwürmer* (1864–68); C. Semper, *Reisen* (1868); Claus, *Die freilebenden Copepoden* (1863); Haeckel, "Familie der Rüsselquallen" (1864, 1866); of this cohort only Weismann never published a classificatory monograph.

9. Frederick Churchill has published extensively on this cluster of topics, with the general aim of showing how they coalesced into concepts of inheritance in the early 1880s (see, e.g., "From Heredity Theory to *Vererbung*" [1987], and "Weismann's Continuity of the Germ-Plasm" [1985]).

sification. A third cluster of problems, again frequently integrated with the others, concerned the effect of the environment on the organism's form and function and was often connected to questions of geographical distribution. While few zoologists conducted research in all these areas, they typically stretched beyond just one.[10]

The felt need to publish on a variety of kinds of problems reflects an important characteristic of the discipline of German zoology in the 1860s: a pervasive commitment to intellectual breadth. One important way of demonstrating breadth was to publish on different problems or topics; another, even more common, was to work on a variety of taxonomic categories. In job searches, for example, faculties expressed a strong preference for candidates who had shown through their publications that their teaching would draw on a research-based knowledge of more than one part of the animal kingdom and that they could handle problems of broad significance. When Leipzig's zoology professorship came open in 1869, the search committee argued for each of the three leading candidates terms of their breadth: the publications of J. V. Carus demonstrated his "fundamental and broad knowledge of specialized zoology" as well as his interest in questions of general theoretical significance; Rudolf Leuckart's numerous publications showed his "comprehensive knowledge of details" in addition to his ability to choose scientifically important problems; Franz Leydig's investigations "stretch over all classes of animals."[11] Similarly, in 1878, the departing professor at Heidelberg argued that the successful candidate should demonstrate that "he commands the entire realm of zoology, through comparative and synthetic works, and through specialized works in at least several major parts [of zoology]."[12] By the same token, those who were lower on a list of candidates were sometimes characterized as "one-sided" or "narrow" researchers.[13]

One might suspect that such comments were simply part of the rhetorical format of job recommendations. After all, wouldn't anyone

10. For example, Ehlers published on the vertical distribution of annelid worms in the ocean; Claus addressed such problems as the border between the animal and plant kingdoms and parthenogenesis in bees; Schneider's study of the flatworms contained extensive histological investigations.

11. "Bericht der philosophische Fakultat, die —— [illegible] Besetzung der zoologische Professur betreffend," 5 December 1868, UA Leipzig, Personalakte Leuckart.

12. "Ursprüngliche Vorlage des Herrn Prof. Pagenstecher in der Berathung der Commission," UA Heidelberg, Doc. 27c, p. 156.

13. In the search at Heidelberg after H. G. Bonn's death in 1862, for example, Bonn's zoologist Franz Troschel was considered but rejected as "too one-sided" (philosophical faculty report, 4 August 1862, BGLA, file 235/19905).

prefer a broadly trained scholar over a narrowly specialized one? But here comparison with searches for anatomists indicates that breadth was not, in fact, a universal feature in hiring preferences: in deliberations over anatomy professorships in the same period, the candidate's philosophical and methodological approach held a much more prominent place than did the breadth of his research.[14] And there is other evidence that breadth was a value taken seriously by zoologists, in scattered references by the youngest cohort of aspiring zoologists to their research strategies. Here the notion of breadth as coverage of different taxa is especially evident. As late as 1878, for example, the young Richard Hertwig wrote to Carl Claus that in choosing to move away from the familiar territory of the medusae for his next project, he was "proceeding from the principle that younger zoologists must use their time to gain an overview over the animal kingdom from their own investigations."[15] The same year, the job-seeking Hubert Ludwig wrote to his mentor, Ernst Ehlers, "My next publication will not consider echinoderms. I already hold in some people's eyes so much the bad reputation of an echinodermist [*eines Stachelhäuters*] that I must take a small break from them."[16] As these quotes suggest, breadth was to be gained not just by reading around in the literature or asking broad questions, but through one's own direct investigations of a wide range of organisms.

What one sees in the 1860s, then, is a broad spectrum of overlapping research activities in a field in which breadth itself was valued. With the exception of the older systematists, who were geographically limited to Prussia and widely perceived to be pursuing an isolated and antiquated variety of zoology, the discipline was not divided among clear-cut methodological or philosophical factions. Zoologists ranged freely across a variety of analytical problems and technical approaches; disagreements tended to be confined to the interpretation of evidence in a particular group of organisms, rather than displaying deep methodological or philosophical divisions. With a commitment to breadth in both coverage and analytical approaches, and with the implications of Darwin's theory just beginning to be explored, Ger-

14. In the 1870s it was generally acknowledged that anatomy had become too large a topic for one professor to cover it all. Search committee recommendations, therefore, focused on how best to divide the subject, and this raised manifold philosophical and methodological arguments in addition to practical ones (see pp. 218–35).

15. Richard Hertwig to Carl Claus, 19 August 1878 (Bll. 11–12), SBPK, Lc 1898 (29): R. Hertwig.

16. Hubert Ludwig to Ernst Ehlers, 12 December 1878, UB Göttingen, Cod. MS E. Ehlers 1158.

man zoologists in the 1860s displayed a flexibility and intellectual openness that would rapidly disappear in the next decade. This characteristic makes it difficult to speak of a "program" of evolutionary morphology in the 1860s and early 1870s; instead, one finds zoologists trying out a variety of linkages among the topics mentioned above, often in connection with evolution, with little differentiation about what was morphology and what was not.

## MAKING EVOLUTIONARY CONNECTIONS: CLASSIFICATION, MORPHOLOGY, AND ADAPTATION

To understand how Haeckel's program of evolutionary morphology fit into the discipline of zoology and how it was received by his contemporaries, we need to understand the ways in which they sought to connect up evolution with their current research problems. Haeckel was by no means alone in seeking to make zoology evolutionary; for most German zoologists, a central task of the 1860s lay in confronting Darwinism and figuring out how it might inform their work. Excepting only Schneider, who was reticent about his views on evolution, all the successful members of Haeckel's cohort were sympathetic to Darwin's theory, as were those of their teachers who counted themselves among the scientific zoologists—Siebold, Leuckart, Leydig, and J. V. Carus. Virtually all of them declared their support in some form during the 1860s; however, most expressed their allegiance cautiously, recognizing that it was probably unprovable in any rigorous sense but that nevertheless, as Franz Leydig wrote, it "is preferable to the assumption of separate acts of creation; it satisfies the intellect the furthest and explains the most."[17]

Consistent with this caution as well as with the open and exploratory character of the discipline, zoologists found a number of different ways to relate evolutionary theory to their research. One response was not to try to unite the older and newer problems, but to treat them as distinct. Siebold and Leuckart, both older men with research programs that were well established, expressed their sympathy with evolutionism[18] but did little to integrate it into their main body of

17. Leydig, *Vom Baue des thierischen Körpers*, p. 6; see also pp. 128, 137–38.
18. Siebold wrote in a letter in 1867, "Indeed, I do believe that Darwin is right, but it can't be proven," Siebold to Ehlers, 17 February 1867, UB Göttingen, Cod. MS E. Ehlers 1815. In a review of Haeckel's *Generelle Morphologie* in 1867, Leuckart wrote that while he couldn't agree with Haeckel in every respect, "on the main issue, on the question of descent," he was "decidedly on [Haeckel's] side" (Leuckart, "Bericht 1866–1867" [1867], p. 167).

research. In Siebold's continuing investigations concerning the existence of true parthenogenesis (the production of offspring by a female with no male contribution), descent had little place, although in a few shorter pieces on other subjects he did address the questions of adaptation and atavism in Darwinian terms.[19] Similarly, in the 1860s Leuckart gained ever more expertise on the subject of intestinal worms, work that drew him into problems of practical interest to physicians, veterinarians, and health officials; although he was sufficiently interested in Darwinism to give a lecture course on it in 1873–74, he was not inclined to modify his research approach or analysis along Darwinian lines.[20]

Leuckart's histologically inclined contemporary Franz Leydig did incorporate Darwin's theory slightly more directly into his work, but used it mainly to justify a line of argument he had long since established rather than changing his outlook to incorporate the theory. In his book of 1864, *Vom Baue des thierischen Körpers,* he took up the Darwinian concept of intermediate forms, applying it not primarily to taxonomic branching points but to the classification of tissues.[21] "There was a time when people drew a sharp boundary between smooth and striated muscles. I was the first to . . . show that both types [*Arten*] of muscles blend into one another in their development and form." Similarly, he wrote, people usually divided nervous tissue into two "species," one containing the white myelin and one lacking it, but in fact there existed intermediate forms linking them. Accepting existing taxonomic divisions, he was concerned to show that histology could be usefully studied through comparison among different invertebrates.[22] In this context, Darwinism offered an opportunity to

19. See, e.g., C. T. von Siebold, "Über die Acclimatisation" (1869), "Über das Anpassungsvermögen" (1875), p. 39, "Haarige Familie" (1877).

20. Leuckart to C. Claus, 24 January 1874, SBPK, Lc 1851: R. Leuckart. See also Wunderlich, *Leuckart,* pp. 73–79.

21. Leydig, *Vom Baue des thierischen Körpers,* p. 113; Leydig did speculate on the phylogenetic connections within the animal kingdom as a whole, but only briefly. Scientists who believed that the vertebrates evolved from the arthropods would soon claim Leydig's support for their view, but I cannot see this as the major argument of the book: he makes the point in one sentence and immediately qualifies it in the next (ibid., p. 114); see, e.g., B. Hatschek, "Entwicklungsgeschichte der Lepidopteren," (1877), p. 141.

22. Leydig, *Vom Baue des thierischen Körpers,* pp. 79, 94. An advertisement for the book stressed that its aim was to unite comparative anatomy and histology (see the copy owned by the Museum of Comparative Zoology at Harvard University; the advertisement appears to have been on the inside of the original front cover, but because of the way it is bound this cannot be certain; the other copy I have seen lacks this page).

explain histological structures by means of an evolutionary continuum.

Younger zoologists from Haeckel's cohort who attempted to integrate evolutionary theory more fully into their work in the 1860s made a variety of linkages. In keeping with the generally cautious tenor of their engagement with Darwinism and the strong commitment to empirical research characteristic of scientific zoology, these efforts tended to be buried in research monographs rather than being announced as programmatic proposals. (As we will see in the next section, here, too, Haeckel would buck the trend.) Three examples by Haeckel's contemporaries illustrate the sorts of connections commonly being made.

In his classificatory/anatomical monograph on the sea cucumbers (Holothuria), Carl Semper explored two different aspects of evolution.[23] First, he made one of the most thoroughgoing early attempts to apply the notion of evolutionary recapitulation expressed in Fritz Müller's *Für Darwin* and Haeckel's *Generelle Morphologie*. Picking one species that seemed to him to show "the most complete primitive history" in its development (without explaining why this was so), he followed the development of its organs step by step.[24] In the individual's development, the water ring developed before the podia (tube feet). This meant that the same was true in the phylogeny of the sea cucumbers: forms without podia had appeared before those with them. The tentacles also appeared in ontogeny before the podia; therefore the tentacles are "phylogenetically more important" (because more ancient) than the presence or absence of podia (fig. 6.1). The logic here is von Baer's with an evolutionary twist: the earliest features to appear in development are the most general ones of the class, which are also the oldest phylogenetically; later ones reflect progressively smaller taxonomic divisions, equivalent to more recent phylogenetic branchings. Following the development of a well-chosen form, then, gave one guidance on how to weigh the various features for classification. Comparing this classification with one based on traditional comparative anatomy, Semper claimed that the two came out nearly the

23. Semper, *Reisen*.

24. Semper quotes the basic principles he draws from Haeckel and Müller on pp. 186–87. These are Haeckel's dictum that ontogeny recapitulates phylogeny and Müller's qualifications of that idea: that the developmental history will be falsified through adaptations gained in the struggle for existence among free-living larvae; that the longer the organism retains its youthful conditions and the less the mode of life of the developing organism differs from that of the adult, the more fully it will reflect its phylogenetic history; the species he used as his key form was *Synapta digitata*.

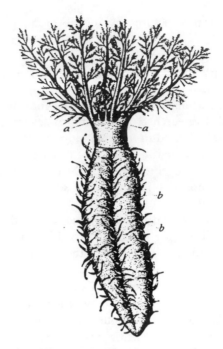

Fig. 6.1. Sea-cucumber (*Cucumaria Planci*), as seen from the underside: *a*, branching tentacles; *b*, tube-feet; the water ring is an internal structure not visible here (from Richard Hertwig, *Lehrbuch der Zoologie*, 9th ed. [1910], p. 333).

same, thereby demonstrating the legitimacy of recapitulation as a tool for evolutionary classification.

The same work contained a substantial separate section on geographical distribution. An important problem for both Darwinians and adherents to the theory of "centers of creation" was establishing the locus from which a particular group of organisms spread. For creationists such information would help pin down the group's point of origin and the conditions of existence for which it had been created; for Darwinians it would help show ancestral connections and the limits of adaptive change. There was considerable debate, however, over the taxonomic level at which geographical distribution provided meaningful information. Did the distribution of an entire family of organisms have any significance, or did one have to confine oneself to lower taxonomic levels, such as the genus or species, to gain information that was meaningful? Although Semper came to no definitive conclusion, he did argue that only Darwin's theory, by offering a causal relationship between distribution and the "laws of [its] dependency on the external conditions of existence," provided a satisfactory accounting of the situation. Semper concluded by stating that a leading task of the future was to investigate, where possible by experiment, "the influence of temperature, light, heat, humidity, nutrition, etc. on the living animal," seeking the "ecological laws" affecting organic forms and their functioning.[25] This work shows Semper integrating evolution with two ongoing areas of German research— embryology and the problem cluster centered on geographical distribution and the effects of the environment—in addition to supplying the expected anatomical descriptions and classificatory suggestions.

Like Semper's study of the sea cucumbers, Carl Claus's monograph on the free-living copepods also reflected an effort to integrate evolutionary thinking into the standard anatomical/classificatory format. Admitting in the preface that Darwin's theory was by no means proven, or even really provable yet, he said that it nevertheless placed "all of morphology in a new light" (p. vi). Like Semper, Claus considered both development and adaptation in justifying his evolutionary classification scheme for the copepods, but the weight he assigned to the two was balanced differently. His classificatory proposal took development into account, but in a less rigorous way than Semper's. In a section comparing the copepods with four other crustacean groups, he pointed to the similarity of their larvae. "From a similar starting-point, each series of forms in its further development departs from the others in a different direction." He emphasized that intermediate forms showed the connections among the different groups, but he did not try to use the details of individual development as a principle for classification. It was the existence of intermediate forms rather than the specific relationships among those forms that interested him.[26]

Far more important to his analysis was the role of the conditions of existence, which in his view persuasively accounted for the divergence of forms into separate varieties and species. Claus argued that by affecting the sources of food and therefore the adaptations necessary to survive, variations in climatic conditions influenced the morphological development of the order. Such variations especially affected the organs of motion, which in turn justified using these as the

25. Semper, *Reisen*, p. 228; Semper explicitly adopts Haeckel's term "ecology" to mean the study of the effects of the external conditions of existence on species, but most researchers of the period used the older term "biology" to mean the same thing. "Ecology" only gained widespread use after 1900, as "biology" took on a different set of associations.

26. Claus, *Die freilebenden Copepoden*, pp. 14–17.

main classificatory criteria: "The copepods are characterized by the possession of paired swimming legs, whose structure is so intimately bound with the conformation of the body, with the type of movement and feeding, with the entire mode of life, that we are justified in viewing those body parts as a general expression of the narrower Type and can use this characteristic as designating the order."[27]

The effect on form of adapting to a particular mode of life came into even sharper focus when one compared the free-living copepods with the parasitic ones. Claus showed that there existed a series of transitional forms going from fully free-living ones, through "guest forms" that lived inside another organism but otherwise maintained their physiological autonomy, to forms that were parasitic for only part of their life cycle, to true parasites that had lost the ability to exist independently of their hosts. Although the presence of these transitional forms meant that there was no hard and fast boundary among the orders of copepods, he found it "the most natural" to divide them into two groups, the true parasites and the free-living ones, including among the latter those partially parasitic forms that retained "all the substantial parts of their chewing apparatus" (pp. 8–9). Through its effect on form, then, the organism's adaptation to different modes of life provided him with the main justification for his classification system.

Yet another route to uniting morphology and evolution was taken by August Weismann, a member of Semper and Claus's cohort who did not use evolutionary theory in the pursuit of classification at all. Alone among his generation in this respect, he concentrated exclusively on the problems of tissue formation, development, and generation that had formed the heart of the program of scientific zoology; with his *Studien zur Descendenz-Theorie* in the middle 1870s he began reinterpreting these problems in evolutionary terms. Thus his paper on the differences between winter and summer forms in butterflies took a question he had long been working on, namely, the process of generation in insects, and addressed it in terms of evolutionary adaptation to two different sets of environmental conditions.[28]

Although they emphasized different links between evolution and morphology, these men remained true to the older program of functional morphology or scientific zoology in one important respect: for all of them, older and younger scientific zoologists alike, an organism's form was intimately dependent on its conditions of existence.

27. Ibid., p. 4.
28. A. Weismann, "Saison-Dimorphismus der Schmetterlinge" (1874).

While Darwin's theory allowed them to replace the much denigrated concept of teleology with the apparently less mysterious, more mechanical vocabulary of adaptation, inheritance, and survival of the fittest, it did not, in their view, change the nature of the project. As Leuckart wrote in 1869, "a main task of scientific zoology" was "to show how the specific conditions of adaptation" affected "the forms of living individuals."[29] It is clear that many of his colleagues agreed. Even Leydig, whose work had the least discussion of adaptation, wrote that it was impossible to produce a purely morphological study, because "the way an animal lives affects its form," just as "the form of the animal determines how it should live."[30] If it had not been for Haeckel, it seems quite possible that evolutionary morphology in the 1870s and 1880s would have looked much more uniformly like the older scientific zoology than turned out to be the case.

## HAECKEL'S GASTRAEA THEORY AMONG HIS CONTEMPORARIES

Haeckel's gastraea theory represented in several respects a departure from the approach of his contemporaries. Whereas nearly all of his colleagues emphasized the extent to which adaptation produced divergence of character and new species, Haeckel's construction of the relationship between evolution and morphology led him to stress the similarities of form produced by common inheritance. But that was just one of the differences. In contrast to their more modest and exploratory attempts to integrate evolution with their ongoing research questions, he set out an ambitious program with clear guidelines about what problems should be pursued. Nor did he leave his proposals as suggestions at the end of a research monograph, although that was where he had originally introduced them; the theory itself became the subject of a series of articles in the early 1870s, and as he relentlessly repeated his position in his numerous publications of the decade, the general outlines of the theory grew more important for him than its empirical foundation.

Each of these features led his contemporaries and older colleagues to look upon his theory with skepticism. His lack of attention to adaptation was viewed by many as a basic weakness in the theory; his dogmatic presentation of his views further contravened the socially accepted openness of the discipline to a variety of approaches. Finally,

29. Leuckart, "Bericht 1866–1867," p. 165.
30. Leydig, *Vom Baue des thierischen Körpers*, p. 2.

to a community in which it was expected that one's breadth would be achieved and demonstrated through empirical research, his penchant for ambitious theorizing was viewed as inappropriate and even dangerous. Thus Haeckel's theory was not greeted by his contemporaries with open arms.

And yet in one important respect, Haeckel was engaged in the same sort of project as his contemporaries, namely, seeking to integrate evolutionary thinking into existing areas of research. It was his contribution to set out a framework that united evolution with one of the most absorbing and significant research problems of the day: the interpretation of the germ layers. Now, the germ layers had long been a subject of interest among vertebrate embryologists, and it had gained in significance in the 1850s, when Robert Remak (1815–65) argued that among vertebrates the "vegetative" organs derived from the internal germ layer and the "animal" or sensory organs derived from the external one. In the 1860s, the Russian embryologist Alexander Kovalevskii (brother of Haeckel's student V. O. Kovalevskii) broadened the discussion by showing that two primitive germ layers existed in many invertebrates as well. This raised numerous questions about whether Remak's physiologically defined division of labor between the layers could be extended to the invertebrates, and in the late 1860s and 1870s embryologists continued to argue over the interpretation of these layers. Was their significance primarily histological—that is, could one trace the different types of tissue to one or the other of the germ layers? Was it morphological—did different organ systems derive from the different layers? What was the source of the late-developing middle layer or mesoderm, which, according to Remak, supplied numerous different types of tissues, including blood vessels, nerves, muscles, and connective tissue: could one trace its cells back to one or the other of the two primary layers? Which cells then produced which organs? Could one find a consistent pattern across different organisms correlating particular organs to specific germ layers? These questions could be studied independently of evolution, and both before and after Haeckel introduced his theory, they often were.[31] But Kovalevskii's claims opened up an opportunity for considering the similarities between invertebrate and vertebrate germ layers as deriv-

31. Note, e.g., Haeckel's distressed acknowledgment that Kölliker found it possible to discuss the significance of the germ layers while deliberately ignoring the gastraea theory (E. Haeckel, "Ursprung und Entwicklung" [1885], pp. 246–49, esp. p. 247 n. 1); for a short historical overview from a perspective close to the action, see Robert Bonnet, "Entwickelungsgeschichte," pt. 1 (1891).

ing from common ancestry, and Haeckel seized it. His interpretation made evolution a central part of explaining the layers.

Haeckel had already set out the general lines of his program for evolutionary morphology in the massive *Generelle Morphologie* in 1866 and his more readable *Natürliche Schöpfungsgeschichte* in 1868. The gastraea theory, however, supplied a more concrete foundation for his claims for the truth and utility of the biogenetic law and for the monophyletic origin of multicellular animals. He first sketched out the theory in his 1872 monograph on the calcareous sponges and then elaborated on it in a series of articles in the *Jenaische Zeitschrift* between September 1873 and November 1876; a summary of it also appeared in his collection of lectures on human phylogeny, *Anthropogenie* (1st ed., 1874). Together with the biogenetic law, the gastraea theory formed the core of Haeckel's evolutionary edifice.

The basic elements of the theory had the kind of simplicity Haeckel found appealing. He began by asserting that the oldest and most fundamental dividing line in the animal kingdom lay between those organisms with germ layers (metazoa) and those without (protozoa). The protozoa might stem from several different origins, but in Haeckel's view the metazoa were monophyletic. The chief evidence for this was that in every phylum excluding the protozoa, one could find representatives with a common developmental stage, consisting of two layers of cells (the primary or primitive germ layers). These two germ layers Haeckel called the "entoderm" (inner layer) and "exoderm" (outer layer, soon renamed "ectoderm"). These germ layers, which could be histologically differentiated by the different shapes, textures, and contents of their cells, took morphological significance from the form in which they first appeared, the two-layered, cuplike gastrula. Its existence in groups as widely divergent as sponges and vertebrates persuaded him that the gastrula was a form that had appeared early in evolution, before all those groups had taken their separate evolutionary paths. Drawing on the biogenetic law, he reasoned that at an early point in evolutionary history, the gastrula must have been the end stage in the metazoans' ontogenetic development. The hypothetical organism taking this adult form he named the "gastraea" by analogy with its parallel developmental stage (fig. 6.2). In his scheme, the gastraea theory provided a morphological justification for claiming the monophyletic origin of all the metazoa, as well as exemplifying the claims that could be founded upon the biogenetic law.

Although this is where most historical discussions of the gastraea

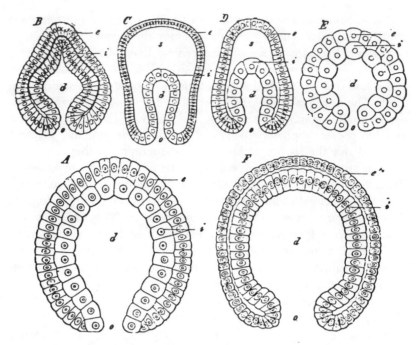

Fig. 6.2. Comparison of gastrulae across the animal kingdom as the basis for the gastraea theory: *A*, gastrula of a zoophyte, *Gastrophysema* (Haeckel); *B*, gastrula of a worm, *Sagitta* (after Kovalevskii); *C*, gastrula of an echinoderm (starfish), *Uraster* (after Alexander Agassiz); *D*, gastrula of an arthropod, *Nauplius*; *E*, gastrula of a mollusc (pond snail), *Limnaeus* (after Karl Rabl); *F*, gastrula of a vertebrate (Lancelet), *Amphioxus* (after Kovalevskii); in all, *d* indicates the primitive intestinal cavity; *o*, the primitive mouth; *s*, the cleavage cavity; *i*, the entoderm, or intestinal layer; *e*, the exoderm, or skin layer (from Ernst Haeckel, *Evolution of Man* [1879], vol. 1, p. 193).

theory end, Haeckel's own analysis did not stop here. Having established to his own satisfaction that the two-layered gastraea was the common ancestral form of the metazoa, he went on to discuss the significance of the germ layers after they emerged from the gastrula stage. Much of his exposition of the gastraea theory concerned whether the products of the germ layers—the tissues, organs, and organ systems deriving from them—can be viewed as truly homologous from phylum to phylum.

Herein came an important blending of concepts with which zoologists were to struggle over the next decades. In Haeckel's language, the "homology of the germ layers" initially meant that the two primitive layers could legitimately be compared in different organisms because they originated from a common ancestor; that was the evolutionary meaning of homology. But in Remak's theory, the germ layers had been less significant in themselves than as the sources of the organs that derived from them. As these two views of the germ layers blended in with one another, it was easy to assume that the evolutionary significance of the germ layers was that their *products* were homologous. That is, one would expect that organs deriving from the exoderm in one species would also derive from the exoderm in another, and so forth.[32] If that were the case, one would have a new and firmer evolutionary explanation for why one could derive true homologies by comparing adult organs: they are homologous because they came out of the same germ layers, which, in turn, reflected their common phylogenetic heritage. This seems to have been Haeckel's basic assumption, but he knew perfectly well that the numerous ontogenetic paths traced by different animals presented a complex picture that did not easily substantiate it. In his presentation of the theory, therefore, Haeckel went through one structure after another, pronouncing on whether each was to be considered homologous in different organisms or not. This aspect of the theory would continue to occupy him in later years, as he sought to reduce the complexity of the ontogenetic evidence to a simple set of rules and exceptions.[33]

What made Haeckel's gastraea theory significant among German zoologists in the 1870s was not its universal acceptance but rather its extremely controversial nature.[34] Haeckel pulled into one theoretical edifice so many different ongoing problems—the relationships of the major taxa to one another, the histological development of the primary germ layers, the morphological and evolutionary significance of the same, and the origin and appropriate interpretation of the mesoderm, not to mention the more philosophical problems of the appropriate balance between theory and empirical evidence, the proper conception of "mechanical cause," and the relative weight to be assigned

32. The logic that blended these two views of the homology of the germ layers was criticized by Hinrich Nitsche, a *Privatdozent* under Leuckart in Leipzig, in "Knospung der Bryozoen" (1875), pp. 390–402; and by Alexander Goette, *Entwickelungsgeschichte der Unke* (1875), pp. 858–59.

33. In the final installment of the theory, dated November 1876, and again in a long essay in 1885, he sought to draw out the implications of his theory for interpreting the germ layers (E. Haeckel, "Nachträge zur Gastraea-Theorie" [1877], "Ursprung und Entwicklung").

34. This statement concerns the German community only. Elsewhere, as Haeckel himself noted, his ideas met with much greater acceptance.

to adaptation and inheritance in interpreting embryological evidence—it is no wonder his theory became the target for criticisms from a multitude of directions.

It is probably fair to say that any such all-encompassing theory would have demanded response from the community, but what is most striking about the earliest reactions is that they were directed at least as much at the dogmatic tone of his presentation as at its content. Eduard von Martens (1831–1904), an old friend from university days and curator of the zoological museum at Berlin, wrote Haeckel upon reading the sketch of the theory in the monograph on sponges, "I'm sorry that the entire presentation diverges more and more from the calm, objective (if also dry and prosaic) [sort] that otherwise rules the descriptive sciences, and becomes a fervent, I want to say almost a preaching, tone, which is from the very beginning directed toward an already established goal and uses everything possible that can be said for it, that sees in every possibility a strict proof, and where it cannot logically lead to the proof, goes around it by saying 'all these phenomena are a tentative proof for my viewpoint.'" Other zoologists also found themselves using similar religious analogies. Haeckel's missionary zeal for the theory of descent had long been noticed: already in 1867 Siebold had written that Haeckel demanded "blind faith" in Darwinism. By 1876, Leuckart was referring (in the same letter) to "Jena's Dalai Lama" and "the Jena Vaticanism" of Haeckel. Semper drew on the latter analogy as well when he huffed in a critique of Haeckel, "fortunately we do not yet have a doctrine of infallibility."[35]

For Haeckel's colleagues the issue was more than just a matter of tone; what made his writing not just offensive but positively pernicious for zoology and perhaps science in general was that his dogmatism was attached to a new and highly speculative theory. As Claus wrote in the 1880 edition of his textbook, Haeckel had lapsed "into the error that also characterized the older *Naturphilosophie*, to maintain as unerring truth an abstraction based on a limited foundation, one that is at best to be viewed as a possibility, and out of attachment to it, to construe dogmatically or even infer a priori the phenomena

35. Eduard von Martens to Haeckel, 12 January 1872, in Jahn, "Ernst Haeckel," p. 95; Siebold to Ehlers, 17 February 1867, UB Göttingen, Cod. MS E. Ehlers 1815; Leuckart to Claus, 1 January 1876, SBPK, Lc 1851: R. Leuckart; C. Semper, "Kritische Gänge" (1874), p. 233; the reference to the doctrine of infallibility refers, of course, to the doctrine of papal infallibility that had just been asserted in the Vatican council of 1869–70.

and processes that are to be observed."[36] Martens had already expressed the same concern somewhat more gently in his 1872 letter: "Science should always candidly separate the factual from the inferred and within the latter, the wholly probable from the somewhat [*ziemlich*] probable, to give the truth its full due. Those scientists who are well acquired with the subject will always be able to distinguish the one from the other in your presentation, but the beginners and those who are interested superficially in science will not, and there are so many of these now. These will then . . . proclaim as the sure result of infallible science much that is still only hypothetical, and that should not happen."[37] Haeckel's presentation of speculation as fact and his drawing of hard and fast generalizations from a few pieces of evidence was especially egregious in his popular work of 1874, *Anthropogenie,* in which the caveats and exceptions acknowledged in the scholarly version of the theory were absent. Weismann was undoubtedly referring to this work when he wrote to Claus late in 1874 that he shared Claus's opinion of the "*naturphilosophisch* swindle" but was persuaded that the author would soon see that he had acted rashly and "make room for more sober *research.*"[38]

Haeckel sought to defend himself against such criticisms by asserting the necessity of balancing observation with reflection (thereby invoking that icon of embryology, Karl Ernst von Baer),[39] but this did not stem the criticism. Semper, who ten years earlier had been a devoted follower of Haeckel's biogenetic law, wrote in an open letter to Haeckel in 1877, "I do not want at all to rob you of your philosophical views. My aim was only to show you that you ignore every-

36. C. Claus, *Grundzüge* (1880), p. 57.

37. E. von Martens to Haeckel, 12 January 1872, quoted in Jahn, "Ernst Haeckel," p. 95.

38. Weismann to Claus, 20 October 1874, SBPK, Lc 1889: A. Weismann. There is no direct reference to Haeckel in the letter, but given that *Anthropogenie* had just appeared, that no other recent publication quite fits the description, and that Weismann went on to say he saw his forthcoming *Studien zur Descendenztheorie* as a contribution to "sober research" on the subject, it seems a plausible inference. Haeckel's idiosyncratic use of illustrations, which included numerous falsifications and shortcuts to make them conform to his vision of things, also came in for damning criticism, dating back to the first edition of the *Naturliche Schöpfungsgeschichte* in 1868 and continuing virtually for the rest of his career (see R. Gursch, *Die Illustrationen Ernst Haeckels* [1981]).

39. Haeckel's often-repeated refrain concerning the necessity of combining "observation and reflection" was a reference to the subtitle of von Baer's famous work *Über die Entwickelungsgeschichte der Thiere: Beobachtung und Reflexion* (see, e.g., Haeckel, *Anthropogenie,* 3d ed. [1877], p. xxii).

thing that you can't use or won't accept." Semper had nothing against "reflection" in its place, but thought Haeckel should have "trusted observation somewhat more, and your reflections somewhat less."[40]

This is hardly the response of a community swept up in admiration of Haeckel's speculative theorizing. One can imagine several reasons why they might have responded this way: a general belief that science properly proceeded through skeptical empiricism, not speculation, and certainly not dogmatic speculation;[41] a heightened commitment to that view in the wake of the pronouncement of papal infallibility in 1870; a sense that Darwin's theory itself was controversial enough not to want to have criticism drawn to it by a too-dogmatic adherence to it; and a sense of frustration, sometimes fanned to indignation, that Haeckel so readily ignored conflicting evidence. All these reactions could only have been intensified by the fact that Haeckel was breaking with the habits of intellectual openness and modesty that had been characteristic of the discipline in the 1860s. All in all, his distinctive combination of speculation and dogmatism did not sit well with many of his colleagues.

Characteristically, Haeckel treated criticisms based on his inattention to evidence as minor quibbles over the evidence itself; looking back in 1884, he wrote that his theory, "like *all* such far-reaching generalizations, naturally contained many mistakes *in its details;* but on the whole it has nevertheless won more and more recognition."[42] Indeed, virtually the only criticisms he viewed as worthy of serious rebuttal were those based on philosophical principles radically opposed to his own. In a 100-page supplement to the *Jenaische Zeitschrift* in 1873, he singled out the anatomist Wilhelm His (1831–1904) and the younger zoologist Alexander Goette (1840–1922) for special consideration.[43] Although in many respects in fundamental

40. Semper, *Offener Brief* (1877), pp. 16, 21; in his textbook Claus reproached Haeckel along similar lines (see Claus, *Grundzüge,* pp. 58–59).

41. The extreme version of this view was expressed by Siebold, coeditor of the *Zeitschrift für wissenschaftliche Zoologie,* the leading German-language zoological journal, in 1875, when he explained his decision to reject the speculative part of an article: "let us leave such far-reaching speculative utterances . . . for other journals. . . . We want to reserve the space in our journal for facts. The more new facts our journal brings, the more it will be necessary for the speculative spirits of our time to change their principles again and again, whereas that which is fact will stand" (Siebold to Ehlers, 23 December 1875, UB Göttingen, Cod. MS E. Ehlers 1815).

42. Haeckel, "Ursprung und Entwicklung," p. 210.

43. E. Haeckel, *Ziele und Wege* (1876); the last twenty pages were devoted to the religiously motivated criticisms of Louis Agassiz, Carus Sterne, Michelis, and Adolf Bastian.

disagreement with one another,[44] His and Goette shared the view that phylogenetic explanations of ontogenetic events and forms were unnecessary and unsatisfactory. Whereas Haeckel found Goette's argument muddy and confused, and therefore treated it only with ridicule, he saw in His's writings a clear alternative worthy of a more sustained attack.

All ontogenetic events, His argued, were the "mechanical" result of differential growth in cells. For example, when more cells grew on one side of the germinal disc [*Keimscheibe*] than the other, that caused the flat form of the immature embryo to bend around and forced the creation of a fold. In His's view, such physical processes supplied the "mechanical causes" of development. But for Haeckel, this explanation solved nothing, because it did not tell why the growth of the cells on one side outpaced those on the other side to begin with.

"No matter how exactly His describes all the individual folds, main folds and subordinate folds and shows these to be the substantial changes in the developing body form, nothing is in the least *explained* by it. For *every one of these simple ontogenetic folding processes is a highly complex historical result, which is causally determined through thousands of phylogenetic changes, of transmission and adaptation processes.*"[45] Haeckel emphasized that, like His, he too sought a mechanical explanation founded on physiological causes. The difference was that His excluded phylogeny from consideration as one of those causes, while Haeckel took it to be essential. Since, in Haeckel's view, phylogeny was nothing other than the dynamic result of heredity and adaptation, and these were themselves important physiological functions ultimately explicable in mechanical terms (though not yet investigated thoroughly by physiologists or zoologists), it was perfectly justified to treat phylogeny as a mechanical cause. Furthermore, Haeckel's mechanical explanation accounted for vestigial organs appearing in development, whereas His could only dismiss them as meaningless scraps of material left over in nature's process of snipping and shaping the developing embryo.

Haeckel acknowledged that a complete mechanical explanation, that is, a perfectly rigorous demonstration of causes and effects, could not be made in most morphological investigations. Because we lack the empirical evidence from the historical record that would be necessary to trace the full sequence of phylogenetic events, "we will never

44. His wrote a highly critical review of Goette's book, "Goettes *Entwicklungsgeschichte der Unke*" (1876).
45. Haeckel, *Ziele und Wege*, p. 24.

be able to explain the ontogeny of even a single organism completely mechanically"; nevertheless, he continued, one is justified in saying that adaptation and heredity, the two processes that have created phylogeny, are indeed wholly mechanical processes that can ultimately be traced back, in theory at least, wholly to chemistry and physics.[46] Just because the tracing of this mechanical chain of events was impossible in practice, that did not justify denying its theoretical significance, as His did.

As Haeckel presented the controversy in this and later publications, there was only one important division: on one side stood those who accepted the entire package of the monophyly of the metazoans, the gastraea theory, and the biogenetic law in its strong form as "phylogeny is the mechanical cause of ontogeny"; on the other side he grouped those who sought the mechanical causes of ontogeny in His's terms along with those who rejected any other parts of the package. Any efforts that some might make to adopt both His's sort of mechanical cause and his own he rejected on the grounds of principle: the two sorts of causes were fundamentally opposed and could not be mediated. As he wrote in 1884, the split between the two approaches, which had grown ever clearer in the last decade, had had the benefit of forcing every embryologist to decide at the outset of his investigations which sort of explanation he was going to seek.[47]

This dichotomy may faithfully reflect Haeckel's view of the situation, but it vastly oversimplifies the actual responses of other zoologists to his theory. Most accepted monophyletic descent as a central contribution of Darwinism, but viewed this as a matter of faith, not something that had been proven. It was further possible to accept in principle that ontogeny often reflected phylogeny while doubting that the biogenetic law could be trusted to provide a reliable guide to an organism's ancestry; similarly one could accept the principle of recapitulation while questioning the evolutionary significance of the gastrula. Claus, for example, objected vociferously to Haeckel's claims for the gastrula but in his textbook presented the similarity of embryos to ancestors as one consequence of evolution.[48] Nor did most zoologists see the argument as revolving around Haeckel's sharp dichotomy between mechanical and phylogenetic causes. That left out the central role of adaptation: surely in order for a feature to become a phylogenetically significant one, it must have been acquired and

46. Ibid., pp. 24–25.
47. Haeckel, "Ursprung und Entwicklung."
48. Claus, *Grundzüge*, pp. 110–13.

preserved at some time in history on the basis of its adaptive benefits. But such an adapative feature might well have emerged in the first place through mechanical means.

Such arguments are especially well illustrated by one of the most thorough critiques of the theory, *Die Typenlehre und E. Haeckel's sogenannte Gastraea Theorie* (1874) by Haeckel's contemporary Carl Claus, a former student of Leuckart by then teaching at the University of Vienna. In this pamphlet Claus mounted a formidable attack on the reasoning Haeckel used in his interpretation of the ontogenetic evidence. For example, Haeckel maintained that all the two-layered, cuplike gastrula forms were homologous. But as he himself acknowledged, these forms did not all originate the same way from the earlier mulberry or "morula" stage. In fact, according to Haeckel there were four main patterns of gastrula formation. In the "primordial" pattern the morula developed into a single spherical layer of cells, which then folded in on itself to create the two-layered "archigastrula" form. But there were three other patterns as well; these he considered to be secondary forms that had evolved from the original archigastrula form.[49]

To Claus and other critics,[50] these multiple modes of formation raised serious problems for Haeckel's theory. What basis did he have for declaring one form "primitive" and the others "secondary"? As Claus wrote, "Clearly it is purely a matter of subjective judgment . . . to set up one single gastraea as the Ur-form." Nor was he enamored of Haeckel's explanation that the other patterns of development "falsified" their phylogenetic history through adaptation—Claus thought this use of "falsification," which Haeckel dubbed "cenogenesis," to be "in its application an exceedingly dangerous hypothesis."[51] The existence of so many different modes of gastrula formation further cast doubt on a central claim of the theory, that all gastrulae were derived from a common ancestor and therefore were homologous. If they came from a common ancestor, then, following Haeckel's logic, they should show the same mode of ontogenetic development. What was common to them, however, was not the pattern of development but the existence of a two-layer stage. To Claus the existence of such a stage without a common developmental path indicated that its significance should be understood not in terms of inheritance from a common ancestor but in terms of adaptation to similar conditions of

49. Haeckel, "Die vier Hauptformen der Eifurchung und Gastrula-bildung," pp. 419–52, sec. 10 of "Die Gastrula und die Eifurchung" (1875).

50. See, e.g., Leuckart, "Bericht 1872–1875" (1873), pp. 419–21.

51. C. Claus, *Typenlehre* (1874), pp. 11–12.

existence.[52] Claus argued along similar lines that the development of the gastrula should be understood in terms of improving the efficiency of the organism. A cell cannot simply continue growing larger and larger, because as the surface-to-volume ratio gets lower the cell cannot continue to supply the interior with nutrition. That is why at a certain point cells divide, and at a later point they differentiate into two layers.

So far, this explanation seems very close to the sort of mechanical explanation His might have provided. But Claus put an adaptationist twist on it: the division of labor between internal and external layers, with the internal layer taking on the nutritive functions and the external layer taking on the sensory and respiratory functions, is an adaptive one that confers on the organism an improved way of dealing with its environment.

He made a similar argument concerning Haeckel's somewhat tortured discussion of the significance of the middle germ layer, the mesoderm. "It is by no means a remarkable phenomenon, somehow explicable only by phylogeny, but a physiologically necessary one, that everywhere the organism achieves a higher organization and stage of life there appears a middle germ layer." Indeed, Claus perceived that the difficulties Haeckel had in determining the phylogenetic significance of the mesoderm derived from his insistence on looking at it as only phylogenetically significant. There were good physiological, adaptive reasons, Claus argued, why such a layer might have held advantages for the organisms that developed it; although it might have originated in different ways in different organisms, it would still confer the same advantages. Rather than selecting one of its several modes of formation as the most primitive and relegating the others to the dustheap of "secondary adaptations" without phylogenetic significance, why not see that the significance lies in recognizing that this structure, however it originates, is advantageous to those that develop it?[53]

For someone taking this stance, morphology as Haeckel defined it was at best a limited enterprise. Claus's whole approach to zoology turned him away from the comparison of embryonic forms and toward the examination of those forms in relation to their conditions of existence. As a result, his work, though Darwinian, was less and less morphological as morphology came to be defined in the 1870s.

52. Ibid., p. 10.
53. Ibid.

Throughout his career, Claus remained committed to a zoology in which the organism's adaptation to its conditions of existence was a main theme; the fact that his textbook went through so many editions suggests that enough people agreed with him to have their students buy it.[54] A number of other researchers, including Siebold, Leuckart, Weismann, and Eimer, pursued this more physiological (and often experimental) line of work in the 1870s and 1880s. Although they considered themselves Darwinians, they saw the processes of adaptation and inheritance and their relationship as matters to be investigated and not, as Haeckel assumed, to be taken as given. Haeckel's approach was oversimplified and logically unrigorous, and it neglected long-standing questions about just how the conditions of existence affected the individual organism and the species as a whole.

If all the responses to Haeckel were like this, his program would hardly have been counted a success. And yet there was a widespread perception by the late 1870s and early 1880s that "phylogenizing" was rampant, even dominating the community. Where did this come from?

The answer is, younger zoologists. The most concerted opposition to Haeckel's approach came from those members of his own generation and older zoologists who had been engaged in or trained in scientific zoology and for whom Darwinism gave a new justification for studying adaptation. Men who came into the field in the 1870s, however, viewed Haeckel's program through quite different eyes. It offered several appeals to an aspiring zoologist that would have been less compelling to his better-established older colleagues.

## THE GERM LAYERS AND THE YOUNGER GENERATION

"Where will all our young zoologists continue?! *Sancte Darwin, ora pro illis!" In* 1876 Emil Selenka was happy to have attracted students at Bavaria's smallest university, Erlangen, but he nevertheless worried over their future. Two weeks earlier his colleague Carl Semper had written with characteristic acerbity, "God knows why every ass now

---

54. Claus's textbook went through four editions between 1868 and 1882; a thoroughly revised version, shortened in text and supplemented with illustrations, appeared in 1880 under the title *Kleines Lehrbuch der Zoologie;* this new version, subsequently retitled *Lehrbuch der Zoologie,* went through six editions by Claus's death in 1897; thereafter it went through several more editions revised by Carl Grobben and a final edition in 1932, revised and edited by Alfred Kühn (E. Korschelt, 'Claus" [1939], pp. 65–66; see also K. Grobben, "Claus" [1899]).

thinks he is fit for zoology; is it because they all think they are related to Haeckel?"[55] These letters expressed with varying degree of sympathy what was becoming evident to everyone: zoology was suddenly becoming disturbingly crowded with aspirants to a career.[56] In zoologists' lecture rooms and labs, in the new journals that were popping up to accommodate dissertations and postdoctoral papers, the swelling numbers were overwhelmingly apparent.

Nine zoologists, six of them in Haeckel's generation, had acquired full professorships between 1865 and 1870, in the process representing a change at nearly half of Germany's universities.[57] Now they were settling into the business of building up their institutes. Although they by no means received especially favored treatment by their university and state administrators, they did benefit from the general university expansion of the period, gradually gaining more space, equipment, and students. While this allowed them to accommodate greater numbers of advanced researchers, those researchers had little hope of gaining university professorships. It took the entire decade of the 1870s for another eight zoology professorships to change hands; the second half of the decade was especially bad, with only two chairs opening up between 1876 and 1880. Reflecting on the situation in 1882 to a cousin living in Holland, Ernst Ehlers wrote that Germany seemed "to labor under an overproduction of zoologists; . . . most here, unfortunately, are forced to seek in their science their living as well."[58] Ehlers would know; a number of his protégés from Göttingen were desperately hanging on in nonuniversity jobs in hopes of returning to academia.[59] By the early 1880s, when a new flurry of

55. E. Selenka to Ehlers 30 May 1876, UB Göttingen, Cod. MS E. Ehlers 1802; Semper to Ehlers, 18 May 1876, UB Göttingen, Cod. MS E. Ehlers 1805.

56. Haeckel himself wrote to Siebold in 1880, "What is to become of all these 'candidates in zoology?'" (Haeckel to Siebold, 14 February 1880, in Uschmann, *Ernst Haeckel,* p. 140).

57. The Haeckel cohort members were Haeckel, Claus, Weismann, Semper, Ehlers, and Schneider. The other three were Karl Möbius (1825–1908) at Kiel; Rudolf Leuckart at Leipzig; and H. A. Pagenstecher (1825–89) at Heidelberg; I have left out Wilhelm Keferstein because he died in 1870.

58. Ehlers to T. W. Englemann, 29 January 1882, SBPK, Lc 1860 (9): E. Ehlers.

59. For example, Hubert Ludwig (1852–1913) directed the Bremen museum for natural history and anthropology from the time of his marriage in 1878 until he landed the Giessen zoology chair in 1881. During those three years he repeatedly wrote Ehlers of his longing to return to academia and anxiously awaited the outcome of the searches at Breslau, Rostock, and Giessen (see Ludwig to Ehlers, 25 July 1879; 29 November 1879; 14 December 1879; and letters of 25 June 1880–17 February 1881, UB Göttingen, Cod. MS E. Ehlers 1158). J. W. Spengel (1852–1921) spent two years as librarian at Naples before succeeding Ludwig in Bremen, where he worked for seven years,

changes took place (nine turnovers between 1881 and 1885), a substantial backlog of experienced zoologists was waiting in the wings.

Thus the men who came into zoology in the 1870s found themselves in a field far more crowded and competitive than their teachers who had entered a decade earlier, and this fact affected their own expectations for their research. Their teachers retained the old expectation that one should demonstrate one's basic skills through empirical studies of the anatomy, development, and classification of a particular group of animals, just as they themselves had. Yet as the field grew more crowded, the demands on younger researchers to publish more rapidly and to offer work of greater significance escalated. The framework set out by Haeckel's gastraea theory offered one way to meet these various demands.

The gastraea theory did not transform these younger researchers into slavish followers of Haeckel, but they found in his program a number of features especially attractive to men in their professional situation. In uniting evolution with the problem of interpreting the germ layers, the gastraea theory gave them a comprehensive and intellectually challenging framework into which to place their empirical findings and, as they developed the program, the possibility of exploiting new laboratory techniques in the pursuit of these findings. As we will see, both the theoretical framework and the new techniques they developed held an added advantage for researchers in a crowded, competitive environment, namely, the ability to produce publishable material relatively quickly. Although they found much to criticize in Haeckel's theory itself, as they followed it up in their investigations they almost inadvertently accepted two aspects of his program that had so inflamed his older contemporaries: the speculative tenor of his theorizing and the dissociation of morphology from the study of the organism's relation to its conditions of existence.

Haeckel's theory and the criticisms of it by his colleagues provided a rich set of problems for investigators to pursue. How did the germ layers develop in the countless organisms that Haeckel had not investigated himself?[60] Were the secondary germ layers homologous—did the same organs derive from the same layers in different

---

ever hoping to gain a university professorship. This he finally did in 1887, again succeeding Ludwig, this time to his chair at Giessen.

60. This was the broadest area for investigation, and the number of works touching on it is far too long to list. Even species closely related to those Haeckel did investigate were subject to careful reexamination (see, e.g., the series of 10 articles on development and classification of sponges, F. E. Schulze, "Untersuchungen über den Bau und Entwicklung der Spongien," 10 pts. [1877–81]).

groups?[61] How should one understand the development of organs that in their adult state appeared homologous, but that took different courses of development?[62] Could some developmental processes be clearly defined as mechanical in His's sense, distinct from others that were phylogenetic"?[63]

In their general structure, the works that considered such questions often looked extremely similar to the earlier studies of anatomy, development, and classification conducted in the 1860s: they went through the animal organ system by organ system, following its development and modes of generation and discussing the significance of the findings for classification. Although some essays still confined themselves to description, more ambitious authors sought to relate their results to larger questions concerning phylogeny and/or the germ layers. In this format such questions were not usually presented as the raison d'être of the research conducted, and their discussion was frequently confined to a section entitled something like "theoretical considerations." Nevertheless the anatomical/developmental studies that had long typified "first works"—dissertations and early postdoctoral papers—now often incorporated discussion of the theory of germ layers.[64]

One might have expected this of Haeckel's students at Jena, and in the 1870s and 1880s the *Jenaische Zeitschrift* contained numerous discussions of germ-layer development that were heavily tilted toward the gastraea theory. But ironically, two of the places where younger researchers showed the most intense engagement with the evolutionary interpretation of the germ layers were Leuckart's institute at Leipzig and Claus's institute in Vienna. Leuckart wrote Claus in some dismay in 1876 of two men working at his institute: "Dr. Hatschek, too, although he has an undeniable technical and scholarly talent, is

61. Oscar and Richard Hertwig, among others, attempted a solution of this problem (*Studien zur Blättertheorie* [1879–80], and "Coelomtheorie" [1882]).

62. H. von Ihering, "Ontogenie von Cyclas" (1876), *Vergleichende Anatomie des Nervensystemes* (1877).

63. For example, Emil Selenka, writing on the sea cucumbers, argued that because a lack of oxygen could lead the form of the cucumber to develop in different ways, one could not simply reject the possibility that "external mechanical stimuli can lead to the construction of new organ systems . . . —as little as such a hypothesis seems to harmonize with the complicated laws of adaptation and inheritance of the Darwinian theory" ("Entwicklung der Holothurien" [1876], p. 175). For another example, see E. Witlaczil, "Entwicklungsgeschichte der Aphiden" (1884), esp. pp. 677ff., "Die mechanischen Vorgänge bei Entwicklung der Insekten: Lageveränderungen des Embryo."

64. An excellent source for titles concerning development and classification in the various invertebrate groups is E. Korschelt and K. Heider, *Lehrbuch* (1890); each chapter ends with a bibliography of works concerning the classificatory group covered; the large majority of the works cited date from the late 1860s and after.

infected by the Jena epidemic, as is Herr Dr. v. Ihering."[65] Leuckart's student Karl Chun (1852–1914), who would later become famous for his hydrobiological work, also took up questions kindled by the gastraea theory in the early 1880s. At Vienna, the pursuit of ontogenetic homologies may have been spurred by Berthold Hatschek (1854–1941), who left Leipzig to work under Claus for much of the 1880s. While he was there, other young scholars, including Karl Heider (1856–1935), Karl Grobben (1854–1945), and J. Ciamician (1857–1922), took up the evolutionary interpretation of the germ layers.

Among these researchers, the stance toward Haeckel's theory itself varied considerably. Haeckel's students Oscar and Richard Hertwig, whose coelom theory attempted to explain the source of the mesoderm and its products in phylogenetic terms, clearly saw themselves as refining and extending his framework.[66] Nevertheless, their conclusions did not favor drawing homologies between the germ layers (especially the mesoderms) across different phyla; like Claus, they considered the germ layers as fairly undifferentiated material from which a variety of tissues and organs could develop through adaptation to different conditions of existence. (Despite this, Haeckel later claimed that the Hertwig's coelom theory proved the utility of the biogenetic law as a "heuristic principle.")[67]

Hermann von Ihering (1850–1930), who published extensively on molluscs in this period, took a radically opposed position. Having followed the ontogenetic development of the germ layers in the snail *Cyclas*, he declared that it did not provide a reliable source of information about the snail's ancestry. Therefore, ontogeny should not be considered the primary criterion of homology; comparative anatomy was a far more reliable source.[68] When Leuckart complained that Ihering had succumbed to the "Jena epidemic," then, he could not have meant that his former student had taken up Haeckel's own ap-

---

65. Leuckart to Claus, 26 November 1876. SBPK, Lc 1851: Leuckart.

66. O. and R. Hertwig, "Coelomtheorie"; on the coelom theory, see J. Oppenheimer, "Non-specificity of the Germ Layers," *Essays* (1967); Weindling, *Darwinism and Social Darwinism*, esp. chap. 4.

67. O. and R. Hertwig, "Coelomtheorie," p. 186; Haeckel, "Ursprung und Entwicklung," p. 211.

68. Ihering, "Ontogenie von Cyclas"; when Ihering published his big work on the comparative anatomy of the nervous system and phylogeny of the molluscs in 1877, he made the point even more strongly (*Vergleichende Anatomie des Nervensystemes*, pp. vi–vii, 11–21); Carl Schmid-Monnard at Würzburg (probably a student of Semper) came to similar conclusions on the greater value of adult comparative anatomy (see "Histogenese des Knochens" [1883]).

proach. Rather, his dissatisfaction came from Ihering's having been drawn onto the territory of the phylogeneticists and perhaps from his adopting their polemical and speculative tone.[69]

Leuckart's comment on Hatschek was more to the point. Beginning with his dissertation under Leuckart, Hatschek took as his own the problem of giving the germ layers a correct evolutionary interpretation. In this first essay, Hatschek strode onto the evolutionary territory with a piece of research on the development of the nervous system in butterflies that was frankly incomplete, but from which he nevertheless thought he could draw some significant conjectures about the evolutionary relationship between the arthropods and the annelids.[70] Although he never mentioned Haeckel's name in this essay, he used the framework of the gastraea theory in his arguments, and a later essay made explicit his view of Haeckel's theory: "As long as we view the construction of phylogenetic stages only as an embodiment of our abstractions and continually remind ourselves that its significance is predominantly heuristic, we can adopt without fear a series of the hypothetical forms set out by Haeckel."[71] He went on to accept all the basic Ur-forms that Haeckel had proposed. Hatschek became one of the leading proponents of the Haeckelian style of morphology; his 1888 textbook of zoology presents a subject that follows Haeckel in all its main features.[72]

The overall impression left by such works is that while in most cases they found something in Haeckel's theory to criticize or modify, certain aspects of his approach gained increasing legitimacy. Even as many younger researchers found their own studies of the germ layers inconclusive for phylogeny, this did not detract from the general goal of seeking phylogenetically valid homologies and using them to make classificatory judgments. Furthermore, while younger researchers were always concerned to contribute new empirical findings to the stock of zoological knowledge, they tended to be more comfortable than many of their elders with offering up speculative comments for discussion. Hatschek, for example, wrote in his first paper, "in our biological science the imperative necessity for theoretical considerations is ever more urgently making itself felt . . . "; he then went on

69. Leuckart to Claus, 26 Nov. 1876, SBPK, Lc 1851: Leuckart.

70. B. Hatschek, "Entwicklungsgeschichte der Lepidopteren."

71. B. Hatschek, "Entwicklungsgeschichte von Teredo" (1881); Hatschek began corresponding with Haeckel in early 1876 and met him on numerous occasions; he considered himself a Haeckel student "at a distance" (Uschmann, *Geschichte der Zoologie*, pp. 133–35).

72. B. Hatschek, *Lehrbuch* (1888).

to invoke Karl Ernst von Baer's call for both observation and reflection.[73] A remark this explicit is unusual, but the presence of new theoretical or general sections in many of these works suggests that speculation and hypothesizing became much more widely accepted among the younger generation.

This generational shift in standards of research and publishing seems a bit odd, given the values predominating among the older generation. After all, the latter ran the laboratories in which the younger scholars worked and controlled the publishing outlets; surely if they disapproved of speculation in general and of Haeckel's gastraea theory in particular, it was strange that the older scholars would permit this new direction to take shape. And yet they did. Although no clear-cut resolutions to this paradox present themselves, we may speculate on some possibilities. To begin with, the line between legitimate theorizing and illegitimate speculation is a subjective one, and although older scholars were not especially happy about speculations, they did approve of theorizing, if properly grounded; after all, one of the features that made scientific zoology scientific was its claim to a higher theoretical status than mere systematics. Haeckel had crossed the line by not distinguishing between what was known and what was conjectured, and by the dogmatic tone of his assertions. Younger scholars avoided these transgressions by keeping their theoretical remarks in a clearly marked section of their work separated off from their empirical results, and by couching the theory in tentative language. Then they could have it both ways: their empirical results could stand as evidence of their research abilities, and their theoretical remarks might gain acceptance, or at least push a problem in a new direction. This role for theory was much more in line with the values of the older generation.

A second possible reason why older scholars did not steer the younger generation more firmly away from Haeckel's theory goes back to the reason Haeckel made a splash in the first place: his theory was magnificently synthetic, covering the entire animal kingdom (except for single-celled creatures) as well as pulling together the problem areas of development, evolution, and systematics. In other words, despite its numerous flaws of evidence and presentation, the theory epitomized the community value of breadth. In may well be that older scholars tolerated their students' grappling with the gastraea theory because it gave them an opportunity to practice big thinking and to learn to place a relatively small-scale research task into a

---

73. Hatschek, "Entwicklungsgeschichte der Lepidopteren," p. 115.

broader comparative anatomical, developmental, and classificatory perspective.

The younger generation's impulse to speculate may also have been fueled by the newly competitive environment in which these men found themselves. The push to publish quickly was widely felt in the discipline, as more and more people began working in closely related areas. The competitive pressure meant that researchers often felt pressed to publish findings, and perhaps to announce their possible significance, before those findings and conclusions were securely established. Indeed, the need for a rapid publication outlet for such preliminary notices was part of the justification for the 1878 founding of the *Zoologischer Anzeiger,* whose issues were published every two weeks, far more often than other journals.[74] Even established scholars felt this pressure: in a letter accompanying a preliminary note for the *Anzeiger* in 1880, August Weismann wrote the editor that he normally liked to make everything absolutely complete before he published on a subject, but "today this is especially dangerous, when such a mass of workers is around that one can hardly work on a subject in peace. I hate this breathless racing very much, but it is also aggravating to sign away to others something of the little bit that one can produce in a lifetime."[75] Similarly, Theodor Eimer wrote in 1877 that certain preliminary results published by the Hertwig brothers duplicated his own, earlier work, which had been awaiting publication until he finished with the "endless physiological experiments" that would round out the morphological results. "I wanted to complete the whole peacefully and fully without 'preliminary notices'—now here come the predators [*Jäger*]."[76] The existence of these many short notices of findings and accompanying speculations suggests that the increased presence of speculation was not entirely due to Haeckel's intellectual influence. Rather, Haeckel provided a rationalization for men who wanted to make their work significant without having sufficient time to check up on all the details.

Although the attention devoted to theoretical considerations grew substantially in the late 1870s and 1880s, that placed on empirical

74. The *Zoologischer Anzeiger* also became a leading forum for priority disputes, which became rampant in the late 1870s, and which may also be seen as a sign of overcrowding (see L. Nyhart, "Writing Zoologically" [1991]).

75. A. Weismann to J. V. Carus, 9 April 1880 (Freiburg) (Bll. 34–35), SBPK, Lc 1889 (23): A. Weismann. Weismann knew what he was talking about, for in the previous two years he had been involved in a protracted priority dispute with his contemporary Carl Claus (see Nyhart, "Writing Zoologically," esp. pp. 55–60).

76. T. Eimer to Ehlers, 14 October 1877, UB Göttingen, Cod. MS E. Ehlers 422.

contributions by no means disappeared. The study of the germ layers allowed researchers a wonderful opportunity to increase the empirical knowledge of ontogeny by exploiting techniques that had not been available before about 1870. In particular, two new technical aids became available at the end of the 1860s that opened up dramatic new possibilities for the microscopic study of invertebrates. One was a new fixative, osmic acid, which allowed fresh specimens to be preserved with little fuss or loss of tissue definition; this was the first of a series of new fixatives and stains that would improve the ability of researchers to identify various embryonic tissues and organs. The other was the microtome.[77] Using these technologies not only meant that researchers could wring new information out of even the best-studied organisms, but it also allowed them to demonstrate that they were on the technical forefront, too.

Section cutters (*Querschnitter*) had been used in the 1860s and earlier, but the most common method for preparing microscopic sections from preserved material was to use a sharp knife and a steady hand. Under such circumstances, making serial sections was an arduous task that few people attempted; most zoologists contented themselves with the occasional cross-section. Then in 1870 the anatomist Wilhelm His described his new microtome, which overcame the disadvantages of the old ones and immensely simplified the preparation of serial sections. Not only did the new device significantly improve the precision of the slicing, but perhaps at least as important, it vastly decreased the time it took to make the sections. Recounting what it had been like earlier to make sections of chick embryos, His observed, "After days of long work with a free hand I had brought matters so far as to achieve a few tolerable sections, but only very unsure ones and with an excessive expenditure of time." In the four years since acquiring the instrument in 1866, however, he had made "probably over 5,000 sections" with it.[78] Even if we allow for rhetorical exaggeration, the contrast must have been dramatic.

77. The first published mentions of osmic acid appeared in 1865, and by the end of the decade it was being widely used (M. Schultze, "Zur Kenntniss der Leuchtorgane" [1865], p. 132; Schultze and M. Rudneff "Weitere Mittheilungen" [1865]); Brian Bracegirdle has written that osmic acid was rarely used before 1882 because it was "expensive and labile," but it is clear from the writings of the German zoologists that they were using it with great regularity from the early 1870s on (Bracegirdle, *History of Microtechnique,* p. 61; this work also provides a useful overview, with pictures, of the development of the microtome). For a contemporary's description of how tissues were actually stained, hardened, and sliced, see A. B. Lee, *Microtomist's Vade-mecum* (1885).

78. W. His, "Beschreibung eines Mikrotoms" (1870), pp. 229–30.

As it gained momentum in the mid-1870s, the use of serial sections changed significantly the practice of studying development. In the 1850s and 1860s, the most common practice was to observe continuously one or a few living individuals as they developed. But in the 1870s more and more zoologists preserved individuals at different stages of development and then sliced them up to reconstruct the position and extent of tissues and organs at each preserved stage. Because the first method was limited to organisms with transparent outer layers, the second one opened up the field of embryological study to a much wider range of organisms. In addition, the new techniques allowed researchers to identify much more readily the size and shape of small organs and to follow the developmental destiny of a given group of cells into tissues and organs. The study of the germ layers (whether considered from an evolutionary perspective or not) was therefore made accessible in a way that had simply not been possible earlier.

At the same time as they opened up a new range of empirical investigation, the changes in technique brought with them a shift in focus from development as a process occurring within a living organism in contact with its environment to development as a series of pictures of tissues and organs. In the 1860s zoologists had invariably made sure to emphasize that their observations of development were made predominantly on living organisms. The practice was so customary that Anton Schneider, whose study of the nematodes drew mainly from the enormous collection of prepared helminthological specimens at the University of Berlin, felt it necessary to defend the relative lack of results on the living worms.[79] By the late 1870s, Leuckart's student Hermann Reichenbach (1848–1921) could state without apology that "the study of prepared [*gehärtete*] embryos" of the river crab produced "far better results" than that of fresh specimens. Many of the papers of the late 1870s and early 1880s do not even bother with that much explanation, simply making such statements as "fresh material was not available to me." The study of the germ layers, then, marked not just a change in technique but a change in the value placed on looking at living organisms.

There is no reason to suppose that this shift was some sort of ideologically motivated turning away from a program focused on adaptation and toward one of "pure" morphology. Nevertheless, as some zoologists at the time saw it, the result was just a shift. While members of the older generation acknowledged the progress brought

79. Schneider, *Monographie der Nematoden* pp. 27–28.

by the new techniques,[80] not all were entirely comfortable with the direction morphological research seemed to be taking. Already in 1876 Siebold was writing a colleague,

> But don't you also fear that the currently ruling phylogenetic orientation, if I may call it that, is making evident a great one-sidedness of the younger generation of zoologists. . . . Section after section is made, as if it held that whoever achieves the most sections for one animal has an advantage over those who have treated the same animal with fewer sections. I also note to my greatest astonishment that such a zoologist, who boasts of having sliced up a minutely small animal into so-and-so many sections, doesn't even know what a Phryganide's casing looks like. Such an artificial product made by an insect, of course, doesn't concern him—it doesn't allow itself be split up into sections.[81]

And later, "We are being literally flooded with these slicing methods. Where is the *observation* of the ways and actions of the *living* animal!"[82] What good does it do, he asked, to know an organism's structures in their minutest details, if you don't know what they do for the living organism?[83]

Even Haeckel seems to have been somewhat dismayed by the direction his program had taken by 1880, when he wrote, "It has unfortunately now become the mode to document one's 'scientific' zoological qualifications primarily through the preparation of cross-sections with the microtome and glossing them over with eosin, methyl green, Bismark brown, etc." He sounded the same theme again the next year: "The next generation of 'scientific zoologists' will only know *cross-sections* and *colored* tissues, but neither the *entire* animal nor its mode of life!"[84]

Siebold and Haeckel were observing one of the most important developments in zoology in the 1870s and 1880s: the creation of a morphology that was divorced from questions of adaptation and the living organism. Haeckel was partly responsible, in that his program played down questions of adaptation, but like his contemporaries he valued the observation of the living object, and his research itself had always been conducted on living and freshly killed specimens. This style of morphology was not entirely of his own making.

---

80. See, e.g., Weismann, "Entwicklungsvorgänge im Insektenei" quoted in K. Sander, "August Weismanns Untersuchungen" (1985), p. 48.

81. Siebold to Ehlers 16 November 1876, UB Göttingen, Cod. MS E. Ehlers 1815.

82. Siebold to Ehlers, 15 November 1879, UB Göttingen, Cod. MS E. Ehlers 1815.

83. Siebold to Ehlers, 18 February 1879, UB Göttingen, Cod. MS E. Ehlers 1815.

84. Haeckel to Wilhelm Peters, 6 March 1880, quoted in Jahn, "Ernst Haeckel," p. 83; Haeckel to Siebold, 5 March 1881, quoted in Uschmann, *Ernst Haeckel*, p. 141.

Siebold himself supplied another possible reason why the stain-and-slice approach was so popular among the younger generation. The observation of the activities of the living animal, he wrote, "requires time, which the current generation of zoologists don't want to spend on such observations, [for] having begun one set of observations, the observer must wait the length of a year or a year and a half or longer before he can publish his results."[85]

But as we have seen, time was in short supply for the younger zoologists, who needed to publish as quickly as possible. Although there is no direct evidence that researchers consciously picked stain-and-slice topics more for this reason than for others, Siebold's observation is quite trenchant. Studying the life cycle of an organism frequently required more than one season—indeed, investigations like that of Weismann on the butterflies could take several years. The microtome had not only made the creation of serial sections much faster in comparison to the old mode of preparation, but had also made studying the germ layers a faster project in comparison to the natural history studies favored by someone like Siebold.

In the 1870s and 1880s, then, the pressures on younger researchers to publish quickly led in two directions, both of which steered them away from the commitments of the older scientific zoologists. By accepting speculation as a legitimate contribution to zoological scholarship, they could claim to publish work of significance even when it was based on small quantities of research. The research that they did conduct was largely based on the production of serial sections, material that could be produced and studied with relative haste, especially as compared with studies of the organism's entire life cycle. A curious and unintended result of this line of research was an increasing neglect of the living organism, which caused a de facto separation of the study of development from the consideration of its relation to the conditions of existence. Here, too, the younger researchers were leading morphology away from their elders' earlier set of associations.

## CONCLUSION

In 1876 Siebold sadly observed, "Biology is not keeping pace with morphology. The biological conditions interest morphologists today not a bit, and yet, one should think that biology is the highest part of zoological science."[86] Siebold's use of the term "biology," meaning

---

85. Siebold to Ehlers, 18 February 1879, UB Göttingen, Cod. MS E. Ehlers 1815.
86. Siebold to Ehlers, 16 November 1876, UB Göttingen, Cod. MS E. Ehlers 1815.

the organism in relation to its conditions of existence, is significant, for it marks a shift in his own thinking about the parts of zoology and suggests a larger shift in the discipline. In the 1850s, morphology and biology had been so closely joined that they were rarely considered separately: the entire program of scientific zoology revolved around a conception of the organism in which the two were united. Although Siebold's own main area of research, the study of parthenogenesis, remained constant from the late 1850s through the end of his life, the sort of investigations he was conducting looked different as the intellectual milieu changed around him. If it might still be called morphological research when judged against that of an older generation of systematists, when compared with the new morphologists of the phylogenetic stripe, it was no longer morphology but "biology." Indeed, as Siebold's own words show, he was shifting his own stance as to what he considered morphology as it developed a new set of techniques and associations.

By the 1880s a new alignment was emerging: for many younger zoologists, morphology was associated on the one hand with the speculative elaboration of phylogenetic relationships and on the other hand with the study of microscopic sections of developmental stages. The study of the organism in its environment and the process by which adaptation changed an organism's form, by contrast, was a separate topic called "biology" or sometimes "ecology" (Haeckel's term for the same thing). In a zoology textbook that ran the subtitle "A morphological overview of the animal kingdom," for example, Hatschek declared that morphology "in its narrower sense" meant "genealogical morphology, . . . the science that concerns itself with the forms in and of themselves, in order to establish their relationship." In this context, "comparative physiology of the animals is a specialty area [*Sondergebiet*]," an "aid" to morphology, but not strictly central to its aims.[87]

This separation is ironic, for in both the program of the scientific zoologists and the logic of Darwin's theory, the organism's relationship to its conditions of existence played a central role. I have tried to suggest that this shift derived from a combination of Haeckel's own relative indifference to adaptation as an object of investigation, the attraction of applying new laboratory techniques that required the animals to be sliced up, and a hurry-up-and-publish mentality that gained momentum from a newly crowded and competitive disciplinary environment. It is true that there remained some important conti-

87. Hatschek, *Lehrbuch*, pp. 19, iii.

nuities. The value placed by the zoological community on intellectual breadth is visible in both Haeckel's generation's efforts to unite anatomy, embryology, and classification under the banner of evolution, and in the first generation of their students, who addressed the same areas in their critical elaborations of Haeckel's gastraea theory. But as developed by the youngest of the three generations of zoological researchers active in the 1870s, evolutionary morphology in German zoology moved in a direction decidedly contrary to the commitments of the older scientific zoologists.

However, that direction was by no means dominant or permanent. The stiff competition facing younger researchers meant that only a few were lucky enough to land university jobs in Germany. Of the Haeckel enthusiasts discussed here only Karl Chun achieved a university professorship in Germany—and Chun moved on to other, decidedly more "biological," research questions by the mid-1880s. Hatschek, Heider, and Grobben did well in their native Austria; Ihering emigrated to Brazil in 1880, where he helped build the state museum in São Paulo; Giacomo Ciamician, who had always devoted only a part of his interests to zoology, developed his other main research area, photochemistry, into a prominent career in his native Italy; and Hermann Reichenbach became an *Oberlehrer* in Frankfurt am Main. The other Haeckel followers are lost to us. Thus it would appear that the vogue for the gastraea theory was intense but, at least in Germany, relatively limited in scope: a cohort of dissertators and young postdoctoral researchers in the late 1870s and early 1880s took up its questions, which enlivened the traditional embryological, anatomical, and classificatory first works with the whiff of grand theory, but they very soon either moved on to other research areas or moved out of Germany. As we will see in chapter 10, Siebold's fears for biology would prove to be largely unfounded.

Carl Theodor von Siebold (from *Zeitschrift für wissenschaftliche Zoologie* 42 [1885] facing p. i).

Carl Gegenbaur in 1861, at age 35 (from Fürbringer and Bluntschli, eds., *Gesammelte Abhandlungen von Carl Gegenbaur*, 1912, frontispiece).

Ernst Haeckel in Tropical Finery (from Haeckel, *Indische Reisebriefe,*
6th ed., Leipzig, 1922, frontispiece).

Haeckel, Fürbringer, and friends, *left to right, standing:* Max Verworn, Heinrich Haeckel, Ernst Stahl; *seated:* Ernst Haeckel, Fanny Fürbringer, Lisbeth Haeckel, Max Fürbringer, Wilhelm Biedermann (from Max Fürbringer Nachlass, Handschriftenabteilung, reprinted, by permission, from Universitätsbibliothek, Heidelberg).

Carl Gegenbaur lecturing in the anatomy hall at Heidelberg, 1892 (from Friedrich Maurer, "Carl Gegenbaur," 1926, facing p. 511).

Carl Semper (reprinted, by permission, from Smithsonian Institution, photo no. 937009).

Rudolf Leuckart, 1891 (reprinted, by permission, from Smithsonian
Institution, photo no. 93-7021).

August Weismann, Winter 1889–90 (reprinted, by permission, from Smithsonian Institution, photo no. 93-7019).

Ernst Ehlers (reprinted, by permission, from Smithsonian Institution, photo no. 93-7015).

## CHAPTER SEVEN

# Evolutionary Morphology in Anatomy

## Carl Gegenbaur and His School

The story of evolutionary morphology within the discipline of anatomy in the 1870s and 1880s has quite a different feel from that in zoology. To be sure, just as the program became identified with Haeckel in zoology, so, too, in anatomy did it have a leading representative in Carl Gegenbaur—though the term used to describe the program in anatomy was more often "comparative anatomy" than "morphology," and the evolutionary aspect was assumed within that term. But the intellectual and social character of Gegenbaur's program differed substantially from Haeckel's. Haeckel inspired big thinking; Gegenbaur demanded detailed empirical research. Haeckel never had a "school" in the strong sense of the word; Gegenbaur did. Conversely, whereas Haeckel's program was taken up with enthusiasm by a number of younger zoological researchers who were not, strictly speaking, his students, Gegenbaur found that his support among anatomists was closely limited to his school.

The differences in the fate of Haeckel's and Gegenbaur's programs in the 1870s and 1880s cannot be explained entirely in terms of the nature of the followings they attracted, however, for those followings were themselves shaped by different pressures within the two disciplines. The discipline of anatomy in this period shows three important contrasts to zoology, differences that help explain the problems Gegenbaur's evolutionary program had fitting into anatomy. To begin with, whereas zoological inquiry in the 1860s and early 1870s was marked by its breadth and fluidity, anatomy was characterized by greater intellectual differentiation, with at least four clearly defined orientations toward research being widely recognized. The comparative-anatomical orientation associated with Gegenbaur was just one of these, often viewed as being in competition with the others. Second, the institutional location and educational function of anatomy differed from that of zoology: whereas zoology was by this time firmly ensconced in university philosophical faculties, anatomy remained just as firmly a part of medical education. Within this institutional framework, evolutionary comparative anatomy as a teaching subject came under considerable pressure from medical educators who found it insufficiently tied to issues of medical practice. Third, the social and demographic structure of the discipline of anatomy that had developed since the early 1850s was quite different from that of zoology, with an internal hierarchy of *ausserordentliche* professors and prosectors and an emerging division into macroscopic and microscopic anatomy as two main subdisciplines of teaching. These features lacked parallels in zoology. In conjunction with the first two factors, anatomy's social structure tended to make it increasingly difficult over the course of the 1870s and 1880s for evolutionary comparative anatomists to find an institutional home there.

## GEGENBAUR'S SCHOOL AT HEIDELBERG

In 1873, Gegenbaur was appointed professor of anatomy at Heidelberg to replace his father-in-law, Friedrich Arnold. This relationship probably accounts in some measure for his decision to leave Jena and Haeckel, for his new wife doubtless preferred to be nearer her relatives. At the same time, Heidelberg held other, more professional attractions: it was a considerably larger university than Jena and was less isolated. Gegenbaur was already forty-eight years old at the time of his appointment, and as Haeckel noted at the time, "in spite of all Jena's advantages, it would not be good to spend one's whole life here."[1]

Any hopes Gegenbaur might have harbored that Haeckel might eventually join him at Heidelberg were quashed in 1878, when Heidelberg's zoologist Heinrich Alexander Pagenstecher retired from teaching. The philosophical faculty explicitly chose to avoid considering scholars "whose *naturphilosophisch* exertions place theory over its foundations in reality in a confusing way, and who thus place in

---

1. Quoted in Uschmann, *Geschichte der Zoologie*, p. 80, n. 384.

serious doubt their teaching success."[2] This can only be a reference to Haeckel, who had recently alienated much of the scholarly community with a notorious speech at the fiftieth annual meeting of the German Society of Naturalists and Physicians in 1877, in which he argued that evolutionary monism ought to replace Christianity as the foundation of state-run education from the elementary years on.[3] In fact, Gegenbaur's hopes for bringing his friend any closer than Jena were fruitless, for after the 1877 speech, Haeckel never received another formal offer for a professorship in Germany.

There were certain advantages to having Haeckel remain in Jena, however. Unencumbered by his dominating presence, Gegenbaur attracted a sizable coterie of students to work in his institute, on projects of his own instigation. Thus the phylogenetic development of the vertebrate body, especially the limbs and the head, formed the starting point for researchers working with him. He also enhanced the possibilities for attracting advanced students by organizing his institute differently than he had at Jena. Most advanced researchers could financially afford to work with a professor for more than a few months only if he employed them as his assistants. At Jena, instead of keeping on one advanced researcher as a long-term prosector, Gegenbaur had chosen to spend the same money to employ two junior assistants from the ranks of the younger medical students.[4] These could devote only part of their time to the anatomy institute, and most moved on to medical careers outside of anatomy. At Heidelberg, where the larger classes placed more demands on an assistant and where there was more money at his disposal, Gegenbaur chose long-term prosectors and assistants who were bent on careers as professional anatomists. In this way, he was able to support a small group of disciples whose work he oversaw and guided for a good number of years.

Five of these students became the core members of his school. Max Fürbringer (1846–1920) and Georg Ruge (1852–1919), who had begun as students with him at Jena, were his first long-term assistants; they were succeeded by Friedrich Maurer (1859–1936), Hermann Klaatsch (1863–1916), and Ernst Göppert (1866–1945) (see table

2. "Gutachten" of the philosophical faculty to Engerer Senat, 22 January 1878, BGLA, file 235/29858. Although *Naturphilosophie* itself was by no means a threat in 1878, it had come to represent the primacy of speculation over empirical research, and so was used as a shorthand for that.

3. E. Haeckel, "Über die heutige Entwicklungslehre" (1877), pp. 14–20.

4. M. Fürbringer, "Carl Gegenbaur" (1903), p. 412.

TABLE 7.1 Gegenbaur's Assistants at Heidelberg

| | 1873 | 1880 | 1890 | 1900 |
|---|---|---|---|---|
| GEGENBAUR | • • • • • • • | • • • • • • • • • • | • • • • • • • • • | • |
| Fürbringer | ┝━━━━━━┥ | | | |
| Ruge | | ┝━━━━━━━━━━━┥ | | |
| Maurer | | | ┝━━━━━━━━━━━━━━━━━┥ | |
| Klaatsch | | | ┝━━━━━━━━━━┥ | |
| Göppert | | | | ┝━━━━━━━━━1912 |

*Note:* Fürbringer: 1873–79; Ruge: 1876–88; Maurer: 1884–1901; Klaatsch: 1888–1901; Göppert: 1892–1912.

7.1). A second group of some dozen disciples worked for lesser periods of time at Heidelberg and then returned to their homes or moved on elsewhere. Especially notable about this group is its international constitution: researchers came from America, Finland, Russia, Sweden, Denmark, Holland, and England to work under Gegenbaur.[5]

All of these men worked on problems extending their teacher's research field in different directions. Some of their research, especially that on the head, gills, and extremities, grew directly out of questions raised by Gegenbaur's theories on the phylogeny of skeletal structures. For example, Emil Rosenberg conducted his early investigations on the development of the spinal column and the hand bones under Gegenbaur, publishing his results in 1875 in the opening volume of Gegenbaur's new journal, the *Morphologisches Jahrbuch*. Rosenberg continued throughout his career to elaborate on the theory of the evolution of the spinal column that came out of that work. And J. E. V. Boas published numerous works on mammalian comparative anatomy, concentrating especially on the head and feet.[6]

More often, Gegenbaur's core students extended his comparative method to other structures besides the skeleton, or to taxa he himself had not worked on in detail. Fürbringer became the world's leading authority on the comparative anatomy of birds and with Ruge developed a line of research demonstrating the intimate connections between musculature and nerves, which allowed each to be used as a signpost for the other in morphological research. Ruge's main area

5. Fürbringer, "Lebenserinnerungen" vol. 1, pp. 354–62, UB Heidelberg, Fürbringer Nachlass. In 1912, Fürbringer wrote up brief resumes of the careers of Ruge, Maurer, and Göppert as part of the deliberations over his successor at Heidelberg (file D-1-f: "Nachfolge in Jena 1900–Heidelberg 1912, II: Rohmaterialien," SB, Fürbringer Nachlass).

6. On student works that followed closely along Gegenbaur's lines, see Fürbringer, "Carl Gegenbaur," pp. 429–38; on Rosenberg, see also A. Hasselwander, "Dem Andenken Emil Rosenbergs" (1930), pp. 81–107.

was the comparative anatomy of the muscles, especially in mammals. Maurer worked on the structure and development of the gills in fishes, embryonic gill structures in higher vertebrates, and the morphology and genesis of epidermal structures, especially hair. During the time that he worked under Gegenbaur and Fürbringer, Göppert concentrated on the morphology of the larynx, the ribs, and the arterial system. Only Klaatsch moved away from the broad comparative anatomy of the vertebrates to focus on human anatomy: he became an anthropologist and ethnographer.[7] Despite differences in the range and subject matter of their research, all of these men were interested in demonstrating the evolution of vertebrate form by means of comparative anatomy and embryology. Even where they moved beyond problems closely allied to Gegenbaur's theories, his disciples still retained the basic methods, aims, and overall style of research their leader had taught them.

Max Fürbringer's first major monograph, published in 1888, exemplifies this Gegenbaurian morphology well. This massive opus, *Untersuchungen zur Morphologie und Systematik der Vögel* (Investigations on the morphology and systematics of the birds), consisted of two huge volumes, a special part and a general part. Fürbringer devoted the "special" volume to a comparative anatomical analysis of the breast, shoulder, and proximal wing region of the bird, a subject on which few synthetic works existed. He set out to create a new synthesis by tying together research on the skeleton, nerves, and musculature. The first section, on the comparative anatomy of the nerves, served as the basis for the second, myological, section, for Fürbringer was convinced that homologies between the musculature of different bird species could be derived correctly only if they took into account the morphological relationship of the muscles to the nerves. And in Fürbringer's eyes these homologies provided the key to establishing the phylogenetic relationships among the birds.

The monograph's general part, he wrote, had the dual task of interpreting the results of the special part morphologically and systematically. The larger portion of the "general" volume was given over to bird systematics, offering a new genealogical system and an examination of the phylogenetic development of warm-blooded animals from their cold-blooded predecessors. The morphological section emphasized the theoretical problems raised by the manifold connections be-

---

7. On Fürbringer, see H. Bluntschli, "Fürbringer" (1922); on Ruge, see F. Maurer, "Ruge," (1921); on Maurer, see E. Göppert, "Friedrich Maurer" (1936–37); on Klaatsch, see R. N. Wegner, "Klaatsch" (1915–16).

tween musculature, nerves, and the bones and tissues that supported them. These issues included the morphological method; disagreements over interpreting forms as cenogenetic or palingenetic (i.e., deviating from or following the developmental path set by the organisms' ancestors), normal or pathological; and the meaning of correlation. Fürbringer also considered the development of the wing area in terms of the larger problem, set by Gegenbaur, of the evolution of the limbs. His book epitomized Gegenbaurian morphology, beginning from a broad empirical base in comparative anatomy to work up a synthetic interpretation of bird form and taxonomy as they could be understood in terms of development, adaptation, inheritance, and phylogenetic history.[8]

The members of the Gegenbaur school did not simply extend their teacher's approach intellectually; they spread it to new institutions as well. The outer circle of foreigners returned to their own countries, bringing Gegenbaur's ideas and style of work with them. After working at Gegenbaur's institute, Emil Rosenberg returned to his prosectorship in anatomy at the Estonian city of Dorpat, later moving on to a professorship at Utrecht, in Holland; Johan Axel Palmen became professor of zoology in his native Helsinki, Finland; William Berryman Scott returned to Princeton to become one of America's foremost paleontologists; A. A. W. Hubrecht later became professor of zoology in Utrecht; Wilhelm Leche returned to Stockholm; J. E. V. Boas represented the Gegenbaur school in Copenhagen as professor of zoology at the veterinary school; Michael v. Davidov would later become the director of the Russian marine station at Villefranche-sur-Mer; Hans Gadow returned to Cambridge, England, to become a leading ornithologist.[9] These men acted as spokesmen for Gegenbaur's morphological approach, bringing his name and his ideas to their native countries and incorporating them into the ongoing biological traditions there.

The core group of Gegenbaur students stayed much closer to the institutional bases of Heidelberg and Jena, and indeed the professorships of anatomy at these two universities were controlled by Gegenbaur and his school well into the twentieth century. After Gegenbaur's immediate successor at Jena, Gustav Schwalbe, left in 1881, the professorship of anatomy there was held in succession by O. Hertwig

---

8. M. Fürbringer, *Untersuchungen zur Morphologie* (1888), vol. 1, pp. ix–xi.

9. Fürbringer, "Lebenserinnerungen" vol. 1, pp. 354–62, UB Heidelberg, Fürbringer Nachlass.

(1881–88), Fürbringer (1888–1901), and Maurer (1901–32). At Heidelberg, Gegenbaur taught until 1901, then passed on his position to Fürbringer. When Fürbringer retired in 1912, he was succeeded by his son-in-law and protégé Hermann Braus, who remained there until 1921.[10] Göppert, who had begun his assistantship under Gegenbaur in 1892, stayed at Heidelberg as Fürbringer's prosector until the latter's retirement. Only in 1912, when his younger colleague Braus received Fürbringer's professorship, did he move on to the University of Marburg as an *ausserordentliche* professor.

As this pattern suggests, the core students of the Gegenbaur school depended very much on each other and on their teacher for institutional support as they moved from assistantships and prosectorships to professorships. As a result, they did not move beyond the range of institutions upon which Gegenbaur could exercise his personal clout. The dominant role of patronage in their careers was established early on, with Fürbringer's appointment in 1879 at the newly founded University of Amsterdam. The Dutch minister of education had been very impressed by Gegenbaur, who had taught his son, and first offered the post to him. He turned it down, suggesting Fürbringer for the position instead. After another, older anatomist had also refused the position, it was duly offered to Fürbringer, who took it.[11] Nine years later, when he moved to Jena, Fürbringer managed to get himself replaced in Amsterdam by fellow Gegenbaur disciple Georg Ruge. Ruge held the post until 1897, when he accepted the anatomy professorship at Zurich (for which Gegenbaur had strongly recommended him). At that point, Ruge recommended Maurer along with Fürbringer's and Haeckel's erstwhile student Richard Semon as possible successors, but the Amsterdam position reverted to Dutch hands.[12]

The story was similar at Heidelberg and Jena. Fürbringer in particular felt it was his duty as the eldest student of the school to look out for his younger "brother" Gegenbaur school colleagues. The tale of Maurer's appointment at Jena in 1900 exemplifies how he did so. Fürbringer was leaving Jena for Heidelberg, and as departing professor of anatomy, he had some say in recommending his successor. A number of faculty members leaned away from continuing the Gegenbaur tradition of morphological anatomists at Jena, preferring to

---

10. Eulner, *Entwicklung der medizinischen Spezialfächer* pp. 548–49.

11. Fürbringer, "Lebenserinnerungen" vol. 1, p. 358, UB Heidelberg, Fürbringer Nachlass.

12. See Letters 176–97 concerning "Ruge's Berufung nach Zürich," file D-1-f, SB, Fürbringer Nachlass.

strike out in a new direction. Fürbringer exerted as much influence as he could in favor of Maurer, especially pressuring the physiologist Biedermann, whom he called "the most influential man in the faculty."[13] When he discovered that Biedermann supported the candidate Carl Rabl, who had had numerous disputes with Gegenbaur and Haeckel, Fürbringer stepped up the campaign for Maurer. He pointed out in a letter to the medical faculty the folly of placing "Pope and anti-Pope" (Haeckel and Rabl) in the same arena.[14] At the same time, he called upon his long-standing friendship with Biedermann to persuade him not to work actively against Maurer. Finally, he badgered the ailing Gegenbaur to get more involved and write a second letter favoring Maurer. Through Fürbringer's efforts, Maurer eventually received the job.[15]

At Heidelberg as well, Fürbringer strove to smooth the path for other Gegenbaur students. Klaatsch wanted to move away from human anatomy to work exclusively in anthropology, and Fürbringer offered to assist him in any way he could.[16] Fürbringer also retained Gegenbaur's assistant Ernst Göppert, even though he had brought his own protégé, Hermann Braus, to assist him. Over the next decade, he scrupulously made sure that Göppert and Braus were always promoted equally. When he retired in 1912 and had to express a preference for his successor, he chose Braus, but it appears that he then worked to find a face-saving way for Göppert to leave Heidelberg.[17]

To the end of his career, then, Fürbringer took seriously his duties as Gegenbaur's successor, showing himself to be a faithful student not only by conducting research in the Gegenbaurian mold but also by taking over his role as patron to the rest of the school. These personal ties were equally important to others in the core group; as Ernst Göppert wrote many years later, "the affiliation with Gegenbaur also formed the bond that united his students to each other."[18] They displayed their mutual devotion in various ways, through their private correspondence, their support of each other at meetings, and espe-

13. Fürbringer to Gegenbaur (draft), 3 November 1900, file D-1-f, SB, Fürbringer Nachlass.

14. Fürbringer to Braus, 2 December 1900, UB Heidelberg, Braus Nachlass.

15. Fürbringer to Gegenbaur (drafts), 3, 4, and 5 November 1900, file D-1-f, SB, Fürbringer Nachlass.

16. Klaatsch to Fürbringer, 10 April 1901; Fürbringer to Klaatsch (draft), 13 April 1901, SB, Fürbringer Nachlass.

17. Fürbringer to Dekan der medizinischen Fakultät, 3 December 1908, documents the parallel promotions and advances to that time, BGLA, file 235/29858.

18. Göppert, "Friedrich Maurer," p. 315.

cially in their efforts to keep each other's names and reputations alive long after the school had faded from the forefront of research.[19] The members of the Gegenbaur school thus formed a familylike unit, devoted lifelong to their fatherly master and to each other. This unity was crucial to the visibility of Gegenbaur's program and magnified its significance in the discipline of anatomy.

## THE INSTITUTIONAL EXPANSION OF GEGENBAUR'S PROGRAM

Gegenbaur's research program in vertebrate morphology was taken up by a particular kind of social group—a school—but it was developed in a particular institutional setting as well: the university anatomy institute. The success of his program may be measured in part by the extent to which it spread to anatomy institutes at other universities. This could happen in two ways: Gegenbaur's students could colonize new universities by receiving professorships of anatomy, or other anatomists could incorporate Gegenbaur's program into their own research efforts.

As we have seen, the pattern presented by Gegenbaur's own research students is lopsided. Outside of Germany, his students were mainly clustered in the states bordering Germany: the Netherlands, Switzerland, Estonia, and the Scandinavian countries. Many of them did not have anatomy appointments but taught zoology instead. Within Germany, Gegenbaur's students were committed to careers as anatomists but did not carry his program beyond Heidelberg and Jena during his lifetime. The roster of former students who contributed to a three-volume *Festschrift* honoring Gegenbaur on his seventieth birthday in 1896 illustrates this pattern nicely. Twelve of the twenty-three contributors were working at institutions outside Germany, five were working in the anatomy institutes of Jena and Heidelberg, and four had zoology positions in Germany. Of the entire group, only two held anatomy appointments at other German universities: Oscar Hertwig, a full professor at Berlin, and Bernhard Solger, an associate professor at the undistinguished University of Greifswald.[20]

19. Thus Fürbringer supervised the publication of Gegenbaur's edited essays; Maurer wrote Ruge's eloge in the *Anatomischer Anzeiger* (1921); similarly, Göppert wrote about Maurer some years later ("Friedrich Maurer"). Göppert also gave a speech at Jena in honor of Gegenbaur's one hundredth birthday in 1926 ("Carl Gegenbaur. Rede zum Gedächtnis seines 100. Geburtsjahres" [1926]).

20. *Festschrift* (1896). The other contributors were all zoologists and included Haeckel, a Heidelberg *Privatdozent* in zoology, and two former students from his Jena days.

Judging entirely by his students' careers, then, it appears that Gegenbaur's influence was greater outside of Germany than within the country. It even seems possible that his program of vertebrate evolutionary morphology lacked influence within German science. This was not entirely the case, however. Gegenbaur spread his program in a variety of ways, and placing his students was only one of them. He published several textbooks of comparative anatomy, which were probably used more by zoologists than anatomists. His textbook of human anatomy, framed on evolutionary principles, went through seven editions between 1883 and 1899, and served as a standard reference work for students and professors alike.[21] In 1875 he founded the *Morphologisches Jahrbuch,* which was devoted solely to morphological studies; he edited it until 1901. And, of course, he published prolifically throughout his career, producing over 120 articles and monographs.[22]

Moreover, even a few anatomists who had not been members of Gegenbaur's immediate circle took up research problems related to his. One was Carl Hasse (1841–1922), who had already been interested in comparative anatomy as a student at Kiel, which brought him to work at Kölliker's institute at Würzburg in 1867. During his six years there, Hasse became an "enthusiastic follower" of Haeckel and Gegenbaur, and their influence is apparent in his research. He published a series of articles in the early and mid-1870s that traced the form and organization of the hearing apparatus through the vertebrates; later he conducted an extensive comparative anatomical study of the spinal cord, aimed at discovering the course of its evolution. In 1873, Hasse accepted a professorship of anatomy at Breslau, where he remained for the rest of his career.[23]

A second morphological anatomist, Robert Wiedersheim (1848–1923), conducted his first important research at Kölliker's institute while Hasse was still there. In fact, Hasse proposed the subject for his doctoral thesis, a histological work on the fine structure of the stomach glands in birds. After Kölliker offered Wiedersheim an assistantship in 1872, Hasse helped him through his difficult first semester of laboratory teaching. A year later, Hasse left Würzburg, and Wiedersheim replaced him as prosector. In 1876, Wiedersheim moved to a prosectorship in anatomy and comparative anatomy at Freiburg, where his brother-in-law August Weismann taught zoology. Upon the

21. Fürbringer, "Carl Gegenbaur," pp. 438–43; see also W. Roux to Fürbringer, 29 April 1910, SB, Fürbringer Nachlass.

22. On Gegenbaur's publications, see Fürbringer, "Carl Gegenbaur."

23. Gräper, "Carl Hasse" (1922).

anatomy professor's retirement several years later, he was promoted to full professor of anatomy. Wiedersheim credited Hasse and Weismann, not Gegenbaur, with piquing and sustaining his interest in evolutionary morphology. Nevertheless, he chose to spend his career working on the problems that Gegenbaur had made his own, namely, the origin of the limbs and the skull. Aside from these studies, Wiedersheim's major contribution was his textbook of vertebrate comparative anatomy, which quickly supplanted Gegenbaur's more difficult textbook for introductory courses.[24]

Although neither Hasse nor Wiedersheim ever studied directly with Gegenbaur, once their work in vertebrate comparative anatomy came to his attention, he took an interest in their careers. He thought sufficiently highly of Hasse to consider nominating him as a potential successor at Jena in 1873, but by then Hasse had already accepted the professorship at Breslau, so nothing further came of it. Gegenbaur also considered suggesting Wiedersheim as a candidate at Amsterdam in 1878, a possibility which Wiedersheim turned down before any formal steps were taken.[25] Thus if neither of these men counted as Gegenbaur's students, they did enter his sphere of influence, both intellectually and professionally.

Wiedersheim and Hasse themselves pursued research problems within the parameters set by Gegenbaur; however, their anatomy institutes did not produce researchers with the same interests. Their advanced students looked instead in new directions to discover the laws of form. As a group, they turned their attentions away from comparative anatomy—which to Gegenbaur was the basis of evolutionary morphology—and toward embryology.[26] Hence the institutes of Hasse and Wiedersheim did not act as new centers for the production of research in Gegenbaurian evolutionary morphology but took on a different character as their members explored new intellectual territory. Belonging to a different generation, these students were affected by a new set of intellectual, disciplinary, and institutional pres-

24. R. Wiedersheim, *Lebenserinnerungen* (1919); see also *Biogr. Lex.* II, s.v. "Wiedersheim, Robert Ernst."

25. Gräper, "Carl Hasse," pp. 211–12; Wiedersheim, *Lebenserinnerungen*, p. 82.

26. Hasse's leading students were Hans Strasser, Wilhelm Roux, and Gustav Born, all of whom went into experimental embryology; Wiedersheim's major students—Ernst Gaupp, Franz Keibel, and Albert Oppel—also inclined toward embryology rather than comparative anatomy. Basic information about these men is given in *Biogr. Lex.* II; more details are provided in C. Wegelin, "Strasser" (1927); F. Churchill, "Wilhelm Roux," and "Roux, Wilhelm"; R. Mocek, *Wilhelm Roux–Hans Driesch* (1974); W. Gebhardt, "Gustav Born" (1900); E. Fischer, "Ernst Gaupp" (1916–17); K. Peter, "Franz Keibel" (1929–30); and W. Roux, "Albert Oppel" (1916–17).

sures that drew them away from the concerns that preoccupied Gegenbaur and his school.

The research school that Gegenbaur forged at Heidelberg in the 1870s and 1880s was a classic of its kind.[27] He set out a fruitful and important set of theoretical questions about vertebrate evolution and provided strict methodological guidelines by which to go about answering them. This made for a vigorous and well-focused research effort that attracted students from all over the world. At the same time, his personality appears to have inspired a strong allegiance that bound his students to him, his program, and each other. These mutual ties in turn played a fundamental role in the institutional survival of the program, through the professional patronage that Gegenbaur offered his students and followers.

A program that relied so heavily on personal ties, however, soon met the limits of expansion. Outside of Gegenbaur's school itself, beyond Jena and Heidelberg, where he could command allegiance to his program, his influence quickly became diluted. Hasse and Wiedersheim took up vertebrate comparative anatomy themselves but did not found schools that would follow the same rigorously single-minded vision that Gegenbaur had. The men who worked in their institutes were simply their students, not their disciples. Thus with regard to his influence on individual researchers, Gegenbaur's reach extended only so far.

The limits of Gegenbaur's program of evolutionary comparative anatomy did not simply reflect the finite power of his ideas and personality, however. To understand just why neither his specific research program nor the broader tenets of evolutionary morphology attracted a wider following among German anatomists, we must place them in the context of the larger changes taking place in anatomy during the 1870s and 1880s.

## APPROACHES TO ANATOMICAL RESEARCH AND TEACHING

When Gegenbaur was hired at Heidelberg in 1873, his move represented not just a separation from his intellectual partner Haeckel, or simply the opportunity for launching his own school, although it was both of those things. It was also one in a flurry of shifts in occupancy

27. On research schools, see J. B. Morrell, "The Chemist Breeders" (1972); G. L. Geison, "Scientific Change" (1981); Fruton, *Contrasts in Scientific Style;* and G. L. Geison and F. L. Holmes, eds., *Research Schools* (1993).

of anatomy chairs that began in 1872, after an almost completely static decade in the 1860s. Between 1872 and 1876, thirteen full professorships of anatomy at eleven universities were newly filled, adding eleven new faces to Germany's roster of anatomy *Ordinarien*.[28] After a few more quiet years another spate of changes between 1880 and 1888 (fifteen turnovers involving ten universities) meant that in the course of less than twenty years, the anatomy professorships at over four-fifths of the German universities changed hands.[29] This protracted activity occasioned a wide variety of public and behind-the-scenes reflections on what really counted in anatomical research and teaching: programmatic statements in the form of articles and pamphlets described and prescribed for the discipline a plethora of ideal scenarios, while faculties and search committees argued at the more practical level of who was affordable, who would blend into their existing structures, and who would best contribute to the mission of the medical faculty as it was conceived of at their own university. Strikingly, the issue of intellectual breadth, so prominent a feature in zoological deliberations, does not emerge from these sources as an important one for anatomists. Instead, it was widely accepted that anatomy was divided into various research and teaching orientations, and arguments centered around which best fulfilled the requirements of anatomy as a medical teaching discipline. As the evaluations of this period show, in this setting, evolutionary comparative anatomy was widely looked upon as a less central aspect of anatomy than many others.

Given the high value placed on research in German academia, it comes as no surprise that in these appointment decisions one of the most important factors was the candidate's research record. But if the quantity and technical quality of an individual's research counted heavily, there was often no single scale by which different researchers could be judged, for they took widely varying approaches to their

28. Of these eleven, seven were vertical moves from *ausserordentliche* professorships or *Privatdozent* positions, two were horizontal moves from outside of Germany proper, and two were from professors formerly in other disciplines—one in pathological anatomy, one in zoology.

29. This does not mean that 80 percent of the faces were new over twenty years, for there was some horizontal recruitment. The changes between 1880 and 1888, for example, brought five men up from lower positions and three full professors from outside of Germany. The rest of the moves—seven involving five people—were horizontal ones. Although no positions were ever explicitly reserved for senior full professors as opposed to new ones, a fairly clear status hierarchy among the universities can be determined by tracing horizontal moves at the level of full professor (see A. Zloczower, *Career Opportunities* [1981], pp. 29–38).

work. Comparative anatomy was only one of these. Whereas in the 1850s only two research orientations could be clearly distinguished— mechanical and microscopic anatomy[30]—by the 1870s and 1880s, these had split and evolved into several different approaches. Search committee records and published programmatic statements in the 1870s and 1880s frequently named four different methodological orientations of research within the discipline. Besides comparative anatomy, the anatomical orientations most often mentioned were described as embryological (or genetic), mechanical (or physiological) and empirical or descriptive.[31]

Each of these orientations signified both a certain cluster of research problems and a more general view of how the structures of the human body would best be explained. Led by Gegenbaur, the comparative anatomists sought to explain the human form by means of evolution; to this end, they compared individual anatomical structures (both macroscopic and microscopic) in different organisms in order to discover the path of evolutionary development. The embryological orientation consisted of those who sought to understand the adult human in terms of its individual development; representatives of this approach, such as Wilhelm His, studied the developing form by itself, without making evolution their primary reference point. The mechanical or physiological anatomists, still led by Georg Hermann von Meyer, sought to explain the human form in terms of the functional coordination of its parts. They focused their attention on the mechanical relations between muscles, joints, and connective tissue and were especially concerned with the body in motion. Finally, adherents to the empirical orientation believed that the key to understanding the human body lay in avoiding theory as much as possible and laying out the facts. They worked in many different areas of human anatomy, but they were especially well represented in the study of cell and tissue structure, areas that were continuing to open up with the steady improvements in microscopic methods in the 1870s and 1880s.[32]

30. See pp. 81–90.

31. Programmatic statements mentioning these four orientations include J. Budge, "Die Aufgaben" (1882); W. Waldeyer, "Wie soll man" (1884); Meyer, "Stellung und Aufgabe der Anatomie"; and W. Roux, "Die Entwicklungsmechanik der Organismen" (1889).

32. On Meyer and the mechanical anatomists, see pp. 81–84. Jakob Henle and Wilhelm Waldeyer both worked on histology and development with a strong empirical bias; on their teaching from this view, see the section, "The Goals of Anatomical Instruction," below; on the development of cytology and fertilization studies, see Coleman, "Cell, Nucleus and Inheritance."

Not all research done by anatomists was encompassed by these orientations; for example, some of the most exciting new investigations in the period concerned fertilization and transmission. Representatives of several orientations worked in these areas.[33] Furthermore, not all anatomists allied themselves with one particular orientation throughout their careers; many tried out different approaches at different times. Nevertheless, the consistency with which these four orientations were named as the major ones in these two decades suggests that in contemporaries' eyes, these were the methodological biases to be reckoned with in the discipline of anatomy.

Each of these research orientations had some affinity with different teaching tasks under the anatomists' purview, but there were no perfect correspondences. Professors of anatomy had to cover a wide range of course subjects treating various different aspects of the human form. *Systematic anatomy* covered the adult human body by tissue types; traditionally, one began with the skeleton, then moved on to the ligaments, muscles, intestines, and blood vessels, and ended with the nervous system. This was often also called "descriptive anatomy" or "special anatomy,"[34] and both descriptive anatomists and mechanical anatomists could claim research expertise in this area. Another teaching subject, *microscopic anatomy* or histology, presented the fine structure of these tissues and sometimes also spilled over into cell physiology; virtually all the orientations claimed some direct connection to this teaching area. Courses in *embryology* covered the fertilization process as well as the development of the human embryo; here the embryological orientation had the strongest ties, but both descriptive anatomists and evolutionary comparative anatomists often touched this area in their research. *Comparative anatomy* taught students to place the human organism in its relation to the rest of the vertebrates, usually from an evolutionary point of view; the comparative anatomists stood nearly alone as experts here, though some embryologists also took a research interest in comparison. Finally, *topographical anatomy* analyzed the human body in terms of the spatial relations between organs. Descriptive and mechanical anatomists held

33. Among German anatomists working on these problems, Waldeyer had a strongly empirical orientation, as did perhaps Walther Flemming; Oscar Hertwig came at the subject with evolutionary concerns; and His's orientation was embryological and mechanical. German physiologists (Pflüger, Hensen), botanists (Strasburger, Naegeli), and zoologists (Bütschli, Schneider, Weismann) also took an active role in research on these topics.

34. See, e.g., J. Henle, *Grundriß der Anatomie* (1880), "Anatomie" (1893).

sway here, although there was little new research to be conveyed to students in this course by the late nineteenth century.

In one form or another, all these subjects were normally taught by members of the anatomy institute's staff. Most German universities had one full professor in charge of the anatomical institute, who was assisted in his teaching by a prosector who might hold the rank of assistant, *Privatdozent, Extraordinarius,* or even, in a few cases, *Ordinarius;* at the larger establishments there was frequently a junior assistant or two in addition to the more senior prosector. Usually, the institute director gave lectures in systematic anatomy one semester and embryology and microscopic anatomy the other. He or his assistant(s) might hold more detailed classes in osteology and "syndesmology" (the studies of bones and ligaments) as well as advanced courses in embryological, histological, or comparative anatomical problems, depending on their interests. They also conducted practical instruction in the dissection and microscopy laboratories, which were usually open daily. Topographical anatomy might be incorporated into systematic anatomy, or it might be taught outside of the anatomy institute by a member of the surgical staff. Although the specific arrangement might differ from university to university, most managed to cover them all.

Of these courses, the most important for the medical student was the introductory two-semester sequence of systematic anatomy and microscopic anatomy. These were normally the first preclinical courses in the medical faculty to be taken by a future physician, and they provided him with the basic knowledge of the human body that he would need throughout his medical education and later in his career. Anatomy also gave him his first exposure to issues in medicine and could shape his entire attitude toward the basic medical sciences and clinical practice. Thus medical educators considered the basic anatomy sequence especially critical for the student's further education.

Since the early nineteenth century, some medical educators had been arguing that a morphological approach to anatomy, emphasizing its comparative and developmental aspects, would benefit medical students by helping them place the myriad details of anatomical knowledge into a coherent, unified whole. Proponents from Burdach in the 1810s through Reichert in the 1850s up to Gegenbaur and his students in the 1870s and 1880s argued that such an approach would provide a sense of the human body as a truly organic structure that displayed lawful relations both internally and in relation to the rest of the living world; following this approach would also help students refine their critical abilities.[35] But increasingly over the 1870s and 1880s, these arguments were dismissed by many medical educators as not answering the practical needs of medical students.

Concerns about the relevance of comparative anatomy were already evident in the deliberations surrounding the 1873 search at Heidelberg that ultimately resulted in Gegenbaur's appointment. Strong arguments were waged against his candidacy and in the deciding vote, the faculty split down the middle. The tie had to be broken by the dean, who chose in Gegenbaur's favor. The search committee's report and the close decision demonstrate just how hard it was for a comparative anatomist—even the best one in Germany—to gain acceptance in a medical faculty.

Teaching was a critical consideration in the Heidelberg decision. The search committee consisted of Willy Kühne, the professor of physiology; the surgeon Gustav Simon; and the professor of ophthalmology, Otto Becker.[36] In their report, they placed their recommendations within a broader statement of the present and future tasks of anatomy. They began by asserting that pure descriptive anatomy was the most important thing for the medical student to learn. Unfortunately, it had recently been subjected to the criticism of being "unscientific." Because of the strong desire among anatomists to be regarded as scientists, and not just knife wielders, there now existed very few descriptive anatomists at all. Instead, three groups now claimed to be doing scientific anatomy: the histologists, the comparative anatomists, and those engaged in studying the mechanics of the human body. The issue at hand was not which of these three groups had the greatest claims to doing science, but which approach was most necessary for anatomical instruction, and which best fit the unique needs of Heidelberg. They then considered each approach in turn, according to these two criteria.

The search committee dismissed the need for an anatomist specially trained in histology, the study of tissues, since at Heidelberg both normal and pathological histology were already taught outside of anatomy (by the physiologist Kühne, who was on the search committee). The situation was similar for comparative anatomy. While of course there were great comparative anatomists currently holding

---

35. On Burdach, see pp. 35–44; on Reichert, see pp. 72–73; see also C. Gegenbaur, "Stellung und Bedeutung der Morphologie" (1875); and O. Hertwig, *Der anatomische Unterricht* (1881).

36. "Commissionsbericht über die Besetzung des an der Universität Heidelberg erledigten Lehrkanzel der Anatomie," 24 April 1873, BGLA, file 235/29858.

chairs of anatomy, the committee argued that it was not central to anatomy proper and, in fact, had much closer ties to zoology than to descriptive anatomy. The committee recognized it was important for the university to be represented in comparative anatomy, but without casting any doubt on the significance of comparative anatomy for medical education (which of course they were), they felt that the appropriate place for it was in the philosophical faculty, as part of zoology.

Having neatly disposed of Gegenbaur's potential candidacy, the committee went on to argue for the importance of the mechanical orientation. Those researchers who studied the mechanical relations of the human body brought anatomy into direct contact with questions of physiology and practical medicine and were the natural heirs to the descriptive anatomists. As such, they should be considered first. Consequently, the search committee nominated three anatomists of the mechanical orientation, Wilhelm Henke of Prague (who would later go to Tübingen), Christian Wilhelm Braune of Leipzig, and Christoph Theodor Aeby of Bern.

It is not surprising that a physiologist, a surgeon, and an ophthalmologist coming out of a surgical tradition should all value research on the mechanical workings of bones and muscles more highly than research on the evolution of the vertebrates. But as the response to the report by the retiring anatomy professor shows, the commission's practical vision of anatomy and its subordinate relation to medical concerns jarred against the morphologist's view of what made anatomy a *Wissenschaft* worthy of the name.[37]

The anatomist Friedrich Arnold was a morphologist of the old school, a contemporary of Johannes Müller who viewed the study of form and development as essential to a complete scientific understanding of the human organism. Although he had taken himself out of the search for the new anatomist because of his familial connection to Gegenbaur, he opposed the commission's report too strongly to remain silent. First, Arnold objected to the commission's cavalier removal of histology from the province of anatomy. Because of the importance of anatomy for all branches of medical science, he argued, the teacher of anatomy must cover both macroscopic (descriptive) anatomy and microscopic anatomy or histology. A physiologist teaching histology necessarily came at it from a different point of view

---

37. Arnold, "Aueßerung über den Commissionsbericht, betr. die Besetzung des an der Universität Heidelberg erledigte Lehrkanzel der Anatomie," 8 May 1873, BGLA, file 235/29858.

from an anatomist, because the physiologist would stress tissue function whereas the anatomist would emphasize form. Furthermore, Arnold continued, the new anatomist at Heidelberg had to be qualified to cover histology in case the next physiologist was not.

Arnold offered further professional reasons why the professor of anatomy should not be a mechanical anatomist. As the search committee had emphasized, the representatives of the mechanical orientation were the heirs to the old descriptive anatomists, whom Arnold equated with topographical or surgical anatomists. But topographical anatomy in the German university was a practical subject designed to teach the medical student where to cut into a patient without damaging anything essential. A good general anatomist would include the basics of topographical anatomy in his courses. The topographical anatomist, on the other hand, was really just a medical technician, untrained in the scientific aspects of embryology and histology. Just as a technical or pharmaceutical chemist would be unqualified to fill a chair of theoretical chemistry, Arnold argued, so was the topographical anatomist unqualified to teach theoretical or general anatomy.

Arnold ended by repeating his plea to consider anatomy not just as a practical discipline, but as a branch of the descriptive sciences. As a *Wissenschaft,* anatomy could aspire to a higher morphological understanding by means of embryology and comparative anatomy. The committee had not even mentioned embryology, and Arnold rebuked them for ignoring its centrality to anatomy, for it was commonly recognized that an adult being could only be fully understood with reference to its development. He also castigated the committee for underestimating comparative anatomy's worth to medical education, "especially in a time when the student of medicine is trying to become something more than just a tradesman [*Gewerbsmann*]." He went on to acknowledge that the mechanical approach might be an excellent one for certain areas of anatomy, but he challenged the commissioners to ask themselves if "the structure of the brain, the organization of the nerves, the outer and inner configuration of many organs of the body can be explained through mechanical relations." On the contrary, it was precisely these essential aspects of human anatomy that could be most satisfactorily explained by embryological and comparative study.[38]

The documents surrounding the 1873 search at Heidelberg illustrate several differences between the morphologist's understanding of anatomical instruction and that of other medical educators. Drawing

38. Ibid.

on an argument that was as old as morphology itself, Arnold suggested that the medical student desired to become not simply a medical craftsman, but a *Wissenschaftler*. Accordingly, he and other morphologists sought to teach systematic knowledge or *Wissenschaft*, which meant conveying a synthetic point of view and organizing the myriad facts of anatomy under a few overarching laws. As a practical pedagogical tool, such a unified framework would enable students to remember the facts more easily. More important, the comparative method associated with the morphological approach would awaken in the student a capacity for critical observation and judgment that would make the future practitioner a true man of science. Although Arnold's arguments could have been made in the 1830s and 1840s, he was able to transpose them with no difficulty to make the case for the evolutionary comparative-anatomical approach of Gegenbaur.

The search committee at Heidelberg expressed different aims for anatomical instruction. In their view, the most important thing the anatomist had to teach was descriptive anatomy. Indeed, in a letter sent a little later to the university senate, half the medical faculty argued specifically against hiring Gegenbaur because he was not a descriptive anatomist.[39] Nor were doubts about evolutionary morphology unique to the Heidelberg medical faculty. By the early 1880s, a similar emphasis on imparting basic knowledge without loading it down with too much evolutionary baggage had gained the support of some of Germany's most distinguished anatomists. Jakob Henle (1809–85), one of the outstanding histologists and anatomy teachers of the century, wrote with some impatience in 1883, "Widespread Darwinism (unfortunately, even over-Darwinized in Germany) is responsible for the fact that the occupation of studying the existing being in itself . . . is often judged unscientific. But the doctor must learn where an artery lies, even if he doesn't learn whether it came there through adaptation or through inheritance. And if human anatomy is not a science, it is indeed an art, in that its task is to build up in the student's imagination the transparent picture of the human form."[40] Wilhelm Waldeyer (1836–1921), who emulated Henle's approach to teaching, echoed his words in 1884 when discussing the necessity for providing the medical student with a good empirical training in anatomy: "It is a question . . . less of a science than of an art, and here the schooling must proceed in a completely simple way; we must be elementary teachers. Here neither a genetic, nor a physio-

39. "Separatvotum," 17 May 1873, BGLA, file 235/29858.
40. Jakob Henle, quoted in H. Hoepke, "Jakob Henles Gutachten" (1967), p. 222.

logical, nor a comparative standpoint may be taken from the start; the young medical man should first learn the naked facts, without any artificial illumination from a special standpoint."[41] Thus Waldeyer criticized the morphologists for investing their anatomical teaching with too much comparative *Wissenschaft* too soon, before the medical student learned the bare facts upon which theories were spun. This reflected his view, similar to Henle's, that the main mission in teaching basic anatomy was to train practitioners, not scientists.[42] The same aim led the Heidelberg search committee to propose hiring a mechanical anatomist: a person of that orientation would presumably organize his teaching to raise clinically directed questions rather than ones with no practical applications. Although there was, of course, no necessary reason why a scholar who focused his research on comparative anatomy could not in theory teach a good introductory course in descriptive anatomy, the assumption seems to have been widespread that one's research orientation would most naturally be reflected in one's teaching.

Both the morphologist's and medical anatomist's views were echoed a number of years later when Gegenbaur's former student Oscar Hertwig expounded on the goals of anatomical education in his inaugural lecture as professor of anatomy of Jena. Like Arnold, he elaborated at great length on the importance of morphological work for *Wissenschaft* but neglected to tie these goals more specifically to medical concerns.[43] A *Privatodozent* at Rostock felt compelled to respond with his own "undelivered address" on the subject in the pages of the *Deutsche Medizinische Wochenschrift*. There he criticized Hertwig's speech partly because it did "not sharply limit the sphere of anatomy, and does not try to answer the question, which parts of the relevant subjects belonged to it, but really speaks exclusively about the value of these subjects for the development of our scientific knowledge in general."[44] The critic went on to assert that anatomy was a

41. Waldeyer, "Wie soll man", p. 593; Waldeyer specifically states that he emulated Henle's style at the blackboard, and it is clear from his autobiography that he considered Henle the mentor who most influenced his life's path (W. Waldeyer, *Lebenserinnerungen* [1920], pp. 77–79); see also J. Sobotta, "Zum Andenken" (1922).

42. This may have much to do with Waldeyer's never founding a school; according to his student, Johannes Sobotta, during Waldeyer's thirty-three years in Berlin, only one of his assistants and prosectors, Sobotta himself, ever became a full professor of anatomy (Sobotta, "Zum Andenken," p. 17). Waldeyer did believe, however, that the anatomist must also be a researcher who extended scientific knowledge; he just saw that as a separate task (Waldeyer, "Wie soll man," p. 593).

43. O. Hertwig, *Der anatomische Unterricht*.

44. P. Schiefferdecker, "Der anatomische Unterricht" (1882), p. 465.

part of medical science, the task of which was the healing of human illnesses. With reference to Hertwig's concern not to dissolve anatomy's links with the broader natural sciences, he wrote that the anatomist "has above all else to see to it that he does not lose the connection with the other *medical* disciplines, and he runs this risk far more than the former, the way the science is going now."[45]

In various forms, all these medical educators were expressing a single dominating criticism of comparative anatomy. Even if one accepted the comparative anatomist's approach as a valid one for increasing general anatomical knowledge, many medical men argued, it had less value than the other orientations for medical teaching, research, and practice. Embryological research had connections to obstetrics and could lead to an improved understanding of birth defects. Mechanical anatomy had direct implications for the treatment of wounds and orthopedic problems. Descriptive anatomy expanded the basic working knowledge of the healthy body, which was necessary for identifying pathological conditions. Comparative anatomy could offer no equivalent: its evolutionary orientation was directed at a general understanding of where humans came from rather than at medically useful information. Most comparative anatomists considered it enough to be pursuing pure scientific research and teaching students to think like a *Wissenschaftler*. But to many medical educators, both within anatomy and in other medical disciplines, comparative anatomy's lack of medical applications made a comparative anatomist a less desirable acquisition than a proponent of another orientation. Thus where medical considerations in research and teaching took precedence over more general contributions to *Wissenschaft,* as it did in most medical faculties, comparative anatomy came under fire.

## BETWEEN TWO CHAIRS: COMPARATIVE ANATOMY IN THE STRUCTURE OF THE UNIVERSITY

As the discussions over hiring a new anatomist at Heidelberg suggest, the criticisms of comparative anatomy were not just a matter of rhetoric. In the ideal world of pure thought that Hertwig seemed to invoke, concerns about anatomy's connections to the medical or nonmedical sciences could be cast aside as unimportant in the face of the larger goal of expanding knowledge. But in the real world such links were

45. Ibid., p. 467 (emphasis added).

critically important, and they were embodied in the structure of the universities. Anatomy was in the medical faculty for a reason: it was supposed to serve the needs of the medical profession. Only by virtue of its history could comparative anatomy claim a place in the medical faculty. And when professorship came open—a rare enough event at most universities—medical faculties could use the occasion to reassess what they valued most about anatomy and to redirect the subject appropriately.

The numerous deaths, retirements, and moves in the German anatomical community during the 1870s and 1880s provided the opportunity for medical faculties to experiment with their institutional structures, as they sought ways to accommodate the growth of anatomical knowledge and the variety of approaches to anatomical teaching. In making these adjustments, a few medical faculties made room for comparative anatomy, but over the course of these two decades that orientation lost out more and more often.

At some universities, an effort was made to encompass within anatomy both the clinician's demand for practical education and the comparative anatomist's emphasis on *Wissenschaft*. This could be achieved by dividing the two tasks between two different anatomy institutes within the university. Such a compromise could be undertaken only where the number of medical students warranted some sort of division of teaching labor for pragmatic reasons, and it appears to have been implemented at just three German universities: Würzburg, Bonn, and Berlin. Since the mid-1860s at Würzburg, Kölliker had run two institutes by himself (with the assistance of various prosectors and assistants), one for human anatomy and another for comparative anatomy, histology, and embryology. This divided practical medical anatomy from the broader concerns of scientific anatomy and satisfied the research requirements of both.[46]

At Bonn the situation was more complicated. There had long been two full professors sharing anatomical teaching duties—Moritz Weber (1795–1875) had been covering both comparative and pathological anatomy since 1830, while the pioneering cytologist Max Schultze (1825–74) had been brought in to head up the anatomy institute in 1859. By the early 1860s it was widely acknowledged that Weber was senile, and without relieving him of his title of *Ordinarius,* a younger man, Adolph Freiherr La Valette St. George (1831–1910), was in-

46. On Kölliker's two institutes at Würzburg, see T. H. Schiebler, "Anatomie in Würzburg" (1982), p. 993; and Feser, "Das anatomische Institut," pp. 44–45.

stalled as *ausserordentliche* professor and prosector to take over Weber's duties under Schultze's direction.[47] When Schultze died in 1874,
it was proposed that La Valette be promoted to *Ordinarius*. At the
same time, however, the medical faculty argued that he was not in
the very top rank of anatomists, and that to preserve the standing of
the subject at their university they should try to acquire a front-
running anatomist to direct the institute. After having been turned
down by their first two choices, Kölliker and Waldeyer, the medical
faculty set forth a new list proposing in first place Nathaniel Lieberkühn, the anatomy professor in Marburg; in second place Gegenbaur;
and in third place a division of anatomy such that La Valette, who
would cover descriptive and microscopic anatomy, would be made
director of the anatomy institute, and Haeckel (!) would be brought in
to teach comparative anatomy and direct the comparative anatomical
museum. In forwarding this proposal to the ministry in Berlin, the
university's chief administrator (*Curator*) noted that the appointment
of Lieberkühn would leave a gap at another Prussian university rather
than bringing outside strength into Prussia. He then objected to Gegenbaur on the grounds that he demanded too much of his students,
covering macroscopic human anatomy in lectures that extended six
hours a week over two semesters, still without covering the ancillary
anatomical subjects; the *Curator* argued that this had contributed
in no small way to Heidelberg's recent precipitous drop in medical
enrollments. The Bonn medical faculty, too, viewed this as far too
much time devoted to basic anatomy, cutting in on the time necessary
to teach chemistry, physics, and other sciences important to the medical curriculum. Thus the third possibility seemed the best. When
Haeckel turned down the comparative anatomy position, it was offered to Franz Leydig, the professor of zoology and comparative anatomy in Tübingen, who accepted. Thus a peculiar combination of contingent events led to the recognition in Bonn of comparative anatomy
as a separate, *ordinierbar* subject on equal standing in the medical
faculty with descriptive and microscopic anatomy.[48]

At Berlin, where Waldeyer taught beginning in 1883, it was decided
in 1888 to appoint as a second anatomist Oscar Hertwig, whose orientation was less strictly medical and more evolutionary. A different
division of labor was originally foreseen by the state administration,
one that would have given the macroscopic subjects (systematic and

47. Curator of Universität Bonn Beseler to Minister der GUMA Falk, 22 June 1873,
GStA Merseburg, Bonn Med. Fak., Bl. 88; M. Nußbaum, "La Valette" (1911).

48. GStA Merseburg, Bonn Med. Fak., Bll. 167–78, 198, 224–25, 227–33,
239–41, 251.

topographic anatomy) to Waldeyer and the microscopic subjects (histology and embryology) to the new professor. But although Waldeyer agreed to give up teaching embryology, he refused to relinquish histology. He was unenthusiastic about the choice of Hertwig, but his own insistence on controlling macroscopic and at least part of microscopic anatomy opened the way for a nonmedically oriented anatomist to be appointed.[49] The situation at Würzburg, Bonn, and Berlin, which put a comparative anatomist on an equal footing with a medical anatomist, was rare; indeed, nowhere else within Germany were two anatomy institutes founded. Nor was it an entirely stable arrangement, at least in one case: upon Leydig's retirement from Bonn in 1887, the two institutes were reunited under La Valette's direction.

This sort of division between comparative and medical anatomy was not more widespread at least in part because at most universities there was another place available for comparative anatomical teaching and research: within zoology in the philosophical faculty. As zoologists had become more firmly established in the 1860s and early 1870s, they began to claim comparative anatomy as part of their territory, often trying to wrest it away from older anatomists. Here more than just intellectual territory was at stake: most universities could only afford one comparative anatomy collection, whose specimens and preparations were used for teaching. At Marburg, for example, the zoologist Carl Claus waged a three-year battle beginning soon after his appointment in 1863, to have the comparative anatomy collection removed from the anatomist's jurisdiction and placed under his own; the final transfer came only in 1867 with the anatomist's retirement.[50] The subsequent professor of anatomy and his successor specialized in embryology, not comparative studies.[51]

Collisions over control of comparative anatomy did not always result in a clear-cut victory for the zoologist, however. Thus at Würzburg the professor of zoology Carl Semper disputed Albert von Kölliker's claim to comparative anatomy. In 1870, Semper was offered an appointment at the University of Göttingen. As one of his conditions for staying at Würzburg, he demanded that part of the comparative anatomy collection be turned over to him. He also demanded

49. See Weindling, *Darwinism and Social Darwinism*, pp. 205–13; on Waldeyer's attitude toward histology, see also Waldeyer, "Wie soll man," pp. 594–96.

50. Staatsarchiv Marburg, Zoologisches Institut, correspondence and documents 1864–67 (unpaginated).

51. These were Nathaniel Lieberkühn (1821–87) and Emil Gasser (1847–1919); on Gasser, see E. Göppert, "Emil Gasser" (1921); on Lieberkühn, see P. Jaensch, "Beiträge zur Geschichte" (1924), pp. 817–18.

that his title be expanded to "professor of zoology and comparative anatomy." Both requests were granted, and he stayed on. However, Kölliker relinquished only the invertebrate collection, thereby diminishing the extent of Semper's victory. Indeed, Kölliker taught and supervised research in vertebrate comparative anatomy for the rest of his career, and he and Semper continued their territorial disputes until Semper's retirement in 1892.[52]

Not every German university had such disputes: at Jena, as we have seen, Haeckel was able to cooperate closely with Gegenbaur and his successors, and comparative anatomy's interdisciplinary status was not a problem. And at Breslau, when the medical faculty was searching for a new anatomist in 1873, they explicitly pointed out that "at our university the professorship of human anatomy is bound to that of comparative anatomy." To fulfill that requirement, they ended up hiring a Gegenbaur sympathizer, Carl Hasse.[53] The zoologist at the time, Adolf Grube, was an older systematist whose approach and concerns did not collide with those of a comparative anatomist.[54]

52. G. Krause, "Aus der Geschichte der Zoologie" (1970), p. 3; the other major outbreak of hostilities occurred in 1883 (see Kölliker, *Die Aufgaben der Anatomischen Institute,* esp. pp. 6–9; and C. Semper, "Zoologie und Anatomie" [1883]; for more detail on the different ways comparative anatomy fit into the university structure, see L. Nyhart, "Disciplinary Breakdown" (1987).

53. GStA Merseburg, Breslau Med. Fak., Bll. 126, 135. That students didn't always appreciate Hasse's approach is suggested by an anonymous "Bummellied" of 1893, evidently written by a medical student planning to leave Breslau because of dissatisfaction with his teaching. The opening verse indicates the student had no trouble with his other professors, but could not bear Hasse. Other stanzas give some of the reasons:

> Von Fröschen und Batrachiern war
> Fast immer nur die Rede,
> Am Ende kam er etwas dann
> Zum Menschen—ziemlich späte.
>
> Die Knochen lehrt er uns genau,
> Die einen Froschkopf zieren
> Als ob ich Hausarzt werden sollt'
> Einmal bei Wirbeltieren. . . .
>
> [The talk was nearly all the time
> Of frogs and batrachians;
> At the end then, finally, he touched upon
> The humans—somewhat late.
>
> He taught us all about the bones
> That decorate the frog's pate,
> As if I should perhaps become
> A doctor of vertebrates. . . . ]

Manuscript in GStA Merseburg, Rep. 92 Althoff A I Nr. 65: Mediziner, Beurtheilungen 2, Bl. 191.

54. On Grube, see Zaddach, "Adolf Eduard Grube."

Although these cases show that it was possible for comparative anatomy to remain at least in part under anatomical jurisdiction, the fact that zoologists were also increasingly claiming it gave ammunition to medical men opposed to hiring a comparative anatomist. At Heidelberg in 1873, as we have seen, members of the medical faculty used such an argument to oppose the hiring of Gegenbaur. Elsewhere, where comparative anatomy was more or less firmly established as part of zoology, the medical faculty could easily afford to leave it out of consideration as part of anatomy, even when they decided to split the anatomical teaching duties among two professors. At Leipzig in 1872 and at Halle in 1876, for instance, the medical faculties received permission to appoint two new full professors of anatomy to replace the one who was retiring. At both universities one of the new professors controlled the macroscopic subjects while the other taught the microscopic ones; comparative anatomy remained the province of the zoologist in the philosophical faculty. In Austria an even sharper macroscopic-microscopic split was made, and a position for histology and embryology, separate from anatomy, was founded at each of the four major German-speaking universities.[55] At all of these universities, the relationship of anatomy to medicine stood in the foreground, while comparative anatomy was relegated to the zoologist.

The issue of comparative anatomy's appropriate location, with its consequences for property and institutional structure, also arose at Munich where, following Theodor Bischoff's death in 1878, the medical faculty decided to add a second chair of anatomy. Here the implications for Gegenbaur's orientation are especially telling, for he was seriously considered as a candidate. The physiologist Carl Voit wrote the major recommendation for the new professor's duties and led the negotiations with candidates.[56] He explicitly recognized "comparative morphology" as well as histology and embryology for having contributed the most to recent progress in anatomical research. He even considered Gegenbaur first on his list in terms of scientific significance

55. On Leipzig and the Austrian universities, see Eulner, *Entwicklung der Spezialfächer*, pp. 39–40, 554–55; he lists the Austrian chairs of histology and embryology but considers the situation at Leipzig and other places where anatomy was divided as short-lived and therefore not really significant. It goes without saying that I have interpreted the evidence quite differently. On Halle, see GStA Merseburg, Halle Med. Fak., pp. 84–95; and B. Solger, "Welcker" (1898); Welcker did value the comparative anatomical approach, but so far as I can tell, he was interested in using comparison only to help explain human structures, and unlike the Gegenbaur school, always kept humans as the central referent in his comparisons.

56. Carl Voit's report, July 1878, UA München, Senatsakten, "Anatomie als Disziplin."

and believed that Gegenbaur had "the widest horizons among the presently living and active anatomists." But then he went on to report that Gegenbaur was said to dislike giving the general anatomy lectures and did not know how to capture the interest of the medical students. Thus his main assets lay in his scientific achievements. But since teaching was crucial at a large medical school like Munich, an overemphasis on research could turn into a liability. However, Voit admitted that he had heard from other reliable sources that Gegenbaur was indeed a first-rate teacher who united his teaching and research in an exemplary fashion, especially in his lectures on comparative anatomy and embryology.

Following this already ambivalent commentary on Gegenbaur's assets, Voit went on to say that for the specific conditions at Munich, he thought that Gegenbaur would demand too many changes in the structure of anatomy for his appointment to be feasible. He suggested that Gegenbaur would want to teach comparative anatomy and run the comparative anatomy collection, and that he would probably want to give up the larger part of descriptive anatomy, for example, the preparation room and topographical anatomy, to leave him more time for scientific work. These putative demands would mean changing the present conditions "from the ground up," which Voit thought was hardly likely. Hugo Ziemssen, the professor of clinical medicine who also reviewed the candidates, agreed that Gegenbaur was "out of the question" because everything would have to be restructured around him.[57]

Neither mentioned that the eminent zoologist Carl Theodor von Siebold was currently lecturing on comparative anatomy in the medical faculty and held a joint appointment in the medical and philosophical faculties, but he was surely an important silent partner in their considerations.[58] Even after Siebold's retirement in 1883, when the zoology appointment was narrowed to the philosophical faculty, zoology's control over comparative anatomy was not to be impinged upon. By the time Munich's other anatomy professorship came open in 1896, Gegenbaur's students Max Fürbringer and Georg Ruge were mentioned but then dismissed because "they are overwhelmingly active" in comparative anatomy, "an area which in Munich is given over to the professorship of zoology."[59]

57. Ziemssen, "Correferat," 9 July 1878, UA München, Senatsakten, "Anatomie als Disziplin."

58. On Siebold's official position, see UA München, Senatsakten, "Zoologie und vergleichende Anatomie"; for courses under his name, see Universität München, *Vorlesungsverzeichnisse*, through the winter semester 1882–83.

59. Dekanat der medizinischen Fakultät an der akademischen Senat, 26 November 1896, UA München, Senatsakten, "Anatomie als Disziplin."

This case, then, strongly resembles that at Heidelberg in 1873. Once again, members of the medical faculty argued against choosing Gegenbaur in terms of institutional imperatives. At Heidelberg, the search committee argued against the necessity of hiring a comparative anatomist into the medical faculty when the subject was already covered by the zoologist. At Munich, the committee expressed their doubts about the feasibility of hiring a comparative anatomist for the same reason. In both cases, the committee deemed that it was easier to preserve the existing structure than to change it. We should note that no one at Munich or Heidelberg attacked comparative anatomy's claims to *Wissenschaft;* on the contrary, Voit was effusive in his praise for Gegenbaur's scientific contributions, and the search committee at Heidelberg acknowledged comparative anatomy's significance for the general advance of knowledge. The point at issue was not whether comparative anatomy was worthy as science, but whether it belonged in the medical faculty as part of anatomy. While strong factional interests at Heidelberg pressed the decision in his favor, at Munich no such lobby existed, and comparative anatomy was definitively excluded from anatomy.

The institutional structure finally arrived at for Munich's anatomy split the teaching between macroscopic and microscopic anatomy, mirroring the case at Leipzig and Halle. Munich's first anatomy chair, it was agreed, would go to Nicolaus Rüdinger, who had already worked there for many years as Bischoff's right-hand man, mainly in descriptive and topographic anatomy. The second chair eventually went to Carl Kupffer, whose strengths lay in histology and embryology. Although Kupffer had done substantial comparative embryological research from a Haeckelian perspective, his interests did not extend far into adult comparative anatomy, and these potentially problematic interests were compensated for, in Voit's eyes at least, by his extensive experience as a histologist in Carl Ludwig's famous physiological laboratory at Leipzig. Kupffer's appointment reinforces the point that it was not Darwinism per se that had made Gegenbaur an unacceptable candidate to Voit and Ziemssen, but rather his demand for institutional territory beyond what they considered to be the appropriate realm of medical education.[60]

Appointment decisions were made university by university, as were decisions about where comparative anatomy should fit in a university's structure. But another kind of structural decision in the early 1880s further pushed comparative anatomy to the fringes of medical

60. Voit, "Gutachten," UA München, Senatsakten, "Anatomie als Disziplin."

education and weakened its justification for inclusion in the medical curriculum. This was the change in examination requirements for medical students. In the perennial controversy over the reform of medical education, a major issue was how to keep from overburdening students with requirements, yet make sure they learned what they needed. The problem was to set minimum standards while avoiding politically undesirable restrictions on *Lernfreiheit;* the solution lay in the regulation of the state examinations. Whatever subjects were required for the exam were sure to draw students, while subjects not required or given minimal importance tended to be ignored by the average medical student.[61] Thus the state examinations assumed prominence in shaping a de facto required curriculum.

The different states had long had different rules governing medical certification. Although most by the 1860s had both a preliminary exam, usually covering basic sciences, and a final qualifying exam covering the more practical aspects of medicine, the coverage in these exams was not uniform. In 1861, Prussia had narrowed its preliminary exam from a "tentamen philosophicum" to a "tentamen physicum," dispensing with the older examination subjects of logic and psychology, increasing the emphasis given to physics and chemistry, and limiting the attention to the "descriptive natural sciences" of zoology, mineralogy, and botany.[62] Other states, however, continued to emphasize the links to the descriptive sciences, along with comparative anatomy, in their exams. At Jena, Haeckel examined medical students in zoology; at the Bavarian universities as well, the descriptive sciences received more attention in the preliminary exam. At Giessen (examining for the grand duchy of Hesse-Darmstadt), not only were the descriptive sciences accorded a significant place in the preliminary exam, but the second examination included comparative anatomy as part of the oral exam in physiology.[63]

Over the course of the 1860s and early 1870s, first in the context

61. On the effects of medical examination regulations on course attendance, see *ADB,* s.v. "Burmeister, Karl Hermann Konrad"; A. Geus, "Zoologie" (1978), p. 175; and A. Weismann, "Bedeutung der Zoologie" (1879), p. 1402.

62. W. Horn, *Das preußische Medicinalwesen* (1863), pp. 130ff.

63. The Würzburg medical faculty's 1871 proposal that the Bavarian universities adopt the Prussian *tentamen physicum,* and the responses to it, make evident the more prominent role played by the descriptive sciences to that point, as compared to the Prussians; see documents between 10 February 1871 and 23 July 1871, BHStA München, file MK 11072; on Giessen's medical exams, see *Ordnung für die medizinische Fakultäts-Prüfungen an der großherzoglichen Hessischen Landes-Universität Giessen,* 3 April 1847, BHStA München, file M. Inn. 61291.

of the North German *Bund* and then within the unified empire, legislators in the various states sought to regularize the state medical exams to make it easier for medical men to move between states for their studies and practice. Baden and Württemberg had agreed to accept the *Bund*'s exam regulations as equivalent to their own in 1869, and as early as 1871 the *Reichstag* had asked Bavaria to accept the exam regulations governing the former North German *Bund* (rules based on the Prussian exam system). Despite some fierce objections, including worries about the power of officials in Prussia to control events in the south, the medical faculties at all three Bavarian universities soon acquiesced.[64] However, accepting the exam regulations in the different states as equivalent was not the same as making the exams themselves uniform, and through the 1870s, the different medical boards continued to examine by their own rules. This was particularly disturbing to the Prussian examiners, who saw students migrating south, where the exams were reputed to be easier and cheaper.[65]

Over the late 1870s, protracted discussion among the medical faculties and various expert commissions led to the adoption of a new set of regulations governing the preliminary and final state medical exams throughout the German nation. Two features of the new preliminary exam stand out: first, comparative anatomy was to be examined under the subject of zoology, and second, the results for zoology and botany were to be combined into a single grade, while anatomy, physiology, chemistry, and physics were each graded separately. This meant that the "descriptive sciences," including comparative anatomy, now held a minor place in the overall exam structure. In the state of Baden, the new regulations also did away with a former requirement that the student show evidence of having attended lectures in comparative anatomy, separate from both anatomy and zoology.[66] By decreasing the exam requirements covering comparative anatomy, medical administrators did not officially exclude it, but did remove much of its importance in the medical curriculum. This, too, both signaled and reinforced the lack of influence that the comparative anatomists had within the academic medical community.

64. See BHStA München, file MK 11072. On worries about Prussianization, see esp. Theodor Bischoff's letter to the medical faculty of Jena, 25 February 1871, in this file, and his printed pamphlet, *Reglement für die Prüfung der Ärzte* (1871).

65. "Aus dem preußischen Abgeordnetenhause" (1877); "Die medicinischen Examina" (1877).

66. *Gesetz- und Verordungsblatt* (1883), p. 175; cf. old Gesetz, 2 December 1871, in ibid. (1871), p. 302.

## THE OUTCOME

The appointment decisions, teaching arrangements, and exam regulations that tended to marginalize evolutionary comparative anatomists had a cumulative effect through the 1870s and 1880s. In 1873, Gegenbaur was hired at Heidelberg and his protégé-at-a-distance, Carl Hasse, went to Breslau. Over the next twenty years (1874–93), twenty-two full professorships in anatomy opened up in German universities, but only two were filled by people whose primary research was in adult comparative anatomy. In 1883 Robert Wiedersheim was named full professor at Freiburg, where he had been working as the anatomical prosector for seven years, and in 1888 Max Fürbringer was chosen as professor of anatomy at Jena, where he continued the line of Gegenbaur disciples.[67] Thus even in a time of relatively steady turnover, comparative anatomists did not gain many professorships, and the new professorships were won at universities where there was already a morphological tradition in the medical faculty.

Course listings also indicate that comparative anatomy was being displaced from the medical faculty. In the summer semester of 1870, courses titled "comparative anatomy" were taught by seven full professors of anatomy but by no full professors of zoology. By the end of the 1880s, however, the medical faction among the anatomists had succeeded in driving courses in comparative anatomy out of basic medical instruction and into the natural science curriculum. Sometimes an assistant or *Privatdozent* in anatomy would offer a course in an aspect of comparative anatomy, but more often the professor of zoology taught comparative anatomy as a major course. Even at Heidelberg, where Gegenbaur taught anatomy from a comparative-anatomical standpoint, the zoology professor Bütschli began teaching his own comparative anatomy course in the mid-1880s. By the winter of 1894–95, nine full professors of zoology out of a total of twenty offered courses with "comparative anatomy" in the title, while only one full professor of anatomy, Oscar Hertwig, did so.[68]

Medical educators were unable to prevent people like Gegenbaur or Fürbringer from teaching systematic anatomy from a comparative

67. These numbers include only Germany proper, not Austria or Switzerland. Some universities had several turnovers and some individuals moved more than once in this period. Thus the total number of universities involved in these changes was sixteen; the total number of individuals was twenty-two.

68. Bütschli's courses are listed for each semester in Universität Heidelberg, *Verzeichnis der Professoren und Privatlehrer;* see also *Vorlesungsverzeichnisse der Universitäten Deutschlands* (1894–95); Brühl, "Zur Statistik" (1870).

perspective, nor was either side likely to change its mind about the importance of comparative anatomy for medical education. But the clinical faction could and did limit the official sanction given to comparative anatomy within medical education, and the reformed preliminary medical exam effectively discouraged interest in comparative anatomy among medical students. The combination of exam requirements and a lack of teachers in turn meant that the number of anatomy students taking up comparative anatomical and phylogenetic problems remained limited as well, so that this part of the research program did not continue to grow.

Practically oriented medical educators did not criticize the comparative anatomists' evolutionary research as bad science. Nor do they appear to have considered evolution to be dangerous or erroneous. Thus one cannot attribute the rejection of comparative anatomy to a deep philosophical opposition to evolution. The arguments against comparative anatomy instead were pragmatic: medical academics tended to view evolution as simply irrelevant to their tasks and an inappropriate burden for medical students.

Given this marginalization, the success of Gegenbaur's school in maintaining the visibility and viability of its evolutionary program at all might be considered remarkable. The devotion of its core members to their teacher and to each other goes a long way in accounting for this, but other factors must be considered as well. The fact that the *Wissenschaftlichkeit* of the program was not seriously called into question in the 1870s and early 1880s meant that members of the school were able to have their work treated as legitimate contributions to anatomical science. Even when criticisms of the specific results of the program increased in the later 1880s and after, its evolutionary aims were not widely attacked by people with the power to be heard until near the turn of the century. Perhaps just as important for the survival of the program was the accepted social hierarchy of anatomy institutes, in which someone who was not the institute director could nevertheless hold a long-term or even permanent position as a salaried prosector. The funds for these positions came from the state, but unlike full professorships the choice of candidates was virtually entirely under the control of the institute director. Whereas at many anatomy institutes the director filled the prosectorship with someone whose skills and interests differed from his own, Gegenbaur and Fürbringer were able to use this structure to support members of the school.

Although a number of historians of science have recently found research schools an instructive unit of analysis,[69] it is important to recognize that this sort of intense cadre, devoted to each other as well as to a particular intellectual program, was rare. Indeed, no other German anatomist of the period had a following of the sort that Gegenbaur did. Placed in the context of the discipline of anatomy, we can see how this unusual social formation served to sustain the ideas and careers of its members over more than a quarter-century, despite the increasingly widespread perception among German medical educators that evolutionary comparative anatomy was irrelevant to medicine.

69. See n. 27, above.

# PART III

Morphology and Biology, 1880–1900

## CHAPTER EIGHT

## The *Kompetenzkonflikt* within the Evolutionary Morphological Program

In the 1880s and 1890s, the fortunes of evolutionary morphology rapidly slid downward. Much has been made of the rise of a rival "new" biology in the decades around the turn of the century, one that came to dominate the life-science disciplines in the way that evolutionary morphology has been said to dominate those disciplines in the preceding decades. Historians have often viewed Wilhelm Roux's *Entwicklungsmechanik,* with its trumpeting of a causal-analytical approach realized through experimentation, as emblematic of this new turn in biology, which has been variously characterized as a "revolt from morphology," a shift from a descriptive style of biology to an experimental approach, or a transition from a museum-based science to a laboratory-based one.[1]

Although there can be little doubt that by the early twentieth century the contours of biology were beginning to look quite different than they had a quarter-century earlier, our understanding of the relationship between the rise of the new biological approach and

---

1. Allen, *Life Science;* W. Coleman, *Biology in the Nineteenth Century* (1971); Maienschein, Rainger, and Benson, eds., "Special Section on American Morphology." Two caveats to this generalization are worth making at the outset. First, those who have developed the characterization most fully have focused on American developments, while European events have often been filtered through an Americanist lens. Second, when historians speak of biology undergoing this change, they nearly always mean zoology. Plant and animal physiology were already flourishing by the 1880s, especially in Germany, where animal (especially vertebrate) physiology was by then a leading discipline in university medical faculties, and plant physiology had rapidly gained intellectual and institutional ground in the 1860s and 1870s.

the presumed decline of morphology still lacks flesh, especially for the German context, where both evolutionary morphology and *Entwicklungsmechanik* originated. We have already seen that in the university setting the Haeckel-Gegenbaur program of evolutionary morphology did not, in fact, dominate the disciplines of zoology and anatomy in the 1870s and 1880s, at least in terms of their institutional success. Haeckel produced relatively few advanced students of his own, and even fewer of them achieved the rank of university professor in universities of the German Empire; the same may be said of his more distant followers. Nor were his theories widely accepted by professional zoologists. The circle of Gegenbaur's students and adherents, too, remained fairly closely circumscribed, their range of influence limited by the professional and institutional role seen for anatomy in the medical schools. Nevertheless, the evolutionary aims of Haeckel and Gegenbaur, if not as monolithically dominant as we have come to view them, were definitely something to be reckoned with in the late nineteenth century. Thus the relationship of this program's decline to the newer directions of the 1880s and 1890s is still worth exploring and delineating.

To gauge the changes that took place, it is useful to disentangle two different sets of events. One, concerning developments we might consider internal to Haeckel and Gegenbaur's program of evolutionary morphology, is the subject of the present chapter. The other side of evolutionary morphology's relative decline concerns the increased attractiveness of other intellectual orientations within anatomy and zoology; the subsequent two chapters are devoted to that side of the picture.

In the last quarter of the nineteenth century, those who remained committed to the morphological orientation of Gegenbaur and Haeckel found their research problems evolving in a way that fractured the program from the inside. These researchers continued to believe that determining the phylogenetic history of large-scale taxonomic groups—phyla or classes—was an achievable goal, and that the comparative method, as applied to embryos, adult animals, and fossils, would yield results to which the community would be able to agree. In the event, however, not everyone weighed the evidence in the same way: those who drew primarily upon the comparative study of adult structures came to clash with those who leaned on embryological evidence for their conclusions. As the split widened, comparative anatomists and embryologists quarreled with increasing bitterness over the relative value of their different kinds of evidence for solving major morphological questions.

Among the most intense of these battles was that over the evolutionary origin and history of the vertebrate limbs. Along with the origin of the vertebral skull, this was a focal point of Gegenbaur's research and theorizing for decades, and the object of a theory whose first version appeared in the mid-1870s. As it evolved over the next quarter-century, the controversy surrounding Gegenbaur's theory drew in younger members of the morphological community, with members of Gegenbaur's school defending their master's theory against the attacks of other morphologically trained investigators. Initial objections, based on embryological evidence, came from a contemporary of Gegenbaur from outside of Germany, the British morphologist (and Darwin critic) St. George Mivart (1827–1900), as well as two younger researchers, the American professor James K. Thacher (1847–91) and the British embryologist Francis Maitland Balfour (1851–82). In the early 1890s, the controversy was taken over by Thacher and Balfour's German contemporaries, primarily men who had taken up the morphological program as students and younger researchers during its headiest days in the 1860s and early 1870s: among them, the most prominent in the debate were Anton Dohrn, Max Fürbringer, and Carl Rabl (1853–1917). As these men argued with each other, the terms of the debate shifted so that it came to signify a division between comparative anatomy on the one hand and embryology on the other. At the same time, the logic of the embryological arguments began to shift as well, reflecting a broader disenchantment with the biogenetic law. These arguments were couched in strident, inflammatory language that reflected the degree to which consensus over the basic assumptions of evolutionary morphological method had disintegrated. The same men who had so enthusiastically taken up the enticing program of evolutionary morphology early in their careers were by the turn of the century at each others' throats.[2]

By 1902 the embryologist Anton Dohrn was writing of the *Kompetenzkonflikt* between embryology and comparative anatomy. This term, literally translated as "conflict of competency," held a double meaning. *Kompetenz* implied not only "having the necessary qualifications," but also jurisdiction. The word *"Kompetenzstreit,"* for example, was a legal term referring to a dispute over administrative authority.[3] Thus the conflict Dohrn wrote about was not simply a

2. This story offers an interesting comparison to David Hull's tale of the evolution of cladistics and the disintegration of its early community into competing camps (D. Hull, *Science as a Process* [1988], esp. pp. 255–76).

3. For a definition of "kompetent" that mentions its legal use, see H. Schulz, *Deutsches Fremdwörterbuch* (1913), s.v. "kompetent."

disagreement over theories or evidence, but one about which group—comparative anatomists or embryologists—had the ultimate authority to decide in evolutionary questions. The *Kompetenzkonflikt,* which peaked in the early years of the twentieth century, marked the extent to which the unity within the evolutionary morphological program had dissolved.

## THE BIOGENETIC LAW AND THE ORIGINS OF THE SCHISM

The split between the comparative anatomists and embryologists was already nascent in the divergent emphasis Haeckel and Gegenbaur placed on the biogenetic law. This was one of evolutionary morphology's most powerful instruments, for it held that one could study embryonic and larval development to uncover an organism's history, thereby opening up a vast field to evolutionary inquiry. Haeckel, who formulated the law and made it the center of his morphological work, stood by embryological evidence throughout his career. Gegenbaur originally subscribed to the law as well, but used it sparingly, and over time grew more doubtful about its trustworthiness, placing ever more of his faith in comparative anatomy.

On the assumption that ontogeny recapitulated phylogeny, one could compare the embryo to fossil ancestors, or to primitive present-day forms, or even to other embryos, and expect to find the same single, true path of evolutionary development. Of course, such comparisons made sense only if there was a true parallelism between the series of embryonic stages and the ancestral line. Even the most ardent supporters of the biogenetic law recognized that this parallelism did not always hold. The historical record revealed by an individual's ontogeny was accurate only to the extent that the embryological process itself was protected from adaptive influences. The true past would be obscured if the developing organism adapted to embryonic or postnatal conditions of life. Development that accurately reflected the history of the species Haeckel named "palingenesis," while that which obscured the past through adaptation he called "cenogenesis."[4] For Haeckel, only palingenetic development provided the homologies necessary to reconstruct phylogenetic trees. Such trees could be reconstructed on the basis of embryological evidence, then, only if palingenesis could be readily distinguished from cenogenesis.

In a revealing analogy first employed in the 1877 edition of his

---

4. Haeckel first referred to these in "Gastrula," p. 402.

book *Anthropogenie,* Haeckel expressed his confidence that one would be able to tell cenogenetic stages of development from palingenetic ones. He likened the ancestral series to a Roman alphabet, which in its compressed palingenetic recapitulation would be missing many of its letters. Cenogenesis was the equivalent of interposing letters from another alphabet amid the existing ones. One had only to find the Greek letters in the Roman alphabet, and then mentally delete them to find the correct (if still incomplete) series.[5] A related philological analogy allowed Haeckel to highlight the distinction between palingenesis and cenogenesis: "It is of the same importance to the student of evolution as is the critical distinction between corrupt and genuine passages in the text of an old writer to the philologist: the separation of the original text from interpolations and corrupt readings."[6] Like the philologist, the evolutionist was able to tell the true past from the corrupted record by critically comparing all the available embryonic "texts." Those forms which were repeated in otherwise widely different types were much more likely to represent the original state than a later interpolation. This is the basis for Haeckel's faith in the existence of the ancient gastraea: since he found the developmental stage of the gastrula to be represented in diverse phyla—vertebrates, mollusks, arthropods, echinoderms, worms, and zoophytes—he could deduce from it the past existence of a now lost evolutionary Ur-text, the primitive common ancestor he dubbed the "gastraea."[7] The fact that only a few representatives of these phyla displayed this form in their development did not disturb him in the least; all other forms that now replaced the gastrula were cenogenetic ones derived secondarily through adaptation.[8] To Haeckel, then, the embryo held the key to uncovering the past history of the organism. To be sure, ontogeny offered only a fragmentary picture. "We are, however, able to bridge over the greater part of these gaps satisfactorily by the help of comparative anatomy."[9]

The comparative anatomist Carl Gegenbaur quite naturally saw the relationship between comparative anatomy and embryology conversely: embryology "introduces us to the organs in their earlier states and thus links them with the persisting states of others, by means of

5. Haeckel, *Anthropogenie,* 3d ed., p. 7, this ed. trans. as *Evolution of Man* (1879), pp. 7–8.
6. Haeckel, *Evolution of Man,* p. 10.
7. See pp. 181–92.
8. On the gastraea and gastraea theory, see E. Haeckel, "Gastraea-Theorie" (1874); see also his *Anthropogenie,* 1st ed., pp. 158–59, 392–94.
9. Haeckel, *Evolution of Man,* p. 9.

which it fills in the evolutionary gaps which confront us in the series of perfected parts of the organism."[10] Gegenbaur and Haeckel thus saw the capabilities of comparative anatomy and embryology to be complementary, each filling in the gaps left by the other.

At least, that was the way it seemed in the early 1870s. But over the course of the next two decades, Gegenbaur gradually modified his view of embryology's capacity for revealing the ancestral past. As the biogenetic law became ever more commonly used, Gegenbaur gave greater attention to it in his textbooks, highlighting both its possibilities and its limitations. In his *Grundzüge der vergleichenden Anatomie* of 1870, his first textbook cast in a Darwinian light, he discussed recapitulation only in a long footnote. However, he did say that it was this connection of embryology with descent, and only this connection, that lent embryology scientific significance.[11]

In 1874 he published a condensed version of the *Grundzüge* titled *Grundriss der vergleichenden Anatomie,* which he then revised in 1878. In the 1878 revision, Gegenbaur elevated the recapitulation doctrine out of its footnote status and gave it a full paragraph in the main text, for the first time presenting the limitations of the embryological evidence: "The significance of these larval arrangements is more obvious in organisms which do not enter immediately into the 'struggle for existence' in the external world, but are developed for a certain time within the coverings of the ovum, and so are less exposed to the moulding influences of the outer world."[12] At the same time, he clarified his stance on the relationship between comparative anatomy and embryology: "Comparative anatomy explains the phenomena of ontogeny. Ontogeny by itself does not rise above the level of a descriptive discipline. . . . The necessity of an exact knowledge of comparative anatomy for ontogeny is sufficiently obvious."[13] Here Gegenbaur avoided a detailed exposition of his doubts about the biogenic law; probably he saw no reason to alienate his friend Haeckel, who often took intellectual disagreements personally.[14]

Over the course of the next decade, Gegenbaur's position changed. Exceptions to the biogenetic law had been piling up rapidly, and by 1888 he apparently felt he had to confront head-on the problems of

10. Gegenbaur, *Grundzüge,* p. 7.

11. Ibid., p. 75.

12. C. Gegenbaur, *Grundriss* (1878), trans. F. J. Bell and E. R. Lankester under the title *Elements of Comparative Anatomy* (1878), p. 6.

13. Gegenbaur, *Elements,* pp. 7–8.

14. On Haeckel's taking intellectual disagreements personally, see Uschmann, *Geschichte der Zoologie,* pp. 67, 82–83, 126.

the biogenetic law, cenogenesis and the relation of embryology to anatomy. At the annual meeting of the Anatomical Society in May 1888, Gegenbaur addressed the subject of cenogenesis.[15] Although he asserted his acceptance of recapitulation, he immediately cautioned that not all ontogenetic processes were recapitulating. The assumption that the order of embryological development could be read as a straightforward ancestral record, Gegenbaur felt, quickly led to gross misrepresentations of evolutionary history. To demonstrate his point, he proceeded to list a number of cenogenetic processes such as heterochronism, in which two sets of organs developed embryologically in an order different from their appearance in evolution. The mammalian lungs and teeth provided a case in point: in ontogeny, the lungs develop much earlier than the teeth, although teeth were indisputably more phylogenetically ancient than lungs.[16] If one took the embryological record at face value, one would naturally mistake the true order.

But Gegenbaur had a more profound objection to the general utility of the biogenetic law. When some adaptive change took place early on in ontogeny, it would give rise to a "whole series of modifications that again are purely cenogenetic."[17] Although Gegenbaur did not spell it out, this statement identified the most fundamental flaw in Haeckel's metaphor of the two alphabets: entwined as it was with the idea of simple deletions and interpolations, it completely neglected the consequences of one adaptive change on the subsequent formation and functioning of the organism.

Apart from the conceptual limitations of the biogenetic law, Gegenbaur objected to the sloppiness with which it was being used. "It is understandable," he wrote, "that the pathbreaking idea of the phylogenetic worth of ontogeny, and the abundance of findings it has brought to light, could produce an overrating of ontogenetic facts. . . . *But through this, false paths [Irrwege] have been opened.*"[18] It was far too easy, he went on, to construct fictitious organisms out of developmental stages and then, assuming them to be real, compare them with existing organisms. The investigation then took on the appearance of comparative anatomy, but since one side of the comparison was imaginary, the entire process was illusory. The first step down this false path was to forget that the embryo developed under different conditions of existence than those in which the adult lived.

15. C. Gegenbaur, "Über Cänogenese" (1888).
16. Ibid., p. 6.
17. Ibid., p. 5.
18. Ibid., p. 7.

Because of this difference, one could never simply assume that an embryonic form had also existed historically as an adult one. Having pointed to the severe limitations of the biogenetic law, Gegenbaur concluded on a more compromising note, saying that there were plenty of palingenetic facts one could study if they were clearly established as such: if one employed them exclusively, the phylogenetic picture would be improved.[19]

Four months later, in September 1888, Gegenbaur wrote about the same problem from a slightly different point of view, considering the relative value of ontogeny and anatomy for uncovering the true phylogenetic pathways. Descriptive anatomy, Gegenbaur stated, was the most basic and least scientific area of morphology, for it simply described the structures found in the adult. Embryology constituted an advance in *Wissenschaftlichkeit* over descriptive anatomy, for it showed where the adult structures came from and how they developed, thus providing a fuller explanation for the facts. However, descriptive anatomy was not to be confused with comparative anatomy, which, like embryology, had a higher critical task. Gegenbaur was distressed by the low regard embryologists held for comparative anatomy: "One reads not infrequently that *only* the knowledge of ontogenetic events can substantiate phylogenetic judgments [*Einsichten*], that no comparative anatomy is necessary for them, yes, even that the latter only stands in the way of the true knowledge of species history."[20] In the search for the facts of phylogeny, embryology and comparative anatomy were now "competitors," not companions.[21] This was not fair to either approach, Gegenbaur objected. It placed demands on embryology that were impossible to meet because of the widespread occurrence of cenogenesis, and it underrated comparative anatomy's worth. Once again he ended on a conciliatory note, asking that the two be reunited with an acknowledgment of the limitations and capacities of both methods. Nevertheless, the overall tenor of the article was to cast doubt on the sufficiency of ontogenetic studies for phylogeny and to promote the explanatory value of comparative anatomy.

Gegenbaur's estrangement from embryological studies did not take place in isolation. As his comments suggest, it was provoked in part by the statements of other morphologists who claimed that the only scientific route to evolutionary knowledge was embryology. But the

19. Ibid., pp. 7–8.
20. C. Gegenbaur, "Ontogenie and Anatomie" (1889), p. 4, essay is dated September 1888.
21. Ibid., pp. 4–5.

methodological stances of both Gegenbaur and his opponents were shaped as well by the particular morphological problems they were trying to solve. One of the most important of these was the origin of the vertebrate limbs. In this debate, which lasted from the late 1870s into the early twentieth century, the methodological commitments of different morphologists took on a more concrete form. And as they argued over the meaning of different kinds of evidence, the assumptions of the comparative anatomists and those committed to embryology came into open conflict.

## THE PROBLEM OF THE LIMBS

Gegenbaur had made his reputation as the leading theorist of vertebrate morphology in part by his work on an essential feature of vertebrate form, the paired limbs. With his controversial account of their evolution, he provided the focus for a considerable amount of the morphological research conducted by anatomists in the last quarter of the century. Gegenbaur's solution to the evolution of the extremities was based on comparative anatomy, while an alternative answer was proposed and defended by morphologists who took embryological evidence more seriously than evidence from comparative anatomy.

The problem was straightforward. Nearly all higher vertebrates have two forelimbs and two hind limbs. How was one to account in evolutionary terms for their appearance? Morphologists of the nineteenth century, like their twentieth-century counterparts, traced the origin of these limbs to a transformation in which the paired fins of fishes evolved into the fore and hind limbs of amphibians and land animals. Some scientists therefore focused their attention on the most primitive amphibians—the newts, salamanders, and other tailed or urodele amphibians. Which forms of fish fins did their limbs resemble most closely? What features in their development might offer clues to their evolutionary history? Other morphologists pushed the question one step farther back: how did paired fish fins evolve at all? After all, the most primitive vertebrates, such as hags, lampreys, and various extinct forms, entirely lacked side appendages. So how did one account for the appearance of side fins in the gnathostomes or jawed fishes?[22]

22. This problem has yet to be solved; even modern comparative anatomy texts will cite the two theories described below as the major, if clearly inadequate, contributions to the question (see, e.g., G. C. Kent, *Comparative Anatomy of the Vertebrates* [1983], p. 240). However, the ontogenetic and phylogenetic origin of the vertebrate limbs has recently resurfaced as a hot issue in debates over a genetically updated version

In a series of monographs beginning in the mid-1860s, Gegenbaur attacked these problems. By the mid-1870s, he had come up with the definitive version of a hypothesis that accounted for the evolution of both the amphibian limb and the fish fin. This became known as the "archipterygium" (early fin) or gill-arch hypothesis. This theory would be extended by Gegenbaur and his students over the next two decades, but only slightly modified in its essentials.[23] Because the fin preceded the amphibian limb in evolution, the theoretical debate centered on the evolution of the fish fin.

The appendicular skeleton of the fish has two anatomical structures that needed to be accounted for: the free-moving fins themselves and the pelvic and pectoral girdles connecting the fins to the vertebral column. Gegenbaur modeled his primitive form on *Ceratodus,* a present-day fish of undeniably ancient lineage, whose fins had a featherlike structure with rows of rays branching off from either side of a central column (see fig. 8.1). Gegenbaur called this column the "archipterygium." The next step was to account for its phylogenetic development. Gegenbaur postulated that in an ancient line of fishes related to *Ceratodus,* the two hindmost gill arches had migrated from the gill region toward the tail and were transformed into the pelvic and pectoral girdles. The freely moving part of the fin he derived from the rays that originally supported the gill arch on each side (see fig. 8.2, *A–C*). Thus both the archipterygium, which formed the most primitive part of the fin, and the rays flanking it emerged from the same structure as the girdles. In the living elasmobranch fishes, another primitive group that included the sharks and torpedo-fish, the fin's rays extend mainly from one side of the base, with only a few rays on the other side: Gegenbaur attributed this difference to adaptive degeneration of the other set of rays (fig. 8.2, *D–F*).

Gegenbaur's theory drew almost immediate opposition and counterproposals from leading morphologists outside of Germany. Over the late 1870s James K. Thacher, a teacher of zoology and physiology at Yale University,[24] and St. George Mivart and Francis Balfour, leading British morphologists, contributed to an alternative solution. This was the "side-fold" or "fin-fold" hypothesis, first proposed in an 1877

---

of the fin-fold theory discussed below. Plus ça change (see *"Hox* Genes, Fin Folds and Symmetry" [1993], and references therein).

23. Gegenbaur's most important works developing the archipterygium theory are "Über das Skelet," "Über das Archipterygium" (1873), and "Zur Morphologie der Gliedmassen" (1876).

24. On Thacher, see the *National Cyclopedia of American Biography,* s.v. "Thacher, James K."

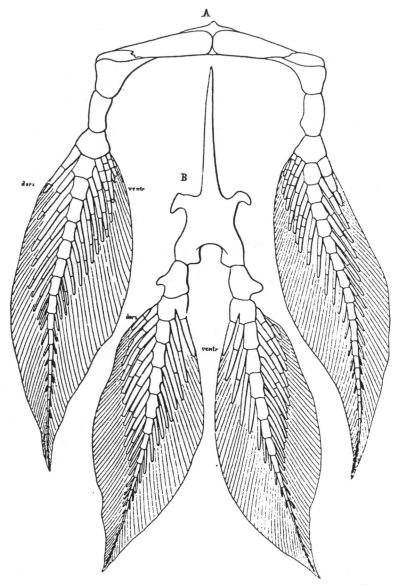

Fig. 8.1. Pectoral and pelvic fins of *Ceratodus forsteri: A*, pectoral girdle; *B*, pelvic girdle; Gegenbaur's archipterygium is the fin's central column (from Rabl, "Gedanken und Studien" [1901], p. 478).

Fig. 8.2. Archipterygium (gill-arch) theory of the origin of the extremities: A, schematic drawing of gill arch and rays; B, C, evolution of featherlike fin (archipterygium) from A; D–F, evolution of elasmobranch fin from archipterygium (from J. S. Kingsley, *Comparative Anatomy of the Vertebrates*, Philadelphia: P. Blakiston's Son [1912], p. 114).

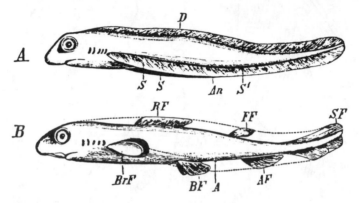

Fig. 8.3. Side-fold hypothesis of the evolution of the paired fins and their skeletal supports: A, primitive stage of continuous fin-folds; B, differentiated fins; AF, anal fin; An, anus; BF, ventral fin; BrF, pectoral fin; D, dorsal folds; FF, adipose dorsal; RF, dorsal fin; S, lateral folds; SF, caudal fin (from J. S. Kingsley, *Text Book of Vertebrate Zoology*, New York: Henry Holt [1899], p. 173).

article by Thacher[25] (see figs. 8.3 and 8.4). Instead of postulating that the fins and girdles had migrated from the head area, Thacher suggested that the fins had developed from lateral folds that had formed lengthwise along the fish's sides, similar to those that produced the unpaired dorsal and ventral fins (fig. 8.3A). Within these folds there developed cartilaginous rays that ran in a series, lined up parallel to each other within the fold (fig. 8.4A). These rays correspond to particular muscles and nerves, each group forming a distinct functional

25. J. K. Thacher, "Median and Paired Fins" (1877), "Ventral Fins" (1877).

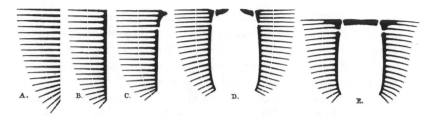

Fig. 8.4. Thacher's side-fold hypothesis: *A*, left ventral fin skeleton seen from beneath in its primitive condition; *B*, the same, with the proximal ends of the radials coalesced into a longitudinal basal cartilage; *C*, the same with the basal cartilage segmented; *D*, both ventral fins, showing the development of a median process from the preaxial basal cartilage of each; *E*, the same, with the preaxial cartilages medianly adherent, and the proximal part of the preaxial radials coalescing (from Mivart, "Notes on the Fins of Elasmobranchs" [1879], p. 468).

segment or metamere. For adaptive reasons, the side folds "became interrupted" into two primitive fins on each side, and the bases of the parallel rays fused to provide stronger support for the fins (Figs. 8.3*B*, 8.4*B, C*). These fused bases then extended inward to link up with the axial or vertebral skeleton, forming the pelvic and pectoral girdles (Figs. 8.4*D, E*).

Independently of Thacher, the prominent British morphologist St. George Mivart had reached a similar conclusion about the evolutionary development of the fins. Mivart particularly objected to Gegenbaur's assertion that the pectoral girdle derived from gill arches. Indeed, he opposed the idea that the girdles were part of the axial skeleton at all, and instead considered them part of a distinct peripheral skeleton that included the girdles and the fins. To Mivart, the evidence suggested that the evolutionary development of the fins proceeded from the periphery inward, rather than from the axial skeleton outward, as Gegenbaur's theory demanded.[26]

Thacher, Mivart, and Gegenbaur all based their views on the comparative anatomy of adult forms, but Thacher and Mivart made reference to embryology as well. Thacher wrote that the lateral folds were "the homologues of the Wolffian ridges in embryos," but he did not elaborate.[27] Mivart had learned of recent embryological work by Francis Maitland Balfour that also appeared to bolster the side-fold

26. St. G. Mivart, "Notes" (1879), esp. pp. 479–82.
27. Thacher, "Median and Paired Fins," p. 298.

theory.[28] Between 1876 and 1878, Balfour had published an extensive embryological study on the shark. In the ontogeny of the shark, there appeared a ridge along the lateral lines of the body, starting behind the gills and extending to the anus. To Balfour, Mivart, and other scientists accustomed to thinking in terms of the biogenetic law, the phylogenetic meaning of this ridge was obvious: it marked an earlier stage in the shark's ancestry, when instead of a fin the ancestor had a lateral fold. The side-fold hypothesis suddenly gained a new source of support.

Not in Gegenbaur's eyes, however. He soon spelled out his objections to the new theory, pointing to ways it did not measure up to his own morphological standards.[29] First, Thacher had based his view on the study of the sturgeon, a fish that, Gegenbaur deemed, unquestionably represented a later form than the shark. But even though the overall form of the sturgeon was more highly differentiated (and therefore more highly evolved), he conceded it was still possible that particular organs preserved a primitive form. One could not a priori rule out the possibility that these fins represented a more primitive form than the shark's fins, and Gegenbaur set out to compare the fins of a variety of jawed fishes (gnathostomes) to find out. This comparison, he claimed, revealed that all of the gnathostomes shared one common feature, the archipterygium. In fact, it was exactly that constancy of structure through a broad range of varying fish forms that demonstrated its primitiveness. Here Gegenbaur was returning to one of the fundamental tenets of evolutionary morphology: that which was common among diverse forms was also most primitive. The archipterygium was arrived at by this criterion; the side-fold theory was not.

Gegenbaur contrasted another part of the archipterygium theory with the side-fold theory. He had derived the archipterygium from the gill skeleton for a good evolutionary reason. The fin skeleton could not evolve out of nothing; it needed a plausible material source. The gill arches provided such a source, since they were cartilaginous structures that could have been transformed through adaptation into a fin skeleton. The side-fold theory provided no such material source; instead, the fin skeleton just appeared. "This must be emphasized," Gegenbaur wrote. In the side-fold theory, "no structure [*Einrichtung*] exists that would be homologous to the appendages."[30] This criticism

28. Mivart, "Notes," pp. 462–63; F. M. Balfour, *Monograph on the Development of Elasmobranch Fishes* (1878), reprinted from *Journal of Anatomy and Physiology* (1876, 1877, 1878).

29. Gegenbaur, "Zur Gliedmassenfrage" (1879).

30. Ibid., p. 522.

was based on the main methodological assumption of evolutionary morphology, that phylogenies could be traced by examining the changes found in homologous parts. But Gegenbaur's critique implied something more: that all complex structures were homologous to more primitive structures, and that therefore homology was the only way to understand a new structure. This also served, of course, to heighten comparative anatomy's importance.

In addition to attacking the assumptions underlying the side-fold theory, Gegenbaur found an ingenious way to accommodate the new embryological evidence to his own theory. The appearance of embryonic ridges preceded the development of the fin, which suggested to the side-fold theorists that ridges also preceded fins phylogenetically. But the ridges could actually indicate something quite different: Gegenbaur suggested that they marked the evolutionary trail of the pelvic fins as they migrated tailwards. Wherever the fins had passed earlier in their phylogenetic history, they could plausibly have produced a thickened epithelial layer. As the fins moved continuously rearward, they could have left an increasingly long ridge in their wake. Gegenbaur even went so far as to say that this supposition fit in better with the evidence than did the side-fold theory. Thus he strove to undermine the side-fold theory in two different ways, first by casting doubt on his opponents' phylogenetic methodology, and second by attempting to show that their evidence conformed to his own theory.

More embryological support for the side-fold theory kept coming, however. In 1881 Balfour published a paper, "On the Development of the Skeleton of Elasmobranchii Considered in Relation to Its Bearings on the Nature of the Limbs of the Vertebrata,"[31] in which he provided a more detailed account of what took place following the appearance of the lateral ridges or folds. Balfour found that in the shark's development, the fin's muscles and cartilaginous skeleton emerged in close conjunction; he also identified a continuous cartilaginous plate from which the skeletal rays extended, like the teeth of a comb. This plate, which Balfour named the "basipterygium" (basal fin or fin base), gradually extended inward and differentiated to form the pelvic and pectoral girdles. Although Balfour also described further differentiations and modifications, these early ones would take on the most significance in subsequent discussions.[32]

Balfour himself was cautious about the significance of his findings for phylogeny. He admitted that his results dovetailed nicely with

31. F. M. Balfour, "On the Development of the Skeleton" (1881).
32. Ibid., pp. 665–67.

the Thacher-Mivart theory while offering no support for Gegenbaur's theory. Nevertheless, he concluded that "the phylogenetic mode of origin of the skeleton . . . cannot . . . be made out without further investigation."[33] Balfour's own intentions to delve further into the problem were cut off by his death in a mountain-climbing accident in 1882, but the problem was quickly taken up by his close friend Anton Dohrn, founding director of the Stazione Zoologica in Naples and a leading player in the morphological community.[34]

A one-time student of Haeckel (one of his very first), Dohrn had long since mapped out a lifelong project to determine the evolutionary origin of the vertebrate body. As part of this project he studied the paired fins of the shark.[35] Although his extensive embryological studies led him to disagree with Balfour on some points, like Balfour, he saw a developmental sequence closely linking the fin's muscle buds and its cartilaginous rays. He reasoned that this ontogenetic relationship indicated that the skeleton and muscles must have evolved together phylogenetically as well. There were two possibilities: either both migrated from the gill region, in accordance with Gegenbaur's theory, or both evolved from the trunk. But if they migrated together from the gill area, then the topographical relationship between muscle and skeleton ought to be recognizably similar in the fin and the gill region. Dohrn's comparison indicated that the relative positions of the muscle and skeleton were so different in the two cases that one could not imagine them to be homologous to each other. Therefore, the fins could not have evolved from the gills, but must have derived from the trunk. Dohrn argued that the ontogenetic record spoke the evolutionary truth, and that the fin skeleton and the fin musculature evolved together from the trunk region.[36]

Dohrn was never a cautious theorizer, and he went beyond Balfour's tentative statements to come out firmly in favor of the side-fold theory. To an outlook conditioned by the biogenetic law, the new evidence clearly lent weight to the side-fold rather than the archipterygium theory. Nowhere in development did one see gill arches migrating toward the tail, or transforming into ribs, girdles, or fins. Indeed, one could not even find homologies between the gill arches and the appendicular skeleton. Nor did the development of the shark bear out Gegenbaur's contention that the skeleton was a unified

33. Ibid., pp. 668–70, quotation on p. 670.
34. On Dohrn and Balfour's friendship, see T. Heuss, *Anton Dohrn* (1940), pp. 191–92.
35. Dohrn, "Studien zur Urgeschichte 6" (1884).
36. Ibid., pp. 166–169, 181–82.

whole, with girdles and fins developing outward from the axial skeleton. In fact, the embryological evidence offered no support for Gegenbaur's theory at all. Instead, Dohrn's results supported Thacher and Mivart's view.

Over the next decade, various aspects of the problem were studied in ever greater detail, as scientists uncovered new comparative anatomical, embryological, and paleontological evidence. Leading morphologists from all over the world contributed to the debate. By 1892 one bibliography could name over one hundred sources relevant to the debate, written by fifty-six authors.[37] Among the Germans, the side-fold theory gained the approval of such respected anatomy professors of the younger generation as Robert Wiedersheim and Carl Rabl, as well as the neophyte researcher Siegfried Mollier, each of whom conducted detailed embryological studies on sharks and other vertebrates.[38]

Their studies, like Dohrn's, stressed the close relationship between the fin's muscle buds and the parallel rays. To these morphologists, it was apparent that each cartilaginous ray, flanked by a pair of muscle buds and their associated nerves, formed a segmented unit or metamere that repeated serially along the body. Metamerism signified more than just a structural arrangement, however: it connoted a particular mode of development as well. Metameric structures arose by a division of one long, relatively undifferentiated structure into numerous smaller, morphologically similar segments. And just as it was an early ontogenetic development, metamerism was believed to be an ancient evolutionary development as well. Thus the metameric structure of the fin muscles in and of itself argued that the development of fins from metameric side-folds was an ancient and generalized occurrence. By demonstrating that the fin muscles and skeleton were not homologous to the head muscles and skeleton, Dohrn only reinforced what he and others already believed: that the evolutionary origin of the fins was to be found in the metamerically structured trunk region.[39]

According to the side-fold theorists, Gegenbaur's theory had to collapse under the weight of the embryological evidence. But the archipterygium theory refused to die. In an 1894 article, Gegenbaur

37. R. Wiedersheim, *Gliedmassenskelet* (1892), pp. 261–66; the bibliography of this work names scientists from Germany, Austria, Britain, Italy, Russia, the Netherlands, Denmark, the United States, Belgium, and Australia.

38. Ibid.; C. Rabl, "Theorie des Mesoderms" (1889, 1893); S. Mollier, "Die paarigen Extremitäten der Wirbeltiere" (1894).

39. On metamerism, see Haeckel, *Generelle Morphologie*, vol. 1, pp. 312–18; Gegenbaur, *Elements*, pp. 61–62.

maintained that the accumulated evidence by no means definitively favored the side-fold theory.[40] Rather than supporting the side-fold theorists' contention that the rays differentiated metamerically from a continuous lateral fold, the most recent ontogenetic evidence actually revealed only that the primordial tissue of the later discrete skeletal parts constituted a single, unified mass of tissue. In a major interpretive leap, Gegenbaur equated this with the statement that the "fin skeleton originates from one spot, that which marks the connection with the shoulder-girdle."[41] This accorded completely with the facts yielded by comparison, and supported Gegenbaur's supposition that the fin skeleton derived phylogenetically from the archipterygium.

For a variety of reasons, Gegenbaur went on, the other ontogenetic evidence for phylogeny was deficient. If one considered the varieties of fin forms in the development of different fish genera and looked for the structures they had in common, one found that the only structure common to all the various forms was the fin base or basipterygium. But in individual development, the fin base simply appeared out of an undifferentiated mass. Thus ontogeny had not provided any clues to its phylogenetic origin, and Gegenbaur did not expect that further embryological research would produce them. Comparison of present-day adult fins and fossil fins, by contrast, yielded results commensurate with the idea that the basipterygium originated out of the archipterygium.[42]

But what about the side fold? What did it signify? Gegenbaur dropped his earlier argument that it represented the trail left by the pelvic fins on their evolutionary migration tailward, but he could not take seriously the possibility that it represented a primitive evolutionary state. This prospect he simply dismissed out of hand. "But if it is not a fin, and if it does not represent a condition that was previously realized, then it can only mark a cenogenesis."[43] As added evidence for this statement, Gegenbaur noted that the side fold failed the test of comparative embryology. If it were indeed an evolutionarily primitive structure, one could expect the side fold to appear at least in the development of more primitive forms. But the development of *Ceratodus,* which Gegenbaur considered unquestionably more primitive than the shark, revealed quite a different ontogenetic history. There were no traces of a sidefold. This, too, indicated that the side fold

40. C. Gegenbaur, "Flossenskelett der Crossopterygier" (1894).
41. Ibid., p. 129.
42. Ibid., pp. 130–31.
43. Ibid., p. 137.

was a cenogenetic development, a structure that appeared in ontogeny to help the embryo adapt to the conditions of embryonic life.[44]

Finally, with regard to the apparent close relationship between the fin skeleton and the muscle buds, Gegenbaur argued that it was also the result of cenogenesis. In phylogeny, he wrote, the metameric muscles and nerves grew out from the trunk, gradually surrounding the fin skeleton, but in ontogeny, this process was so compacted that the differentiation of muscles, nerves, and skeleton all appeared virtually simultaneously.[45] In Gegenbaur's view, the side-fold theorists caused their own main difficulty by insisting that metamerism had to apply to both the muscles and the skeleton of the fins. For him it was not a problem that the fin skeleton evolved in a different manner than did the muscles and nerves; the fin's skeleton could have evolved by migration, while its musculature developed metamerically.

Gegenbaur's arguments in favor of his theory grew somewhat byzantine, but they were guided by a few simple beliefs that remained constant through the 1880s and 1890s. First, a structure could safely be defined as primitive if it was found in a wide variety of taxonomic categories. In fact, commonality could be used as a trustworthy test of a structure's primitiveness. According to Gegenbaur, the archipterygium passed this test, but the side fold did not. Second, a new structure could not emerge out of nothing. A skeletal structure had to evolve from another skeletal structure. "Like all other organs, when they fall out of use, skeletal parts become reduced, lose their function and take on a new function. That is the law of nature."[46] Finally, as he stressed in his general programmatic statements of 1888, one must take cenogenesis seriously. Ontogeny was not an accurate repetition of phylogeny. This fact weakened ontogeny's utility for discovering the true phylogenetic history of a form.

The 1894 article was Gegenbaur's last direct contribution to the debate over the origin of the extremities. He retired from the controversy to revise his textbooks of human and comparative anatomy.[47] By that time, main terms of the debate had been set: the origin of the

---

44. Ibid.

45. Ibid., pp. 138–39.

46. C. Gegenbaur, review of "Morphologische Untersuchungen," by F. Rautenfeld (1884), p. 326.

47. Although his textbook of comparative anatomy contained a lengthy section detailing the evolution of the extremities, it was a summation of his old arguments rather than a new contribution to the debate (C. Gegenbaur, *Vergleichende Anatomie* [1898]).

fin musculature and rays together as a metameric unit versus the unitary origin of the skeletal base exclusive of muscles; the primacy of ontogenetic or comparative anatomical evidence; and the nature and extent of cenogenesis. The controversy itself, however, did not die down with Gegenbaur's departure. Instead, it was continued by Dohrn and two anatomists of the now dominant generation of Gegenbaur and Haeckel students, Max Fürbringer and Carl Rabl.

Fürbringer picked up where his teacher had left off and became the main advocate of the archipterygium theory. Meanwhile, Carl Rabl took over as the leading defender of the side-fold theory. Rabl, too, had strong credentials in evolutionary morphology. Having been inspired by Haeckel's *Natürliche Schöpfungsgeschichte* as a seventeen-year-old, Rabl went to study with him in 1874–75, soon after Gegenbaur had left Jena. His first published works stood entirely within the framework of the gastraea theory. But by the early 1890s, when he became more deeply involved in the debate over the limbs, he had long since expanded his research to other leading embryological questions of the day, concerning cell division, fertilization, and the germ layers after gastrulation. Although phylogenetic questions continued to spark his interest, they comprised just one aspect of his predominantly embryological research.[48]

As the debate entered this new stage with these players, it changed its character as well. Increasingly, the problem of the limbs was perceived to be part of a larger crisis over morphological methodology. In cases of disagreement over the evidence, which sort should take priority—adult comparative anatomy or embryology? By what criteria should one judge the evidence? This was not just a matter of intellectual priorities. As the increasingly polemical tone of the debate suggests, this methodological question was heavily laden with personal and professional stakes as well.

## KOMPETENZKONFLIKT

In early 1893, Carl Rabl wrote a long letter to his former teacher Ernst Haeckel. Amid the news of Austrian and German politics, the birth of his first child, and his relations with his father-in-law, Rudolf Virchow, Rabl mentioned his disagreements with Gegenbaur. "I am sorry that in the most important questions I keep coming into collision with Gegenbaur. This happened with the question of the metamerism

---

48. On Rabl, see H. Held, "Nekrolog auf Carl Rabl" (1918); and A. Fischel, "Carl Rabl" (1918–19).

of the vertebrate head, with the morphological significance of the ribs, the question of the origin of the extremities, and now I find myself once again in conflict with him over the problem of the structure of the vertebral column. Gegenbaur will say that that comes from a one-sided consideration of development; I cannot admit that."[49] In another letter toward the end of the year, he articulated the source of these differences: "The antagonism between my standpoint and Gegenbaur's is very sharply expressed in Klaatsch's words: 'Without comparative anatomy, ontogeny cannot make the simplest process comprehensible.' I turn this sentence (which characterizes the Gegenbaur school splendidly) on its head and say: without comparative embryology, anatomy cannot make the simplest fact comprehensible. Gegenbaur proceeds from the finished form; I proceed from the growing form. I thus stand much closer to your viewpoint than to Gegenbaur's."[50]

These two excerpts disclose a widening rift between the comparative anatomists and the embryologists, one that went beyond the problem of the extremities to a variety of issues in vertebrate morphology. Morphologists had long differed on the weight they gave to various kinds of evidence. But beginning in the mid-1890s, this issue appeared ever more prominently in studies of vertebrate morphology. At the same time, the language in which it was couched became increasingly strident, making reconciliation and compromise less and less likely. Thus in 1897 Fürbringer wrote a gigantic monograph on the comparative morphology of the spino-occipital nerve, which included a statement on the appropriate relations between comparative anatomy (or "comparative morphology," as he called it) and embryology. Embryological studies, he wrote, had produced a mass of confusing and conflicting facts which revealed few constant, phylogenetically meaningful patterns in and of themselves. But in comparative anatomy, "the older, deeper, and more far-sighted sister," the investigator would discover "the proper device by which to measure the worth of particular ontogenetic results, the hand that can sift out the cenogenetic from the palingenetic" like the chaff from the wheat.[51]

These proclamations could not be left unanswered. Anton Dohrn matched Fürbringer's rhetoric in the twenty-first installment of his "Studies on the Primitive History of the Vertebrate Body." He sarcas-

---

49. Rabl to Haeckel, 29 January 1893, in M. Stürzbecher, *Deutsche Ärztebriefe* (1975), pp. 158–60, quotation on p. 159.

50. Rabl to Haeckel, 5 November 1893, quoted in Uschmann, *Geschichte der Zoologie*, p. 131.

51. Fürbringer, "Über die spino-occipitalen Nerven" (1897), p. 712.

tically challenged Fürbringer to answer a series of questions regarding the morphology of the nerves that appeared unsolved to his "merely ontogenetically trained eyes," which "assuredly will be effortlessly solved from a comparative morphological standpoint."[52] With these questions, Dohrn demonstrated inconsistencies between the results of comparative anatomy and those of embryology, while casting Fürbringer's position in an absurd light. The solution to these apparent contradictions, Dohrn wrote, "remains closed to the limited, serflike intellect [*Unterthanenverstande*] of the descriptive ontogeneticist, and must be clarified by the reigning authorities of comparative morphology."[53]

Dohrn went on to enumerate more examples exposing the differences between the phylogenetic reconstructions based on embryology and those based on comparative anatomy. In these cases, the comparative anatomists denied the meaning of the embryological evidence, instead offering elaborate cenogetic processes that would account for the evidence while negating its value for phylogeny. Only confirmatory evidence from embryology would be admitted by the comparative anatomists; contradictory evidence could always be explained away by cenogenesis. This was the most irksome of all to the embryologist, for in Dohrn's view comparative anatomy was in no position to judge or predict which explanations the ontogeny of this or that form might afford for this or that question. Even less did comparative anatomy have the competence to present one group or species as a priori more derivative or more primitive, and so give it special trustworthiness for the transmission of palingenetic or cenogenetic embryonic conditions.[54] It was understandable, Dohrn continued, that Gegenbaur, as one of the founders of modern evolutionary morphology, might set himself up as the highest judge of morphological studies, but this did not make any more defensible the numerous places where he exceeded the bounds of his competence. It was intolerable, however, that Fürbringer could "consider himself Minos, Aeakos and Rhadamanthos in one person regarding morphological questions, and from his elevated position (in his own eyes) . . . allow himself to issue censures for ontogeny and the ontogeneticists."[55]

So that no one would miss what he was driving at, Dohrn had named the issue in the title of his study: "Competenzconflict zwischen

---

52. A. Dohrn, "Studien zur Urgeschichte 21" (1902), p. 236.

53. Ibid., p. 239.

54. Ibid., p. 262.

55. Ibid., p. 263; Minos, Aeakos, and Rhadamanthos were the three judges of Hades in Greek mythology.

Ontogenie und vergleichenden Anatomie." When Dohrn complained of Gegenbaur "overstepping his competence," he meant not simply that he was reaching beyond his abilities but also that he was trespassing onto the jurisdiction of the ontogeneticists. If a signal was needed to show just how tense the situation between the embryologists and comparative anatomists had grown, this was it.

Dohrn's strongest ally in the conflict of competence was Carl Rabl, one of the most powerful and controversial anatomists of his generation. In an 1898 essay, Rabl summarized the history of comparative anatomy and embryology over the previous twenty years. Formerly, he wrote, each side had recognized the value of the other. But over the years, embryological studies had expanded and had shown that many theories based on the comparison of adult forms could not be justified in embryological terms. As the amount of this evidence increased, the opposition between the two approaches sharpened. Rabl expressed the divergence in vividly mixed metaphors: "While comparative anatomical thought became increasingly impoverished and finally confined itself to contriving new variations on well-known melodies, embryology marched vigorously onward with the freshness of youth."[56]

In claiming priority for developmental studies, Rabl proceeded from a standpoint not terribly far removed from Gegenbaur's own. Like Gegenbaur, Rabl had doubts about the biogenetic law. Adaptations took place at every stage of development, and changes at early stages often affected the form at later stages. Gegenbaur had used this observation to caution against an overly simple view of cenogenesis, and to argue that cenogenesis provided a serious limitation to the biogenetic law. The conclusion he drew was that embryological studies were often suspect, while comparative anatomical ones were more secure. Rabl, however, drew a more radical conclusion: the biogenetic law had no basis at all in reality. A given developmental stage did not correspond to an adult ancestor, or even to a stage of development of an ancestor. It represented nothing beyond its own development.[57]

How, then, could embryology hold any information at all for phylogeny? Wouldn't this conclusion imply that the only secure source of phylogenetic information was comparative anatomy? Yes and no, answered Rabl. Comparing adult forms was useful but very limited, since the adult was but one stage in the life of an organism. Why should not all stages be equally worthy of comparison? If one allowed

56. C. Rabl, "Über den Bau und die Entwicklung der Linse" (1898), p. 80.
57. Ibid., pp. 83–85.

this to be the case, then one had a vastly expanded field for comparison.[58] The basic rules of evaluation for phylogeny would remain the
same: a feature common to many different forms was likely to be
primitive, while one that appeared only here and there was a later
adaptation. Only now instead of just comparing adults to other
adults, one could compare equivalent embryonic stages to sift the
common, ancient features from the more varied, recent features. One
had to compare equivalent stages, however: adults and adults, earliest
stage of development to earliest stage, and so forth. Without the assumption of recapitulation, it made no sense to compare embryonic
stages of more complex forms with the adult stages of more primitive
forms.

Just as before, one could still compare two organisms' ontogenies
to determine their degree of relatedness: the longer their ontogenies
ran parallel, the more closely related they were. But, Rabl cautioned,
this parallelism did not result from both individuals recapitulating the
development of the same lower form.[59] He did not offer an alternative
explanation for why parallel individual developments should signify
common ancestry rather than common conditions of existence; it appears to have been an article of faith. In practice, then, the main
difference between operating on the assumption of recapitulation and
operating on Rabl's principles lay in the prohibition of comparing
embryonic stages to adult forms.

Rabl demonstrated his evolutionary use of comparative embryology in "Thoughts and Studies on the Origin of the Extremities"
(1901).[60] To take just one example, he disputed Gegenbaur's contention that the shark was a more primitive form than the sturgeon,
and that the sturgeon's fin could be derived from the shark's fin.
As a starting point, he took an observation from the comparative
embryology of the two forms. Close comparison indicated a lack of
homology between the muscle-cartilage relationship in the sturgeon's
fin and that in the shark's.[61] Consequently one could not derive the
sturgeon from the shark phylogenetically. His alternative solution was
based on the assertion that all jawed fishes displayed a lateral fold
along each side early on in their development—precisely the point
that Gegenbaur had denied—and this indicated that the common ancestor had it as well. One had only to suppose that their ancestry
diverged at a time before the side folds had evolved a musculature

58. Ibid., p. 78.
59. Ibid., pp. 97–98.
60. C. Rabl, "Gedanken und Studien" (1901).
61. Ibid., p. 490.

and skeleton. Thus the evolution of the side fold into a true fin took diverging paths in different fishes, an assumption that fit in with the embryology of the sturgeon and shark, in which the subsequent mode of muscle development differed substantially.[62]

In these examples, Rabl used the standard critical method of comparative anatomy to determine ancestral unity and divergence; the only difference was that he applied the method to embryonic forms rather than adult ones. On the basis of the embryological evidence, he was prepared to assert the existence of an ancestor with primitive side folds instead of fins. He likewise adapted Gegenbaur's method to embryology when he agreed with Gegenbaur that an anatomical structure could lose its old function and take on a new one—a key argument in Gegenbaur's derivation of the fin skeleton from the gill arch. But such a change in function and structure, Rabl asserted, invariably affects development. The burden therefore fell on Gegenbaur and his school to demonstrate the legitimacy of the archipterygium theory using developmental evidence.[63]

Here Rabl was asserting new ground rules for morphology. In his system of reasoning, ontogeny provided the surer standard for phylogenetic research. It was up to comparative anatomists to see that their theories conformed to the evidence produced by embryologists, not the other way around. Nor could comparative anatomists fall back on cenogenesis when ontogenetic evidence failed to conform to their assumptions. In a system in which the biogenetic law was not accepted, palingenesis and cenogenesis had no meaning. (Here, despite his earlier assertions of allegiance to Haeckel, Rabl was clearly attacking his former teacher's most cherished ideas as well.) Adaptations were simply adaptations, whether they occurred at an embryonic stage or later.[64] Rabl's standards for morphology were incommensurable with those of Gegenbaur, for no matter how pessimistic Gegenbaur was about the methodological utility of the biogenetic law, he never denied its reality. In fact, the biogenetic law was an essential prop allowing him to assert the primacy of comparative anatomy over embryology, for without cenogenesis, Gegenbaur's main argument for embryology's untrustworthiness fell away.

62. Ibid., p. 491ff.

63. Ibid., p. 528; Rabl also advocated increased use of functional argumentation and evidence, to demonstrate why a change should have occurred, in addition to formal arguments demonstrating that it could have done so.

64. Dohrn expressed similar views in arguing for the relativity of cenogenesis and palingenesis, but he never entirely gave up the biogenetic law (see Kühn, *Anton Dohrn*, esp. p. 152).

As well as setting new ground rules that Gegenbaur and his school failed to follow, Rabl's 1901 study also tried to show how Gegenbaur failed to live up to more general scientific standards. Nowhere did this appear more clearly than in the twenty-page appendix to the article, "Historical-Critical Remarks on the Archipterygium Theory," in which he accused Gegenbaur of all manner of professional sins. First, he was guilty of bad scientific method. For example, he employed cenogenesis in an irresponsible and arbitrary way; so far as Rabl could make out, Gegenbaur claimed that everything in fin development after the initial form of the basipterygium was cenogenetic. But to invoke cenogenesis was to evade any real explanation at all.[65]

The kind of explanation Gegenbaur did provide for the evolution of the fins was also based on poor method, since it was entirely hypothetical. It started from the presupposition that the skeleton was the phylogenetically most ancient part of the fin, and that this skeleton later acquired muscles and nerves. Gegenbaur could point to no positive support from embryology or physiology for this; on the contrary, the embryological evidence showed that the muscles always preceded the skeleton. Gegenbaur's main way out, aside from cenogenesis, was simply to assert as "facts" unproven statements. In fact, like the *Naturphilosophen* who had derived the fins from the ribs, Gegenbaur's entire theoretical structure consisted of one unsubstantiated hypothesis built upon another.[66] Here Rabl intensified his accusation with guilt by association, for the *Naturphilosophen,* though dead for over two generations, had come to symbolize the worst of pre-Darwinian morphological speculation—just the sort of unacceptable theorizing that the post-1859 morphologists had thought Darwin's theory would overcome.

This kind of unsubstantiated claim, Rabl intimated, was typical of Gegenbaur's irresponsible rhetorical method. Rabl traced Gegenbaur's language concerning the archipterygium over the previous decades: although he had first spoken of it as a "conjecture" [*Vermuthung*], it soon became a "hypothesis" and eventually a "theory," all without any increase in positive evidence to support it.[67] Gegenbaur had also misrepresented Rabl's own findings, a further breach of professional conduct.[68] Despite Gegenbaur's misguided and possibly dishonest attempts to keep this theory afloat and to discredit the embryologists, however, the truer, surer way of embryology would win out.

65. Rabl, "Gedanken und Studien," p. 538.
66. Ibid., pp. 524–37.
67. Ibid., p. 532.
68. Ibid., p. 536.

As Rabl grandiloquently finished, "Embryology will go its own way, unhindered by whether or not its teachings are convenient to the guildlike wisdom of any school. Its past is the guarantor of its future."[69]

Rabl's pugnacious accusations and Dohrn's assertions of the *Kompetenzkonflikt* brought the long-running tension between the embryologists and the comparative anatomists to a head. If comparative anatomy was to hold any sway in the judgment of phylogenetic questions, their two essays demanded a response. Max Fürbringer accepted the challenge in 1902, within a year of these two essays, he published his two-hundred-page answer, "Morphologische Streitfragen" (Morphological points of controversy).[70]

Most of the piece was directed at Rabl. Dohrn, for all his contentious rhetoric, was led to his disagreements with Gegenbaur and Fürbringer in a "natural way," as a result of his intimate acquaintance with shark embryology and anatomy. In his response to Dohrn, therefore, Fürbringer discussed specific differences he and Dohrn had over the interpretation of embryological evidence. Rabl, by contrast, seemed only to want to put an end to the Gegenbaur school.[71] Fürbringer responded in kind, allowing himself to comment not only on Rabl's scientific method and results but also on his polemical approach.

Fürbringer sought to derail both Rabl's claim that embryology was a surer route to evolutionary knowledge than comparative anatomy and his accusation that Gegenbaur was conducting poor science. Fürbringer's approach in both cases was to stress the moderation and openness of the Gegenbaur school, implicitly contrasting it with Rabl's own extreme approach. According to Fürbringer, the problem was not that embryology and comparative anatomy as a whole conflicted with each other; neither was it the case that embryology was entirely worthless for uncovering an organism's ancestry. The problem was that Rabl and a few others had made exaggerated claims for embryology's capabilities, and had made bad use of embryological evidence.[72] Thus Fürbringer's main aim was to show where Rabl as an individual scientist went wrong, while simultaneously maintaining an overall stance of reconciliation between comparative anatomy and embryology. These aims sometimes conflicted with one another, lending a peculiarly equivocal tone to Fürbringer's arguments.

69. Ibid., p. 540.
70. M. Fürbringer, "Morphologische Streitfragen" (1902).
71. Ibid., pp. 90, 85.
72. Ibid., pp. 238–40.

He first concentrated on Rabl's polemical attack on Gegenbaur. He cited one passage after another in which Rabl had quoted Gegenbaur out of context or otherwise misrepresented him. But his rebuttals were directed at curious parts of Rabl's arguments. For example, in response to Rabl's passage describing the elevation of the archipterygium from supposition to hypothesis to theory, Fürbringer did not defend the status of the archipterygium as a theory. Instead, he pointed out that Gegenbaur had not called it a theory since 1876. Since that time, he had always been cautious about how he termed it. Fürbringer had no doubts that the archipterygium was the true source of the fin, but he, like Gegenbaur, remained suitably cautious about its scientific status.[73] Here and elsewhere Fürbringer's need to present his mentor Gegenbaur as a good scientist overrode his need to defend the validity of the archipterygium.[74] He took pains to portray Gegenbaur as a thinker open to evidence of all kinds, a scientist cautious about speculation, who never overextended his claims beyond their basis in fact. To maintain this claim, it was important to avoid any statement that might justify Rabl's opinion that the archipterygium was simply an "article of faith," or the "dogma" of a school.[75] It was the burden of another part of the essay to demonstrate that the archipterygium met the demands of a scientific theory better than did the side fold.

At the same time, by refuting Rabl's characterization of Gegenbaur, Fürbringer cast doubt on his scientific judgment. Indeed, if Gegenbaur was the hand-waving charlatan that Rabl indicated, one could only wonder that the rest of the scientific community had been duped by him for so long. "What a light this casts on the judgment of his contemporaries!"[76] In Fürbringer's portrayal, Gegenbaur was a paragon of scientific moderation, who stood in sharp contrast to the overzealous and one-sided ontogeneticists. Such a man would never stoop to the kind of unscientific, ad hominem attack and low polemical methods that Rabl had used.

Having dispensed with Rabl's polemics, Fürbringer moved on to consider his science. He first paused to examine Rabl's overall methodological assumptions. Rabl's rejection of cenogenesis especially troubled him. If cenogenesis did not exist, then obviously all individual developments were palingenetic. According to Fürbringer, this was how Rabl could claim that a series of embryological preparations

73. Ibid., p. 156n.
74. For example, pp. 153–54.
75. Rabl, "Gedanken und Studien," p. 499.
76. Fürbringer, "Morphologische Streitfragen," p. 160.

would suffice to uncover an organism's true phylogenetic history. But even if this were the case, which everyone knew was false, the ontogenetic evidence still did not support the side-fold theory.[77] Here Fürbringer turned to his main argument against the side-fold theory.

In his detailed examination, Fürbringer attacked Rabl's position from a number of angles. His most important objection to the side-fold theory was the same one Gegenbaur had voiced several years earlier: its advocates mistakenly believed that embryology demonstrated the metameric origin of the fin skeleton. Rabl had even gone so far as to claim that the segmented muscles actually caused the rays to develop.[78] Fürbringer claimed that the most recent embryological evidence in fact opposed the side-fold theory. This evidence, produced by two of Fürbringer's protégés, Richard Semon and Hermann Braus, demonstrated that the base of the fin skeleton (which represented the evolved form of the archipterygium) developed before the musculature, and therefore was independent of it. This lent support to Gegenbaur's approach of considering the skeleton separately from the musculature.

On this subject, as well as others, Fürbringer attempted to demonstrate the poor quality of Rabl's theorizing. First, Rabl had not tried to account for the new findings of Semon and Braus but had simply ignored them.[79] Second, his theory did not even have the evidence he claimed for it: the side fold was not a widespread occurrence in the ontogeny of the fishes. Since it was not a common feature in either ontogeny or primitive adult forms, any phylogenetic significance claimed for it was "fictitious."[80] Finally, what facts he did have actually spoke against his theoretical premises rather than for them.[81]

Fürbringer blasted his way through the rest of Rabl's paper with similar tactics, and ended some thirty pages later with a lengthy summary of Rabl's transgressions. However, when he finally came to the true conclusion of the paper, which addressed both Dohrn and Rabl, he sought to strike a more appeasing note. Ideally, he wrote, all approaches to a problem deserve equal respect, "when they are earnestly, carefully, and thoughtfully carried out, and when they yield good results."[82] Perfect results from different methods would con-

77. Ibid., pp. 160–63.

78. Ibid., p. 175, referring to Rabl's comments at the 1898 meeting of the *Anatomische Gesellschaft* (see *Anatomischer Anzeiger*, supp. 17 [1898]: 179).

79. Fürbringer, "Morphologische Streitfragen" p. 170.

80. Ibid., pp. 183–84.

81. Ibid., pp. 188–89.

82. Ibid., p. 239.

verge on a single truth. But scientists are only human; therefore they make mistakes, which result in contradictions. The problem was that they did not always admit their mistakes or recognize the limits of their particular approach. The wise and experienced Gegenbaur was aware of the strengths of many approaches—and also of their limitations. His criticisms of embryologists applied not to embryology in general, whose capacities he recognized, but to "the pretensions of certain representatives of the new embryological approach."[83] Fürbringer called for a return to Gegenbaur's morphology, a morphology in which comparative embryologists and comparative anatomists would each have their say and would each be conscious of where they needed to depend on each other.

Despite this conciliatory stance, Fürbringer could not keep his prejudices entirely hidden. They emerged when he reviewed his own statements of 1897 that had so aroused Dohrn. At first playing down the leading role he had claimed for comparative anatomy, he stressed that he had called embryology and comparative anatomy "sisters," "the two guides" of morphological research. Anyone who had perused his writings could see the high regard in which he held ontogenetic research in general. That he could not say the same of Dohrn's study of the development of the eye was not his fault—it was a self-contradictory and unclear piece of work. This was not surprising, Fürbringer continued, since the state of ontogenetic knowledge in many areas was similarly labyrinthine. Here Fürbringer gave way to temptation and maintained that, like his honored teacher, he still found comparative anatomy the trustworthy standard by which to judge ontogenetic evidence. This was especially true in the matter of cenogenesis. Despite the "trumpet-duet" of Dohrn and Rabl, he wrote, the existence of a real and regular cenogenetic element in ontogeny could not be denied.[84] Recovering from this lapse, Fürbringer ended on a blandly reconciling note: in the long run, there was no essential *Kompetenzkonflikt*. As knowledge increased, embryology and comparative anatomy would work hand in hand to reach the common goal.[85]

The 1901 and 1902 publications of Dohrn, Rabl, and Fürbringer marked the peak of the *Kompetenzkonflikt*. Although they were the most active players in the conflict, it was followed closely by other morphologists. In 1902 and 1903, Fürbringer received numerous let-

83. Ibid., p. 240.
84. Ibid., p. 251.
85. Ibid., pp. 253–54.

ters from Gegenbaur's students and allies applauding his actions. Rudolph Burkhardt, *ausserordentliche* professor of zoology at the University of Basel, wrote Fürbringer, "The '*Morphologische Streitfragen*' had to come sometime, for the persistent botching of morphology by these one-sided embryologists and their insane pretensions had to be set in their true light. . . . Rabl . . . I consider nothing short of pernicious and inwardly untruthful."[86] Hermann Klaatsch wrote, "Hopefully Rabl, whom I consider in certain respects pathological, will now become somewhat more careful. It is inconceivable to me that some older and younger colleagues consider him a genius."[87] And Richard Semon thanked Fürbringer for undertaking "to put the entire nonsense of the ontogeneticists . . . in the proper light."[88]

Fürbringer attracted even more sympathy in 1903, when he hosted the annual meeting of the Anatomical Society at Heidelberg. Speaking on the first day of the meeting, Rabl ended his paper on the homologies of the tortoise limbs with a short, hurt reply to Fürbringer's attack on him. He claimed to be interested only in the truth and innocent of the pugnaciousness and dishonesty Fürbringer imputed to him. He regretted that Fürbringer had stooped to such a low level. "A fight with weapons like Fürbringer's is infrequent even in public life; in science, happily, it is an extremely rare, entirely isolated exception."[89] Fürbringer's ad hoc response was not recorded. A written reply, published as part of the proceedings of the meeting, disclaimed ever having accused Rabl of deliberate dishonesty, only of being overzealous and sometimes careless. Otherwise it neither added to his old arguments nor backed down on any of them.[90]

Fürbringer received more letters congratulating him on his conduct at the meeting and decrying Rabl's behavior. The zoologist Richard Hertwig, for example, had not read the polemical exchange, but he wrote Fürbringer that it was "not comme il faut" for Rabl to attack the man hosting the meeting.[91] Rudolph Burkhardt had heard from

86. R. Burkhardt to Fürbringer, 20 November 1902, SB, Fürbringer Nachlass.

87. Klaatsch to Fürbringer, 29 March 1902, SB, Fürbringer Nachlass; Fürbringer himself applied a similar character assessment to Dohrn in his unpublished autobiography: although Dohrn had performed an important service for science in founding the Naples Station, Fürbringer condemned his ungrateful, even "hateful" behavior toward Haeckel. This he then excused with the comment that Dohrn had a "very nervous, even psychopathic" character, requiring him to repair frequently to sanatoria (Fürbringer, *Lebenserinnerungen*, vol. 1, p. 287, UB Heidelberg, Fürbringer Nachlass).

88. Semon to Fürbringer, 17 March 1902, SB, Fürbringer Nachlass.

89. C. Rabl, "Über einige Probleme" (1903), p. 190.

90. M. Fürbringer, "Erwiderung" (1903), pp. 190–97.

91. R. Hertwig to Fürbringer, 15 November 1903, SB, Fürbringer Nachlass.

someone who had attended the meeting "that your victory over Rabl was sweeping. I only hope that this is a sign that people are fed up with the encroachments of the embryologists more generally as well."[92]

But Burkhardt's hopes were not borne out. Widespread sympathy for Fürbringer's difficult situation at the meeting did not change anyone's position in the *Kompetenzkonflikt*. In fact, after this point the fortunes of the comparative anatomists slid rapidly. With Gegenbaur's death in June 1903, less than a month after the meeting of the Anatomical Society, evolutionary comparative anatomy lost its greatest advocate and practitioner. Fürbringer continued to teach and conduct research for another decade, but he was well aware that his style of science was considered passé by most younger researchers, and he believed that his influence within the anatomical community had diminished to almost nothing.[93]

During the same period, the controversy over the extremities continued, but it took a direction that also marked the waning of the old evolutionary program. After 1903, the debate came to center on Rabl's claim that in ontogeny the fin muscles actually caused the fin skeleton to form. As Rabl battled with Fürbringer's protégé and son-in-law Hermann Braus over this issue, the crux of the debate shifted to the causes of development within the embryo. These two men became increasingly caught up in the problems and methods of *Entwicklungsmechanik,* and comparative anatomy, the biogenetic law, and phylogeny drifted into the background. As a result, the old fight over methodological jurisdiction became irrelevant and the *Kompetenzkonflikt* faded away.

## CONCLUSION

The debate over the limbs both signaled and hastened the internal dissolution of Haeckel and Gegenbaur's program of evolutionary morphology. The debate itself went through two phases. In the first, between the early 1870s and the late 1880s, Gegenbaur himself was the prime defender of his theory; the course of the debate during this phase ran parallel to Gegenbaur's growing skepticism toward the biogenetic law. In the second phase, from the 1890s to the early 1900s, the debate moved into the hands of the next generation of

92. Burkhardt to Fürbringer, 28 June 1903, SB, Fürbringer Nachlass.

93. Fürbringer to Semon, 17 September 1903, UB Heidelberg, Fürbringer Nachlass; Fürbringer to Oppel (draft), 6 January 1910, SB, Fürbringer Nachlass.

morphologists. Among these researchers the issue became more clearly one of morphological methodology.

The oldest and firmest assumption of vertebrate morphologists, dating from even before the appearance of Darwin's theory, was that comparative anatomy and embryology would work naturally together to yield the laws of form. In the biogenetic law, Haeckel had linked the two more closely while giving their use a new, evolutionary justification. The biogenetic law in its turn became the most prominent tool of evolutionary morphology. But when the results from embryology and those from comparative anatomy conflicted, the weaknesses of the biogenetic law became apparent even to those inside the program, and it was attacked from both sides. Gegenbaur and his school found a loophole in cenogenesis, which they used to explain away embryological results inconsistent with their own. Rabl and others claimed that all development was adaptive, and that therefore hard and fast distinctions like cenogenesis and palingenesis were meaningless. Not only did both sides find the biogenetic law much less lawful than Haeckel had claimed, but they could not even agree on the reasons for its lack of validity. In the long run, this was even more damaging than the attack on the biogenetic law itself, for it signified the disintegration of consensus about the basic methods of evolutionary morphology.

In its second phase, other stakes came into play as well, as signaled by the highly polemical and personal tone it took on. It is evident that by 1903 Fürbringer had come to view the controversy over the limbs not just as a scientific problem but as one involving the defense of his revered teacher and an entire way of doing science, one which he believed was on the decline. Reasons for the degree of personal investment in the controversy by Rabl and Dohrn are somewhat less clear, but one may speculate that they viewed their own orientation as much more on the cutting edge of morphology and had an intellectual and professional duty to put the older approach in its place. Indeed, one might even view the debate as reflecting a clash between two different personal and professional styles, with Rabl and Dohrn represented an aggressively forward-looking style of professional consciousness in which staying on the cutting edge of research (in Dohrn's case, through his promotion of new approaches at the Naples Zoological Station; in Rabl's case, through his pursuit of a variety of hot research problems) was the main thing, while Fürbringer identified himself entirely with a school and the pursuit of its agenda.

Whichever of these layers may have been evident to contemporaries, the debate over the limbs amply demonstrated to newcomers the hopelessness of determining large-scale phylogenies by means of comparative anatomy or embryology. Each side in the debate directed powerful arguments at the other's poor use of evidence, unjustified and forced interpretations, inconsistencies, lapses in argumentation, and inability to account satisfactorily for contradictory evidence. To anyone who stood outside the debate, it was far easier to see the weaknesses of both sides than the strengths of either one.

Adding to the sense of evolutionary morphology as an increasingly barren intellectual field was the concurrent emergence of numerous exciting new problem areas within anatomy and zoology that—in contrast to the Haeckel-Gegenbaur program—were producing dramatic conceptual yields. Discoveries concerning fertilization and the early stages of cell development, begun in the 1870s, provided considerable fodder for novel theorizing over the next two decades, as research on inheritance and subcellular structures intensified. Numerous researchers busied themselves studying the effects of the physical environment—light, pressure, temperature, and so forth—on organic form and function. At the same time, the phenomenon of symbiosis, established by botanists in the 1870s, invited zoologists to move beyond the physical environment to look more closely at the effects of organisms on each other. Other zoologists and anatomists pursued the puzzling phenomena surrounding regeneration. Still others devoted themselves to studying chemical and physical processes in hopes of finding exact models for organic ones. Many of these problem areas touched on the nature of animal form and form change, but they did so by developing methods and questions quite distinct from those of Haeckel and Gegenbaur.

The combination of the conceptual weaknesses of the Haeckel-Gegenbaur program with the excitement generated by these alternative problems and approaches makes it easy to assume that the community concerned with animal form underwent a wholesale shift away from an evolutionary-morphological approach to a "modern" causal-mechanical, experimental approach. As we will see in the following chapters, however, in fact neither the phylogenetic enterprise nor the comparative method were completely repudiated. Instead, a broader generational division took place: those in Rabl and Fürbringer's generation, including such powerful innovators as Wilhelm Roux and Oscar Hertwig, tended to try to find other ways to pursue questions about the causes of form without abandoning either an evolutionary framework or a belief that comparative anatomy and embryology were useful methods. Researchers entering the professional community in the 1890s, however—men such as Hans Driesch—were more likely to view the controversy as indicative of morphology's bank-ruptcy, a reason to give up on evolutionary morphology altogether. But they still wanted to understand the nature and significance of animal form. Indeed, toward the end of the nineteenth century it seems that practically everyone was looking for the key to understanding animal form. They were just looking in different places.

CHAPTER NINE

---

# New Approaches to Form, 1880–1900

---

## Rhetoric, Research, and Rewards

D espite its importance within the narrower circle of evolutionary morphologists, the *Kompetenzkonflikt* hardly held the
center of action within the wider communities of anatomy
and zoology toward the end of the century. For researchers
interested in animal form who did not identify themselves with the
evolutionary goals of Haeckel and Gegenbaur, the period from the
late 1870s to the turn of the century was an extraordinarily tumultu-
ous one. As the Haeckel-Gegenbaur program began to unravel, alternative projects that placed greater emphasis on experimental and
"causal-analytical" approaches to form were capturing researchers'
attention. Concurrent with these intellectual developments was a massive turnover in the makeup of the anatomy and zoology communities: between 1880 and 1900 about two-thirds of the full professors
in each field retired or died, opening up both disciplines to younger
men. This conjunction of events might lead one to think that the shift
from the old evolutionary morphology to a new experimental biology
was a simple matter of generational change, in which young experimentalists finally gained the positions that would allow them to replace the old morphologists and go their own way.

As this chapter and the next show, however, the course of change
was not that straightforward. Neither the intellectual situation nor the
institutional one displayed a clear dichotomy between the concerns of
evolutionary morphologists and the newer ones of causal-analytical
researchers. To complicate matters further, anatomy and zoology differed both in their intellectual environments and in their fine social

278

structure. For example, both anatomists and zoologists took up questions about the proximate causes of form. In zoology, however, causal questions tended to be taken up in quite a different fashion than among anatomists. Whereas the anatomists Roux, Born, Rabl, and O. Hertwig looked to the cell and the earliest stages of individual development for the causes of individual form, many zoologists sought their causes by means of "biological" research on the organism (developing or adult) in relation to its environment. This is not to say that zoologists shunned cellular research—far from it—but that they often looked elsewhere as well, seeking causes of differences among species, varieties within a species, and generations within the same variety in the effects of such external factors as temperature, pressure, and nutrition on the organism as a whole.

The structures, values, and demography of anatomy and zoology also made different agents of change relatively more important in each discipline. In zoology the overabundance of young zoologists first noted in the mid to late 1870s led to a rapid diversification of zoological problem areas among younger zoologists, in part pursuing lines set out by some older, firmly established zoologists. At the same time, however, the opportunities for academic employment in zoology remained limited, with the result that not all the new lines of research were able to gain the institutional security necessary to take root and grow. Thus the criteria by which successful zoologists were selected from the sizable pool of aspirants is an especially important factor in illuminating the direction that zoology took by the turn of the century and should help inform our understanding of both the decline of Haeckelian evolutionary morphology and the rise of the newer approaches. To detail that story, however, requires a chapter in itself; it will therefore be held off to chapter 10.

In anatomy, researchers interested in problems of morphology and evolution continued to face pressures that directed them away from comparative anatomy and toward problems and methods viewed as more appropriate to a basic medical science. It should not be surprising, therefore, that Wilhelm Roux (1850–1924) argued for *Entwicklungsmechanik* as a new anatomical orientation that would be of medical use as well as general scientific interest. Yet Roux's rhetorical stance does not by itself sufficiently clarify the relationship between evolutionary morphology and causal-mechanical approaches to form within anatomy. To do so means inquiring into the development, reception, and institutional fate of his program predominantly within his own disciplinary community.

This chapter therefore begins by looking at the evolution of Roux's program in its immediate context of the anatomy institute at Breslau during the late 1870s and early 1880s. Although it has been common to regard Roux's program as marking a sharp break with evolutionary morphology,[1] the research actually undertaken by Roux and some of his contemporaries during the period suggests that the relationship was far more ambiguous. The second section turns from research to the rhetorical plane and considers the programmatic and philosophical claims Roux made for *Entwicklungsmechanik,* along with some prominent reactions to it. To what degree did he and his respondents view his program as an alternative to evolutionary morphology, to what degree an extension of it? To what extent was the focus of discussion experiment—intervening in the normal functioning or development of the organism—as distinct from Roux's claims for a new, causal-analytical method that went further than mere experimental manipulation to seek the causes of form? The chapter ends by considering the fortunes of *Entwicklungsmechanik* as a new orientation within anatomy, as represented by its ability to attract adherents and by their success in gaining professorships.

## ENTWICKLUNGSMECHANIK AND EVOLUTIONARY MORPHOLOGY

On 12 November 1889, the newly appointed professor of anatomy at Innsbruck, Wilhelm Roux, gave the inaugural lecture traditional for a professor just arriving at a new university. The lecture itself, though in its form also traditional, proclaimed its novel content in the title: "Die Entwickelungsmechanik der Organismen, eine anatomische Wissenschaft der Zukunft" (The developmental mechanics of the organism, an anatomical science of the future). Long interested in morphological problems, Roux hoped with his new program to succeed where the evolutionary morphologists had failed, and to satisfy both intellectual and professional demands that previous approaches to understanding animal form had left unanswered.

He began by naming four orientations that had held sway in anatomy until now: descriptive, physiological ("mechanical," in Hermann von Meyer's terms), embryological (*"Entstehungsweise"*), and comparative anatomical. He went on to say that even if these four were perfected, morphology would still not be complete, because "we lack

---

1. The most influential proponent of this view is Allen, *Life Science,* p. 25; for an assessment that takes into account Roux's evolutionary concerns, see F. Churchill, "Wilhelm Roux" (1966), and "Roux, Wilhelm."

the knowledge of the *direct causes of origination* of these forms."[2] Comparative anatomy, he declared, could only partly satisfy the "inborn need for causality" that every person possesses. That need would be fulfilled only when one understood what powers existed in the fertilized egg to build a complex organism out of an undifferentiated mass. The main path to discovering these causes was experiment.

Roux had already conducted some embryological experiments that indicate what sort of causal analysis he had in mind. Beginning in the early 1880s, he had undertaken a series of experiments on the earliest stages of frog development, seeking to discover the causes determining the cleavage patterns of the fertilized egg and their relation to the later-developing central axis of the frog's body. Following up on some experiments by the Bonn physiologist Eduard Pflüger, who had argued that gravity affected the cleavage planes, Roux devised an apparatus that would rotate the eggs so that the force of gravity would be negated. He argued from his results that Pflüger was wrong, and that internal causes, not the external force of gravity, determined the planes of cleavage. In another, soon-to-become-famous set of experiments, he used a hot needle to destroy half the cells resulting from cleavage (blastomeres), and then observed what happened to the remaining part of the egg as it developed to the gastrula stage. The result was a well formed half-embryo, which suggested to Roux that the remaining cells had not been affected by the destruction of the others. These experiments and many others that Roux performed during the 1880s combined controlled manipulation of the organism with intellectual analysis to discover the mechanical causes of form.[3]

Roux was never one to play up his intellectual antecedents or his debts to other researchers. In his relentless quest to gain intellectual elbowroom for *Entwicklungsmechanik,* he was certainly not disposed to emphasize its connections to evolutionary morphology. And yet his own history suggests that these debts were not inconsiderable. In particular, his experience at Breslau, along with those of his colleagues there, suggests some of the ways that evolutionary concerns could be connected with problems of more proximate causation in the late 1870s and 1880s.

Roux had originally studied at Jena, where he heard lectures from Haeckel, Gegenbaur, and the physiologist Wilhelm Preyer as a student in the philosophical faculty, and later as a medical student learned

---

2. "Es fehlt die Kenntnis der *directen Ursachen des Entstehens* dieser Gebilde." W. Roux, "Entwickelungsmechanik" (1889), p. 27.

3. Churchill, "Roux, Wilhelm."

anatomy from Gegenbaur's successor Gustav Schwalbe. Following his promotion to doctor in 1878, he worked for a year and a half under Franz Hofmann in Leipzig's hygiene institute and then moved to Breslau in the fall of 1879 to work as second assistant in the anatomy institute under Carl Hasse, an enthusiastic follower of Gegenbaur's evolutionary comparative anatomy. Roux's earliest publications derive from research he conducted for his Jena dissertation, out of which he developed his notion of "functional adaptation." In retrospect, Roux considered this early work as merely a preliminary step in his progress toward formulating a truly causal-analytical, experimental program for studying development, but in its time it represented something else. In his earliest years of research, Roux was concerned with problems that came straight out of mechanical anatomy, the orientation of Hermann von Meyer and others who had been inspired by Carl Ludwig's new physiology in the 1850s. Roux's publications of the late 1870s and early 1880s mark one of the earliest efforts within the German anatomy community to unite this mechanical orientation with the morphologists' concern to demonstrate the reality of organic evolution. His first major innovation, therefore, was not his championing of experiment, or even his stress on causal analysis, but his attempt to synthesize analytical approaches from the two disparate orientations of mechanical anatomy and evolutionary morphology.

Roux's first two publications, on the role of blood flow in determining the angle of branching in blood vessels, took on a subject with a solid history in mechanical anatomy, namely hemodynamics.[4] Although he implied in his autobiography that he had thought up this first causal-mechanical problem on his own, it seems unlikely. The structure of blood vessels had been a topic of considerable interest among mechanical anatomists since the work of Ernst Heinrich Weber in the 1850s, and was being actively pursued in the mid- and late 1870s by anatomists well known to Roux at Leipzig and Jena.[5] Roux's

4. W. Roux, "Über die Verzweigungen" (1878), and "Über die Bedeutung der Ablenkung des Arterienstammes" (1879).

5. This problem area had been taken up by Wilhelm His and Wilhelm Braune in the mid-1870s; and even if Roux was unaware of these works, he surely knew of the research undertaken by Jena's prosector of anatomy Karl Bardeleben, who was working in the same institute as Roux, and who in 1877 and 1878 published studies on the structure of blood-vessel walls and the effects of blood pressure and other mechanical relations on their thickness ("Über den Bau der Venenwandung" [1877], "Über den Bau der Arterienwand" [1878]). Gustav Schwalbe, the head of the institute, also took enough interest in Roux's problem to publish an article, "Über Wachstumsverschiebungen und ihren Einfluß auf die Gestaltung des Arteriensystems" (On displacements

work on the blood vessels may therefore be viewed as a contribution to an established problem area in which anatomists were already applying one sort of causal-mechanical analysis—the kind that examined the physical effects of pressure on the shape of anatomical structures.

What Roux added that made his work different was not causal or mechanical analysis, but Darwinian adaptation. Instead of simply ascertaining the mechanical effects of blood flow on vascular structure, he placed this phenomenon into the evolutionary context of adaptation and inheritance, trying to establish which structural aspects were inherited and which were adaptive. By 1881 this work had grown into his first major theoretical contribution, *Die Kampf der Teile im Organismus. Ein Beitrag zur Vervollständigung der mechanistischen Zweckmässigkeitslehre* (The struggle of parts within the organism, a contribution toward the completion of the mechanistic theory of purpose).[6] He pursued a similar goal in another investigation published in 1883. His study of the structure of the dolphin's tail, though almost unreadably theoretical, had as its basis a kind of problem that would be familiar to the older mechanical anatomists, namely, the functional correlation and cooperation of locomotor structures. This paper dealt essentially with the effect of function on form, in this case on the formation of the dolphin's tail. Once again, he differed from the older mechanical anatomists in seeing this problem not simply as a physiological-architectonic one, but as a problem of evolutionary adaptation. In seeking to demonstrate the reality of functional adaptation, he was bringing together mechanical considerations and the traditionally morphological concerns of descent.[7]

Between his early work on blood vessels and his 1881 publication, Roux moved to Breslau, where he would remain for eight years. At

---

of growth and their effect on the formation of the arterial system) in 1878, which immediately followed Roux's dissertation in the *Jenaische Zeitschrift* and served as commentary on and extension of Roux's article. Thus Roux was hardly developing his problem and his approach in isolation.

6. W. Roux, *Kampf der Teile* (1881); Jane Oppenheimer has translated "Zweckmässigkeit" in this title as "teleology," but I believe "purpose" is a slightly more suitable translation (see Oppenheimer, *Essays*, p. 66).

7. W. Roux, [Autobiography] (1923), p. 155; Roux also wrote several other papers on functional adaptation, collected in his *Gesammelte Abhandlungen*, vol. 1; a caution to the reader of the collected works: although Roux uses brackets to show where he has made additions or corrections to his essays, he also makes subtler changes in typeface to emphasize phrases that were not always emphasized in the original; this can result in somewhat different readings of the material.

Hasse's anatomy institute he met two other anatomists his age who shared his interests in combining broad evolutionary questions and other anatomical approaches. When he arrived in 1879 as second assistant in anatomy, his colleagues in the institute were the prosector Gustav Born (1851–1900) and the first assistant Hans Strasser (1852–1927). Both men offered him stimulating intellectual company, working in just the areas that Roux would claim as his own.[8]

Strasser, who overlapped with Roux during the latter's first two years in Breslau, was first assistant in Hasse's institute from 1876 to 1881.[9] Like Roux, he was interested in incorporating both mechanical and evolutionary morphological considerations into his work. In his 1877 dissertation, for example, he analyzed the air sacs in different classes of birds, opening with a descriptive section and then continuing with ontogenetic and phylogenetic analyses of their origin and importance. The structure of the argument places it superficially, at least, in the realm of an evolutionary morphological study. But Strasser was not especially interested in either establishing a phylogenetic tree or using the biogenetic law. He consistently argued instead in terms of the mechanical and physiological significance of the air sacs as they developed, and focused his argument on the functional advantages they brought to the bird.[10] Although Strasser left Hasse's institute before Roux developed his *Entwicklungsmechanik*, his interests in combining evolutionary questions and mechanical ones surely reinforced those of Roux; it also demonstrates that the latter was not alone in his quest to bring together the two areas.[11]

Gustav Born followed a somewhat different path, although, like Roux, he also had links to the Haeckel-Gegenbaur program of evolutionary morphology. Having attained his medical education by touring through Breslau, Bonn, Strassburg, and Berlin and graduating from Berlin in 1873, he went to work the next year in Gegenbaur's

8. Both Strasser and Born were undoubtedly important influences on Roux, but Born probably contributed less to Roux's general mechanical approach and more to his specific experimental technique; on Born's life and work, see Gebhardt, "Gustav Born."

9. On Strasser's life, see Wegelin, "Hans Strasser."

10. H. Strasser, *Über die Luftsäcke* (1877).

11. Strasser's published use of the term "functional adaptation" in 1883 brought a wounded cry from Roux, who felt his toes had been stepped on; see Roux to Strasser, 13 May 1883, Medhist. Inst. Zürich, Strasser Nachlass; H. Strasser, *Zur Kenntniss der funktionellen Anpassung* (1883). In this monograph Strasser suggested that Roux had not always given others due credit for undertaking research of a like-minded orientation: "So have perhaps other workers also built up quietly the area that W. Roux, with an auspicious and daring beginning, has made his own" (p. 18).

Heidelberg institute. Beginning in 1875 and continuing through the early 1880s, he published a number of papers in Gegenbaur's *Morphologisches Jahrbuch*. Couched in the broad framework of the evolutionary morphological program, these works began to reflect some of his doubts about some of its central tenets: in 1879, for example, objecting to Haeckel's assumption that new anatomical adaptations are added on to the end of development, he concluded in one article that "ontogenesis is . . . a highly imperfect recapitulation of phylogenesis, because species formation, that is, the new formation and change in anatomical characters, mostly are already put into place during ontogenesis and not onto the grown animal, even if the fixation of the changes ensues only in a later age of life."[12]

Even before his morphological articles were published, Born had returned to Breslau to work under the morphologically inclined anatomist Hasse. There, beginning in the late 1870s, his research expanded to problems of breeding and development—problems still loosely connected to evolutionary questions but increasingly taking on their own significance. They were also questions amenable to an experimental approach. In 1881, for example, he discussed experiments he had conducted in the previous year on sex determination in frogs. This was a classic effort at trying to decide between alternative hypotheses by varying one factor at a time while controlling all the others. Based on his efforts to do this, Born argued that in his frogs, sex was not determined strictly by the interaction of sperm and egg but could, in experimental situations, be affected by factors such as nutrition that impinged on the fertilized egg during its development.[13] Although he continued also to publish on problems oriented toward the ontogeny and phylogeny of the vertebrate skeleton, these publications were increasingly interspersed with ones reflecting his newer interests. By 1884 Born's publications had turned almost exclusively to experimental investigations concerning breeding and early development of amphibian eggs, and to new techniques and instruments that would facilitate these investigations.[14]

Thus during the first three years or so of Roux's tenure at Breslau,

12. G. Born, "Die Nasenhöhlen und der Thränennasengang" (1879), p. 428.

13. G. Born, "Experimentelle Untersuchungen" (1881); Born explicitly noted that he believed nutritionally induced sexual alteration to be the result of the abnormal experimental conditions, and that in nature such other factors as the age of the parents were probably more important in determining sex. Nevertheless, the main result was that sex in frogs could change during development.

14. A bibliography of Born's works is available in Gebhardt, "Gustav Born," pp. 140–43.

he and his contemporaries in the institute were trying to combine the logic and arguments of German evolutionary morphology with other ways of thinking and investigating form. Beginning in 1881, however, the year Strasser left, Born's research took a new turn, and Roux soon followed him onto the fertile field of experimental embryology, an area in which both gained acclaim as pioneers through the rest of the 1880s. The story of their famous embryological experiments, along with those of the physiologist Eduard Pflüger and the anatomist Oscar Hertwig, has been told elsewhere and is not the burden of these pages.[15] The point here is that the research of Roux, Strasser, and Born in the late 1870s and early 1880s shared a similar pattern: all three displayed an early exposure to and interest in the Haeckel-Gegenbaur style of evolutionary morphology (which their director, Hasse, also pursued), but none was satisfied with staying strictly within an explanatory framework that used comparative anatomy and descriptive embryology as the primary means of understanding evolutionary continuity. At first they mixed their evolutionary questions with ideas and questions from other orientations, but for both Born and Roux, at least, the excitement and sense of novelty attached to their new experimental research soon overtook their earlier questions.

For these men, the early 1880s marked a rapid transition in questions and techniques; nevertheless the evolutionary framework in which they began did not entirely drop away. Thus when Born wrote Gegenbaur's leading disciple Max Fürbringer in 1888 to say that he still considered himself a Gegenbaur student, it was not solely a fawning remark.[16] Roux also continued to take seriously both mechanical and historical components in the production of form. In fact, one of his major concerns of the 1880s was to develop a way of distinguishing between aspects of development due to inheritance, which preserved the past, and those resulting from adaptation, which were assumed to be acquired through mechanical means. This problem was implicit in Haeckel's formulation of the biogenetic law, but no viable approach to solving it had previously been worked out. Roux's eventual solution to the problem was to break down the great dilemma of understanding inherited versus acquired characters into a set of tasks more readily approachable by experiment.[17] Although this solu-

15. See Churchill, "Wilhelm Roux"; Weindling, *Darwinism and Social Darwinism* esp. chap. 5; for an especially succinct overview, see J. Maienschein, "Origins of *Entwicklungsmechanik*" (1991).

16. Born to Fürbringer, 6 July 1888, SB, Fürbringer Nachlass, no. 281.

17. Churchill, "Wilhelm Roux," chap. 5, esp. pp. 155ff.

tion moved him away from direct engagement with evolutionary issues, he never discounted the importance of inheritance and the organism's links to the past the way some other embryological researchers
did.[18] Thus although Roux's program (for which Born, and to a lesser
extent Strasser, was an important catalyst and contributor) was taken
by later commentators to have freed embryology from its ties to evolutionary morphology, the intention of all three men appears instead to
have been to reform morphology, not to reject it.

Among Roux's contemporaries who would become prominent in
anatomy, a few others shared this pattern of an early orientation that
combined the Haeckel-Gegenbaur framework of evolutionary morphology with a mechanical and experimental approach, which then
gradually took over as the predominating one. As we will see in the
next section, Oscar Hertwig exhibited just such a pattern, and the
similarity of personal history may have increased the deep tensions
between him and Roux. The same pattern was followed by the anatomist Carl Rabl, who studied under Haeckel in the mid-1870s but then
went on to work under the physicalist physiologist Ernst Brücke in
Vienna, later becoming prosector in anatomy under Brücke's colleague Carl Langer, and eventually being named successor to His's
anatomy chair at the prestigious University of Leipzig. Rabl's works,
too, show a similar mix of interest in the specific problem areas set
out by Haeckel and Gegenbaur together with cell division, gastrulation, and germ-layer development; by the mid-1890s Rabl was also
caught up in pursuing a causal-mechanical approach to some of these
questions, occasionally using experiments.[19]

Roux, Hertwig, and Rabl came to be among the most innovative
anatomical researchers in Germany in the 1880s, and among the most
influential in the 1890s and after; their mix of frameworks, entailing
a gradual but not absolute shift away from phylogenetic concerns, is
therefore worthy of historical note. Although this bridging position
was not terribly common among anatomists, it does indicate that
not everyone viewed evolutionary and causal-mechanical research as
incompatible alternatives.

18. In fact, Roux considered inheritance a sufficiently important issue to criticize
the physiologist Eduard Pflüger, who conducted embryological experiments in the early
1880s, for not paying attention to it (see W. Roux, [Autobiography] [1923], p. 159).
By the early 1900s the discounting of the past in shaping the organism's present had
grown sufficiently widespread that Theodor Boveri, himself an experimental embryologist, felt compelled to lodge a public protest (see T. Boveri, *Die Organismen als historische Wesen* [1906]).

19. On Rabl, see Fischel, "Rabl"; and Held, "Nekrolog." For a more detailed
discussion of one area of Rabl's research, see chap. 8.

## RHETORIC: ENTWICKLUNGSMECHANIK AND THE
## MECHANICAL CAUSES OF FORM

The reception of Roux's program by the community of researchers concerned with form provides another vantage point from which to assess the relationship between *Entwicklungsmechanik* and the other research orientations taken by researchers at the time. It would almost seem as if his colleagues would have to respond, for Roux never missed a chance from the mid-1880s on to proselytize for his new program. Having introduced the term *Entwicklungsmechanik* into print in 1884 in a journal local to Breslau, he reached a broader audience the next year with an article laying out his principles in the recently restructured *Zeitschrift für Biologie,* a journal edited by two physiologists explicitly interested in bringing morphological questions under their purview. (Although the article in the *Zeitschrift für Biologie* was labelled as the first contribution, the second one, published in the *Breslauer ärztliche Zeitschrift,* appeared first.) Taking advantage of the German genre of review journals, in 1888 he introduced *Entwicklungsmechanik* as a subcategory of his assigned review area (*Entwicklungsgeschichte*) in Hofmann and Schwalbe's widely read *Jahresberichte über die Fortschritte der Anatomie und Physiologie.* He continued these reviews until 1890. In 1892, he switched over to a new review journal, Merkel and Bonnet's *Ergebnisse der Anatomie und Entwicklungsgeschichte,* for which he wrote a lengthy overview of his program as a preliminary to its being made a regular category of review. Two years later he began publication of his own journal, the *Archiv für Entwickelungsmechanik,* which opened, of course, with another essay recapitulating his program. The following year, in 1895, he had all his previous programmatic and research articles gathered into a massive two-volume work titled *Gesammelte Abhandlungen über Entwickelungsmechanik der Organismen.* In 1897 he wrote a lengthy new essay in his journal reasserting his goals and methods, "Für unser Programm und seine Verwirklichung." And in 1905 yet another programmatic essay appeared at the launching of a new series he was editing, *Vorträge und Aufsätze zur Entwickelungsmechanik der Organismen.*

Even the choices of publishing venue of his substantive research articles show him taking every opportunity to blanket the field of his prospective audience with his ideas: his famous series of seven fundamental "Contributions on the *Entwicklungsmechanik* of the embryo," published between 1884 and 1893, appeared spread over six different journals. His other publications displayed a similar strategy. Until he founded his *Archiv* he rarely published in the same journal twice, and even after that, he often sent articles to places that might extend his audience.[20]

Despite the impressive range of audiences to whom he outlined his program, direct responses were not overwhelming in either their quantity or their grasp of his aims. Much to his frustration,[21] only a small number of the older German morphologists responded to it. Haeckel mentioned it only to dismiss it; Gegenbaur appears to have taken even less notice of it. The first scholars to take *Entwicklungsmechanik* seriously appear to have been Haeckel's recent students Hans Driesch (1867–1941) and Friedrich Dreyer (1866–1903?), who gave it a close reading in the early 1890s.[22] A few members of Roux's own generation soon became engaged with it as well, with Oscar Hertwig responding most fully to Roux on his own programmatic and philosophical plane. Although the differences in response reflect very much the personalities and individual battles of the men involved, they also suggest the different challenges, threats, and opportunities Roux's program held for the three generational cohorts represented, respectively, by Haeckel, Hertwig, and Driesch and Dreyer.

In the foreword to the fourth edition of his *Anthropogenie* (1891), Haeckel attacked the "basic errors" of what he called "*Entwicke-lungs-Mechanik*." Even though Roux viewed this work as evidence of Haeckel's rejection of his program,[23] what Haeckel himself meant by *Entwickelungs-Mechanik* referred not to Roux's program alone, but primarily to that of his own longtime opponent Wilhelm His. Noting that he had already set out his objections to this approach in 1875 and again in 1884, he repeated some of the main ones here: that this "pseudo-mechanical school," as he called it, unnecessarily and unscientifically limited its empirical study to ontogeny, ignoring paleontology and comparative anatomy; that it limited itself methodologically to "exact" descriptions using "compass, measuring stick and weight"; and that it was limited philosophically by excluding comparison with other organisms and the study of the relation of the

20. W. Roux, "Beiträge zur Entwickelungsmechanik des Embryo. Nr. 2" (1884), "Beiträge zur Entwickelungsmechanik des Embryo. Nr. 1" (1885); for the aims of the *Zeitschrift für Biologie*, see W. Kühne and C. Voit, "An unsere Leser!" (1883), p. 3; a full bibliography of Roux's works appears in Mocek, *Wilhelm Roux–Hans Driesch*.

21. Roux complained in "Ziele und Wege" (1892) that the comparative anatomists either "completely ignored" his program "or like Haeckel declare directly that it is superfluous," pp. 426–27.

22. One might also include Curt Herbst (1866–1946), a third member of their circle, who took up *Entwicklungsmechanik* as a method but did not engage in programmatics about it.

23. Roux, "Ziele und Wege," p. 426.

parts to the whole. All of these, in Haeckel's view, were signs of "the most narrow-minded specialism" threatening the study of development.[24] None of these objections was directed particularly at Roux; indeed, Roux was not even mentioned by name. Nevertheless, by referring to the program of *Entwicklungsmechanik,* a term that was clearly Roux's, Haeckel placed him in the same camp as His. Roux's own version of what set him apart from Haeckel was somewhat different: he argued that Haeckel "and some of his students" viewed the "so-called biogenetic law alone as a sufficient explanation of embryonic formation," whereas he viewed the biogenetic law as merely an expression of typical events that themselves required an explanation in causal terms.[25] Thus between Roux and Haeckel there was considerable cross-talk, as Haeckel folded Roux's *Entwicklungsmechanik* into the His camp while Roux, who actually regarded His's ideas as somewhat crude, focused on the deficiencies of the biogenetic law.

For Haeckel's rebellious students Hans Driesch and Friedrich Dreyer, who entered the discussions as postdoctoral researchers in the early 1890s, Roux's program resonated in a different way. Having completed his doctorate under Haeckel in the spring of 1889 and spent most of the next year traveling, Driesch had spent the early summer of 1890 reading up on works by the embryologists His and Goette, which had brought home to him "the question of *how,* through what means and forces, the individual organism actually developed out of the egg."[26] With this preparation, when Roux's 1889 Innsbruck address fell into his hands, it seemed quite naturally to him to belong in the same framework, only with the improvement that Roux was conducting experiments to support his hypotheses. Thus in his first philosophical-programmatic publication of 1891, Driesch grouped Roux's ideas about the mechanics of organic development together with those of His, Goette, and other mechanical thinkers such as the anatomist Hermann von Meyer and the botanist Simon Schwendener. But Driesch took his opposition to the Haeckelian evolutionary approach a giant step farther than any of these men, arguing on epistemological and methodological grounds that neither the selection theory nor the broader theory of descent had been—or could be—in any rigorous sense "proven." Given this situation, Driesch argued, it would be best to follow the model of the inorganic sciences,

24. E. Haeckel, *Anthropogenie,* 4th ed. (1891), pp. xxi–xxii.

25. Roux, "Ziele und Wege," pp. 426–27; Roux had already listed his objections to the biogenetic law in an 1886 review (reprinted in Roux, *Gesammelte Abhandlungen,* vol. 1, pp. 438–57, on pp. 444–47).

26. Driesch, *Lebenserinnerungen,* p. 67.

in which the historical science of geology "applies the theories that are offered by [nonhistorical physics,] the sister that stands so infinitely much higher in philosophical value." Thus the historical aspects of biology, which are necessarily hypothetical and inexact, should defer to the exact mathematical-mechanical approach now being developed.[27] Driesch would continue to find historical explanations inadequate even as he grew increasingly skeptical of the value of mechanical explanations of the organism and began developing a neovitalist stance.[28]

Friedrich Dreyer, another Jena graduate and a friend of Driesch,[29] also coupled his advocacy of a mechanical approach to biology with a rejection of the historical-morphological approach epitomized by the descent theory. In his 1892 pamphlet *Ziele und Wege biologischer Forschung* (Goals and paths of biological research), Dreyer dispensed with specific references to the recent literature, painting instead a broad programmatic picture of the past, present, and future of biology. Dreyer, whose scientific worldview appears to have been shaped by Haeckel more than he might have admitted, drew a sharp contrast between the historical-morphological approach and what he called the "etiological-mechanical" approach, arguing that even if we had constructed a true and complete phylogenetic tree, "our causal needs would remain as unsatisfied as before. . . . The theory of descent can give us an external view of the paths of development but not insight into the causal chain of life processes that . . . produces [these paths] as its visible products."[30] Nevertheless, Darwin's theory was not entirely without merit: by making the mistaken claim that the chief mechanical cause of descent was selection, it called biologists' attention to the need to examine mechanical causes. Dreyer himself thought that of the physical sciences that produced mechanical explanations, chemistry would probably play the most important role in the biology of the future.[31]

The early writings of Driesch and Dreyer reflect their view of a polar opposition between the morphological approach associated with Haeckel and the mechanical approach they associated with both His and Roux. One might argue here that the dominance of Haeckel in their educations had something to do with this. They clearly absorbed their teacher's notion that there were only two choices: either

27. H. Driesch, *Mathematisch-mechanische Betrachtung* (1891), p. 57.
28. H. Driesch, *Biologie als selbständige Grundwissenschaft* (1893).
29. Driesch, *Lebenserinnerungen*, p. 39.
30. F. Dreyer, *Ziele und Wege biologischer Forschung* (1892), p. 75.
31. Ibid., pp. 78, vii n.

to swallow Haeckel's historical approach to morphology, accepting the biogenetic law as true and the search for phylogenetic trees as both scientifically legitimate and achievable, or to reject it in favor of another. They chose to reject it, revolting from the style of morphology set by their teacher. This explanation of their behavior, however, must be incomplete, for Roux and Oscar Hertwig, who both took a more reformist stance, had also been Haeckel's students once but did not accept the limited choices he offered. Although this problem cannot yet be definitively solved, two partial explanations suggest themselves. First, Roux's own programmatic writings could be read selectively to support a more extreme position than he himself did; in this way, he may have created the bridge to a position that he himself did not accept. A second factor may lie in differences of age and experience: whereas Roux and Hertwig had long since worked their way beyond Haeckel's frame of reference through their own research, Driesch and Dreyer at this point were just beginning scholars and may have known no other options than to accept the framework of possibilities set by their teacher.

Roux himself thought these two had gone too far, arguing that they "condemn the theory of selection too easily; they raise once again the old ostensible objection that selection is not a principle of active formation, but simply a sorting principle [*Ausleseprincip*] and thereby undervalue the additive effects of this sorting of variations that are brought forth by principles of formation belonging to *Entwicklungs-mechanik* (which will be ascertained only gradually through long research)." He went on to point out that much comparative anatomical work is based on unconscious developmental-mechanical assumptions, implying that *Entwicklungsmechanik* would contribute to evolutionary theorizing, rather than lead away from it.[32]

The response by the other leading participant in the programmatic and philosophical discussions of Roux's program is somewhat more complex. Oscar Hertwig, a member of Roux's own generation, mounted a sustained, direct, and increasingly vehement attack on Roux's program in the later 1890s. Hertwig and Roux had come into repeated conflict since the mid-1880s over interpretations of their experiments on the causes of development, and by 1897, Hertwig's ire had overflowed beyond the research journals into a pamphlet of some two hundred pages attacking Roux's program.

In this pamphlet Hertwig presented Roux as claiming necessary, unique, and novel links between causal analysis, mechanics, and ex-

32. Roux, "Ziele und Wege," p. 425.

periment. He found flaws in Roux's articulation of both these concepts themselves and their relationships. Claiming that Roux attributed causal thinking only to *Entwicklungsmechaniker,* Hertwig argued by contrast that all biologists think causally and are looking for causal laws. "Where, then, are the researchers to be found, who have concerned themselves with the theory of development and who have *not* proceeded from the assumption that . . . the investigation of the causes of form-building is one of their main tasks?"[33] Hertwig also argued that experiment was less reliable than simple observation; he thought it was at best used in conjunction and comparison with observation of normal processes. "There are no logical grounds" for ascribing to questions answerable only by experiment "a higher knowledge value than questions for which the observation of nature by other methods gives information. The kind of resources [*Hilfsmittel*] with which a discovery is made does not decide its greater or lesser epistemological value [*Erkenntnisswerth*]."[34] As to the sort of mechanical thinking that would reduce all processes, including living ones, to atoms subject to physical forces or "energies" (which Hertwig took to be Roux's idea of mechanics), Hertwig believed that the results of biological research, including those from experiment, had deepened rather than narrowed the gap between the living and nonliving worlds.[35]

For Roux, Hertwig's arguments were deeply frustrating. It seemed to him that Hertwig had willfully misunderstood many of his points, quoting him out of context and sometimes stooping to petty and ridiculous criticisms of Roux's turns of phrase. Especially galling was Hertwig's claim that Roux said only *Entwicklungsmechaniker* thought causally. Roux had, in fact, devoted several pages of the 1894 introduction to the *Archiv* to detailing the relationship between comparative anatomy and *Entwicklungsmechanik,* which he viewed as continuous and mutually supporting. "Since comparative anatomy seeks to determine the genetic relationship, the family tree [*Stammbaum*] of organisms," Roux had written, "it is by its nature a causal science. It analyzes forms according to the two components of variation and inheritance." Variation (or adaptation, which Roux used as an equivalent term) and inheritance were general principles that "themselves urgently require causal explanation, that is, analysis into their constantly effective [*wirkungsbeständigen*] components; this

33. Ibid.
34. O. Hertwig, *Zeit- und Streitfragen* (1897), p. 80.
35. Ibid., p. 85; for a more extended discussion of this text, see Weindling, *Darwinism and Social Darwinism,* pp. 129–34.

analysis is a task of *Entwickelungsmechanik*." He went on to note that many hypotheses already developed by comparative anatomists were *entwicklungsmechanisch* in character, and he gave three examples of mechanical reasoning used in comparative-morphological context. Such chains of reasoning, Roux contended, could benefit from further study from a mechanical point of view. Thus comparative anatomy "continually sets new tasks for *Entwickelungsmechanik*, . . . just as for its part *Entwickelungsmechanik* . . . is the extension and simultaneously the support of comparative anatomy." To be sure, it was the experimental testing of the hypotheses of comparative anatomists that would lend them a secure footing, but here as elsewhere Roux presented the two as going hand in hand. "Both orientations pursue the same goal, which will be neared more quickly through working together."[36] Hertwig's deliberate ignoring of this important four-page section seemed to Roux typical of the sort of distortions Hertwig regularly employed.

Concerning the reliability of experiment as opposed to other methods, Roux defended his view that only experiment could, in principle, yield true certainty but softened this declaration with explanations and caveats, repeating earlier statements he had made to the effect that not all experiments were "artificial experiments." "Natural experiments" in the form of "variations, malformations or other pathological events" could also yield valid causal knowledge; the same passage acknowledged the difficulty of achieving such knowledge even with the help of experiment.[37] Furthermore, Hertwig seemed to Roux to miss the emphasis on the analytical aspect of the causal-analytical experimental method, the effort to tease apart the complex causes of form into their simpler components.[38] In response to Hertwig's admonition against overvaluing experimental results, Roux replied that insofar as conclusions based on other methods were really certain, Hertwig's caution was justified. But the problem was exactly that the observation of "the normal events of formation give no absolutely sure information," whereas experiments (often requiring a series of variations on the same theme) could.[39]

Finally, Roux defended his claims for the theoretical reducibility of the causes of form to physical forces acting on brute matter. Hertwig had not understood the most important part of this discussion, Roux griped, which acknowledged the difficulty of breaking down

36. W. Roux, "Einleitung" (1895), pp. 25–29.
37. W. Roux, "Für unser Programm" (1897), p. 238.
38. Ibid., p. 247.
39. Ibid., p. 249.

the intricate processes of development directly into atoms in motion obeying physical forces, and which pointed to the importance and necessity of looking to higher-level combinations of these atoms and forces. When Hertwig dismissed Roux's discussion of these "complex components" as "words, empty words, and nothing more!" he missed the point. Roux answered his criticisms by saying that without such complex components, Hertwig was left with only the most general notion of mechanical cause, which was useless for guiding research. No wonder Hertwig found the task Roux had set out an impossible one![40]

This group of philosophical and polemical writings surrounding the causal-mechanical approach to development in the 1890s reveals several important points. To begin with, although experiments came up for debate, they arose as a methodological subtopic of a deeper philosophical issue: what counts as a mechanical cause? Researchers concerned with form in this period used the term in a bewildering variety of ways, and their discussions suggest the penumbra of confusion it wore at the time. Roux himself ascribed much of the resistance to his program to this confusion:

> First [there is] the opposition of those who understood "mechanics" from their school physics as only the theory of matter in motion, and who furthermore have not read my works enough to understand the sense in which I have used the word. These believe I wanted, like His, to trace back developmental events solely to the pressure and stress of growing parts on the neighboring areas; they did not recognize that it was a question of ascertaining *all* the mechanisms operating in development as well as the combinations of these factors and the amount of effect of [these combinations].[41]

But the situation was even more complicated than Roux suggested. His's emphasis on uneven cell growth was actually an instance of his broader view of mechanical cause as the physical pressure of bodies in contact. Haeckel had said that the mechanical cause of individual form was the biogenetic law; he had also talked about mechanical causes in a broader sense as those which were nonmetaphysical. Both Driesch and Roux claimed to understand the term "mechanical" in the Kantian sense of law-bound causation, but they took different Kantians (or perhaps different Kants) as their source. For Roux, introduced to philosophy by Rudolf Eucken, "law-bound causation" meant both the general exclusion of supernatural or metaphysical

40. Ibid., pp. 49–73.
41. Roux [Autobiography], p. 146.

agents (similar to Haeckel) and a stricter determinism in which true causes yield consistent effects with absolute predictability;[42] however, he also wrote sometimes of seeking the direct causes of form, which appears to have held a narrower meaning more like that of His. For Driesch, who drew his Kantianism from the idealistic neo-Kantian philosopher Otto Liebmann, mechanism as law-bound causation was a limited concept when applied to the organism; it could not explain development without the complementary concept of teleology.[43] Hertwig, for his part, threw up his hands and suggested abandoning the term altogether.

A second important feature of these discussions is that the participants hailed from the three different intellectual generations that dominated the community of scientists concerned with form in the 1890s: Haeckel's cohort, now composed of the senior professors still active; Roux and Hertwig's cohort, most of whom had settled into their full professorships in the 1880s and early 1890s and who were clearly in ascendence; and Driesch and Dreyer's group, born in the mid to late 1860s, who were just entering the fray. Although one cannot say the participants in this debate were representative of their respective generations, their interpretations of *Entwicklungsmechanik* do reveal distinct underlying concerns connected with their experiences, which reflect the concerns of different times. Haeckel associated Roux's term *Entwicklungsmechanik* with His's somewhat different program of mechanical embryology, carrying his old disagreements with His from the mid-1870s into the 1890s and enlarging his arguments to encompass anyone who criticized his own evolutionary approach to morphology. Driesch and Dreyer picked up on the same dichotomy from the other end, rejecting the evolutionary-morphological approach as at best passé, and turning wholeheartedly to modeling their study of the organism along the lines of physics and chemistry. Driesch soon found limits to that study and became a vitalist, but even so, for him, historical, evolutionary explanations of the organism's form held little scientific value.[44]

What is so interesting about the writings of both Roux and his

42. Roux, "Für unser Programm," p. 43.

43. Driesch, *Biologie als selbständige Grundwissenschaft*, pp. 56–58; on the differences between Driesch's and Roux's Kantianism, see Mocek, *Wilhelm Roux–Hans Driesch*, pp. 92–94.

44. In the mature statement of his vitalism, Driesch wrote, "For there is no sound and rational principle underlying phylogeny; there is mere fantastic speculation. How could it be otherwise where all is based upon suppositions which themselves have no leading principle at present?" (Driesch, *Science and Philosophy* [1929], p. 163).

contemporary antagonist Hertwig is their refusal to abide by this sharp dichotomy between historical and mechanical-causal explanations. Despite the hostility and depth of the disagreements between Roux and Hertwig, both operated in a framework in which certain aspects of evolutionary morphology were still embedded.[45] To be sure, the terrain of their discussion was largely shifted from the postgastrula stages of development back to the properties of the cell—a monumental shift the significance of which has recently been highlighted by Paul Weindling (especially for Hertwig).[46] But this newer object of investigation still allowed for continuity with older assumptions. Although both men rejected the biogenetic law as an adequate account of the causes of ontogeny, for neither of them was the understanding of individual development entirely divorced from an understanding of descent. For both, the causes of development had to include at some level the factors that were transmitted from parent to offspring; because of this, the organism's ancestral history could not be completely ignored as inconsequential or merely descriptive. While it was possible to study individual development by itself, this was not yet for them a wholly independent field of investigation. The connection between descent and the development of the individual, now usually mediated by the cell, persisted.

Roux argued, for example, that the study of ontogenetic causes would ultimately yield information about phylogenetic ones, positing a "phylogenetic *Entwicklungsmechanik*" that would eventually follow upon the more accessible "ontogenetic *Entwicklungsmechanik*."[47] Similarly, the entire flow of Hertwig's multifaceted research, from his work on the middle germ layer in the late 1870s and early 1880s to his epoch-making findings on fertilization and cell division in the 1880s and 1890s, was motivated by fundamental questions raised jointly by Darwinism and cell theory: the nature of inheritance, the role of external versus internal factors in determining heritable forms and form change, and the relationship between individual development and phylogenetic history. Although Hertwig came to reject the biogenetic law and, by the early twentieth century, to express

45. This framework was shared by at least one other respondent to Roux from his own generation, namely, Otto Bütschli (1848–1920); for the dialogue between Bütschli and Roux, see O. Bütschli, "Betrachtungen über Hypothese und Beobachtung" (1896), p. 13, "Bemerkungen über die Anwendbarkeit des Experiments" (1897), p. 591; Roux, "Für unser Programm," p. 250; on Bütschli's evolutionary framework and its relation to his protozoological work, see N. X. Jacobs, "From Unit to Unity" (1989).

46. Weindling, *Darwinism and Social Darwinism*.

47. Roux, "Ziele und Wege," pp. 418–29.

larger doubts about monophyletic descent and natural selection, there seems little doubt that Darwin's theory, and especially Haeckel's version of it, remained a touchstone of his intellectual development.[48]

A final and related point about these discussions surrounding *Entwicklungsmechanik* is obvious but important: although Roux first began advertising his program in the mid-1880s, it was not until the early 1890s and after that it caught the attention of other researchers. Only then did mechanical causes and experiment come to be the objects of widespread rhetorical pronouncements, and only then did the term *Entwicklungsmechanik* come to be a synonym for modernity among a group of younger enthusiasts. It is important to recognize this timing, because beginning in the 1890s, the perceived opposition between causal-mechanical and descriptive-historical-morphological approaches to form was often projected back into the previous decades and linked up to earlier debates, often by people who, like Driesch and Dreyer, had not lived through them as researchers. However, as is evident even in the writings of Roux and Hertwig, this opposition crystallized out of a much more complicated relationship. The rhetoric of dichotomy of the 1890s has helped to obscure the more complex and continuous interactions between these research approaches in the earlier period.

## THE PLACE OF ENTWICKLUNGSMECHANIK IN ANATOMY

The question remains, however, whether Roux's causal-mechanical, experimental approach to embryology gained ground relative to other orientations within anatomy, either by attracting adherents or by being promoted at higher administrative levels. Here it is useful to recall the emphasis on the practical interests of medical students that held sway among most university medical faculties in the 1880s, which resulted in the devaluation of comparative anatomical research and made it difficult for Gegenbaur's school to gain a foothold outside of its bases at Jena and Heidelberg. This emphasis led faculties to place great weight on the anatomist's teaching skills, and to value especially those anatomy teachers who directed their efforts toward training future medical practitioners (often at the expense of developing research schools). Indeed, innovative research itself appears to have been a less important consideration in recommendations at the faculty

---

48. On Hertwig's research career, see Weindling, *Darwinism and Social Darwinism.*

level than teaching promise, where "teaching" was understood to include not only lecturing, but even more important, conducting laboratory exercises in dissection and topographic anatomy. In the numerous professorial appointments of the early 1880s, just when Roux was beginning to develop his program, positions tended to go to men who would make their reputations primarily as teachers, and who knew how to make anatomy speak to clinical and pathological issues. Thus two of the most important anatomical positions in Germany, at Berlin and Halle, went to Wilhelm Waldeyer and Karl Eberth, men who had previously been employed as pathological anatomists; Waldeyer's 1883 appointment was especially significant because he rapidly became a close, informal advisor to Friedrich Althoff, the new Prussian overseer of university appointments, and in this capacity Waldeyer was in a position to lobby for appointments of more men like himself.

At Göttingen, where another major position opened up in 1885, the importance of practical (i.e. laboratory) teaching was also evident. There the faculty was split between Wilhelm Henke, a topographical anatomist, and Friedrich Merkel (1845–1918), a histologist, both known for their excellent practical teaching. The main concern of the faculty was not to gain a spectacularly innovative researcher, but rather to acquire someone with sufficient expertise in both microscopic and macroscopic anatomy to cover the two areas proficiently in the teaching laboratory. In the event, the deciding vote appears to have been cast by the deceased professor of anatomy, Jakob Henle, who, the university's *Kurator* informed the cultural minister, had often expressed his view that Merkel (Henle's son-in-law) would be "his most suitable successor."[49] Once at Göttingen, Merkel did indeed become a beloved teacher; much like Waldeyer, he never developed a research school, apparently preferring to put his time into a variety of different research projects, critical overviews, and a textbook that connected anatomy to practical clinical areas.[50]

49. Medical faculty to U. Göttingen Curator Warnstedt, 9 June 1885; Warnstedt to Minister Gossler, 10 June 1885, UA Göttingen, Personalakte Merkel.

50. E. Kallius, "Friedrich Merkel" (1921); Henle, who was a pioneer in anatomical teaching methods, and who, near the end of his life, reasserted the importance of teaching aimed at future practitioners, appears to have been a marked influence on a number among this group of anatomists: besides Merkel, Waldeyer, Albert von Brunn (1849–1895), and August Froriep (1849–1917) had studied or worked under Henle, and some of these men explicitly took his approach to teaching as a model (Sobotta, "Zum Andenken an Wilhelm von Waldeyer-Hartz"; W. Waldeyer, "Albert von Brunn" [1896]; M. Heidenhain, "August Froriep" [1917–18]); see also the discussion on Waldeyer in chap. 7, pp. 226–27, 230–31.

Given the situation, it is little wonder that when Wilhelm Roux announced his "anatomical branch of the future" in 1889, he, too, was careful to emphasize its practical possibilities. Although he devoted most of his inaugural address at Innsbruck to situating *Entwicklungsmechanik* within the traditional and newer aims of anatomical research, explaining the need to understand causes and his new experimental method for doing so, he also asserted the connections of *Entwicklungsmechanik* to pathology and clinical medicine. He noted that much of the existing causal knowledge of form derived from research conducted in these fields, and that in the future, his orientation could be useful not only to these but also to fields such as orthopedics as well. Whereas researchers in these other disciplines dealt with the human form in a pathological state, it was up to the anatomist to determine the laws and causes of form under normal conditions, from which the pathological deviated. At the end he called for those schooled in the exact mathematical methods of physiology to take up his program, but reasserted that his approach was properly an anatomical one.

It was important for Roux to assert these professional links to medical practice, while at the same time justifying why anatomists should be the ones to take up *Entwicklungsmechanik*. For all his rhetoric of novelty and change, Roux was after scientific goals that distinctly resembled those of the evolutionary morphologists; in fact, he took *Entwicklungsmechanik* to be the next step in the perfection of morphology.[51] But it was dangerous for an anatomist to ally himself too closely to evolutionary morphology in the late 1880s. The comparative anatomy side of the program no longer had much of a place in anatomy, and the criticisms aimed at its lack of medical connections could easily be directed toward those researchers hoping to understand evolution by means of experimental embryology as well. If he wanted to gain support for his new morphological orientation, Roux had to take care to avoid the problems of the comparative anatomists. One way that he did this in his speech was to play down his interest in evolutionary comparative anatomy (which he later reasserted only on informal occasions)[52] and stress the appropriateness of his approach to anatomy as a basic medical science.

Despite these claims to both intellectual innovation and medical

51. Roux, "Entwicklungsmechanik," pp. 27–28.
52. See, e.g., his comments at the meeting of the Anatomical Society in 1902 (*Anatomischer Anzeiger,* supp. 16 [1902]; 162); and Roux to Max Fürbringer, 13 December 1893, SB, Fürbringer Nachlass; both of these express Roux's interest in morphology, which by this time was identified with evolution.

relevance, Roux's efforts to gain support for his program met remarkably little welcome in anatomy.[53] As Roux himself acknowledged repeatedly, if privately, he owed his first break to the powerful Prussian university administrator Friedrich Althoff. During eight years at Breslau's anatomy institute under the *Ordinarius* Carl Hasse, he had moved up from assistant to *Privatdozent* in 1880 and *ausserordentliche* professor in 1886, events that were controlled mainly from within the university. Just after Christmas in 1887, though, he had occasion to thank Althoff for his "lovely Christmas present," the funding of a separate small institute for *Entwicklungsmechanik* at Breslau. This had come in spite of skepticism from Althoff's frequent consultants in Berlin, the anatomists Wilhelm Waldeyer, whom Roux described as "not in any way for mechanics and causality," and Oscar Hertwig, of whom Roux wrote that he "believes he has accomplished excellent things in this field, whereas in my view this is exactly his Achilles' heel."[54] (One might add that Roux had not shone in the opinion of Merkel either, who in 1885 had declined to recommend Roux for the anatomy professorship at Greifswald; he characterized Roux's work to Althoff as "occupied substantially in philosophically illuminating the results of other scholars.")[55]

Despite his gratitude for Althoff's support, Roux could not remain content for long with his still-subordinate status as *ausserordentliche* professor. The partial independence that came with the new institute led to conflicts with Hasse, who viewed *Entwicklungsmechanik* and embryology as anatomical subdisciplines that should continue to be subject to the supervision and control of the professor of anatomy. By March 1888 Roux was pressing Althoff to create a second full professorship in anatomy at Breslau, to specialize in embryology and *Entwicklungsmechanik* and to be filled, of course, by Roux himself.[56] Althoff had been pushed hard enough, however, and replied curtly that he didn't have time for this sort of conflict, advising Roux to find ways to get along with Hasse.[57]

53. The lack of interest shown by his anatomist colleagues in Germany contrasts sharply with the historical assessment that Roux's program "gained almost immediate support" (see Allen, *Life Science,* p. 34).

54. Roux to Althoff, 28 December 1887, GStA Merseburg, Rep. 92 Althoff B, Nr. 156, Bd. 2, Bl. 12.

55. Merkel to Althoff, 25 July 1885, GStA Merseburg, Rep. 92 Althoff B, Nr. 130, Bd. 1, Bl. 133.

56. Roux to Althoff, 21 March 1888, 26 August 1888, GStA Merseburg, Rep 92 Althoff B, Nr. 156, Bd. 2, Bll. 14–15, 20–23.

57. A copy of Althoff's reply to Roux's complaints, dated 28 August 1888, is on Bl. 23 of Roux's letter, 26 August 1888.

By the summer of 1889 Roux had found a way out with his appointment to Innsbruck, Austria, as professor of anatomy. In writing to Althoff about the probable Innsbruck offer, he expressed regret that his career strategy hadn't succeeded in gaining him a professorship in his own country: "I had hoped on the basis of my work, which also was planned to encompass all of anatomy and to serve surgical and pathological interests, to make myself eligible for advancement." As long as he held a position that covered only a part of anatomy, however, he felt (probably accurately) that he would be stigmatized as a specialist, and that he would be in a better position to gain a German *Ordinariat* from an equivalent position outside the country.[58] He proved to be right, for in 1895, Althoff engineered Roux's return to Prussia as anatomy professor at Halle, arguing to his minister that Roux was "one of the most outstanding anatomists of the present," whose work was "pathbreaking" and "groundlaying."[59]

Althoff's enthusiasm for Roux's new orientation has been interpreted as part of a broader program on Althoff's part to develop experimental biology, and the same has been held for his appointment of Oscar Hertwig in Berlin.[60] Although it is true that Althoff had been in communication with both men long before their Prussian professorial appointments, and that Hertwig's installation came without support from Berlin's medical faculty, it is difficult to find direct evidence that Althoff was interested in promoting experimental biology in particular, rather than innovative research in general. In Halle, to be sure, Althoff bypassed the rank-ordered list of candidates sent up by the faculty; they had placed Roux behind Zurich's Philipp Stöhr, regarding them equal in scientific merit but stressing Stöhr's greater teaching talents.[61] This suggests that whereas the faculty was somewhat more concerned with the anatomist's teaching ability, Althoff himself may have been more inclined to value his scientific research—but this does not necessarily imply he specifically supported an experimental approach. For Halle, in fact, Althoff's own first choice appears not to have been Roux, but someone else entirely—Walther Flemming, the anatomist at Kiel, who was a great contributor

58. Roux to Althoff, 14 August 1889, in GStA Merseburg, Rep 92 Althoff B, Nr. 156, Bd. 2, Bll. 27–30.

59. Roux was the second choice of the medical faculty at Halle, but Althoff recommended him without even mentioning the faculty's first candidate (GStA Merseburg, Halle Med. Fak., Bd. 7 [April 1894–September 1900], Bll. 115–20).

60. See Weindling, *Darwinism and Social Darwinism*, p. 213; G. Ziernstein, "Friedrich Althoffs Wirken" (1991).

61. UA Halle, Nr. 220, Dekanatsbuch 12 July 1895–12 January 1896, item 33, Medizinische Fakultät to Minister Bosse, 27 July 1895.

to the theory of cell division but not an "experimental biologist" in the manner of either Roux or O. Hertwig.[62]

Even if it had been Althoff's intention to promote an experimental approach in anatomy, this criterion had to be balanced against the faculties' criteria of teaching ability and experience running a laboratory. Roux, at least, was clearly aware of the importance of both to his career prospects: in the Halle negotiations, he had reminded Althoff of his desire for sole control of "human anatomy (systematic and topographic anatomy). . . . Systematic anatomy and the dissection exercises are my intellectual life-element and my pleasure."[63] The criterion of experience in running a laboratory also effectively limited Althoff's choices. The usual way to gain such experience was to hold a prosectorship, a salaried position devoted to running the laboratory exercises and the practical business associated with them. This post was generally sufficiently well paid that someone could support a family on it, and if one was not successful in moving on to a full professorship, it was possible to retain it for an entire career; even most of those who succeeded in moving up from this post spent a good number of years in it.[64] In the late 1880s and through the 1890s, when Roux's program was becoming well known and beginning to gain some adherents, the pool of candidates in these ranks comprised his contemporaries in prosectorships or those holding professorships at smaller universities, born between about 1846 and 1852.

For an administrator seeking to promote the experimenters from this pool, the pickings were modest and made even more so by various disqualifying factors. Other than Roux and O. Hertwig, few experimentally inclined anatomists were eligible for the positions he had to offer. Gustav Born, as a Jew, had no hope of a full professorship at a smaller provincial university, and as he himself recognized, "one *has* to start with the little ones!"[65] Strasser appears not to have come

---

62. Althoff had asked Flemming if he was interested in the position, but since Flemming was not even on the faculty's list of candidates, he declined to be considered (GStA Merseburg, Halle Med. Fak., Bd. 7, Bll. 127–30).

63. Roux to Althoff, 31 July 1895, GStA Merseburg, Halle Med. Fak., Bd. 7, Bl. 131.

64. The status of prosector had risen considerably over time from that of a dissecting assistant, and by the end of the eighteenth century had already been used as a stepping-stone position for a number of professors of anatomy (see H. Schierhorn, "Der Prosektor und seine Stellung" [1985]).

65. Born to M. Fürbringer, 3 July 1888, SB, Fürbringer Nachlass, A. 1. 280; when Waldeyer recommended Born as the best young anatomist for the Halle position in 1894, Althoff penciled "*Jude*" (Jew) in the margin (Waldeyer to Althoff, 8 January 1894, GStA Merseburg, Rep. 92 Althoff B, Nr. 192, vol. 1, Bll. 113–14).

under consideration, perhaps because he was Swiss and not a German national, perhaps because he was considered neither an especially gifted teacher nor a very broad researcher.[66] Moritz Nußbaum (1850–1915), who coined the term "experimental morphology," and who conducted outstanding experimental research on sex determination, regeneration, and inheritance while moving up the ranks at Bonn to the level of *ausserordentliche* professor, was known as a poor teacher who regarded lecturing and administration as "unpleasant interruptions of his new experiments."[67] Leo Gerlach (1851–1918), another experimental researcher in anatomy, was well placed at Erlangen, where he first worked under and then succeeded his father as professor of anatomy.

Nor did the choices widen very rapidly in the 1890s and after, even if one looked farther down the ranks: when Roux was considering candidates for *ausserordentliche* professors to assist him soon after his arrival in Halle, he complained to Althoff that "the number of causally striving anatomists is currently still very small; most causally researching biologists are [physiologists?—illegible] and zoologists." In 1897 Roux was already worried that "more foreigners—Americans and Italians—than Germans take an interest in *Entwickelungsmechanik*." By 1905 the complaint must have begun to sound familiar: the Americans, Roux wrote Althoff, had numerous representatives of this orientation, some in zoology professorships, some in positions specially designated for this approach; in Germany, there was more interest in *Entwicklungsmechanik* among zoologists than among anatomists, where "'descriptive' anatomists still reign."[68]

Roux's lament was justified. In the last two decades of the century, the priority placed on teaching future medical practitioners and on the need for administrative laboratory experience most benefited those members of the anatomical community who had made their reputations as teachers and researchers in descriptive anatomy, rather than

66. Wegelin, "Hans Strasser."

67. R. Bonnet, "Moritz Nußbaum" (1915–16), p. 491; when the anatomy *Ordinariat* at Bonn came open in 1907, Nußbaum was placed out of the running by his colleagues because of his age and his lack of enthusiasm for teaching. It seems probable that his receipt in early 1907 of a "personal *Ordinariat*" for biology and a distinct "biological laboratory" within the anatomical institute were meant as a consolation prize in recognition of his research achievements; on the 1907 situation, see Schultze to Fürbringer, 8 November 1906, and Fürbringer to Schultze, "Gutachten" [n.d.], Section D.2.k., "Bonner Berufungsangelegenheit November 1906," items 204, 206, SB, Fürbringer Nachlass.

68. Roux to Althoff, 10 November 1895, 14 Sept. 1897, 12 August 1905, GStA Merseburg, Rep. 92 Althoff B, Nr. 156, Bd. 2, Bll. 50–52, 75–76, 89.

as especially innovative researchers. One final case is especially telling: in 1895 the anatomy professorship at Prussia's smallest university, Greifswald, came open. The faculty's list included Roux as well as his contemporaries the descriptive embryologist and histologist Robert Bonnet (1851–1921) and the Gegenbaur student Georg Ruge. As Waldeyer wrote to Althoff, these three were definitely the best of the pool; the hardest part was to choose from among them. "Roux and Ruge I would hold to be the more important scientifically; Bonnet is certainly, so far as I know, the more outstanding teacher." After this comment, Waldeyer's next statement is startling: "I would recommend in the first place the appointment of Ruge," because he wanted to see "an outstanding comparative anatomist in a Prussian chair"—a reflection, perhaps, of Waldeyer's recognition that comparative anatomy had been squeezed almost entirely out of anatomy.[69] Despite the recommendation, however, it was Bonnet, the descriptive researcher and outstanding teacher, who won the post.

This anecdote sums up the institutional mission of anatomy in its late nineteenth-century German university context and suggests its consequences. Just as Gegenbaur's school of comparative anatomists found little room for expansion in this climate, neither did experimental morphology succeed in becoming a prominent orientation in the population of anatomy professors. Roux's hopes that *Entwicklungsmechanik* would indeed become an "anatomical science of the future" were not destined to be borne out by the early twentieth century.

69. Waldeyer to Althoff, 7 July 1895, GStA Merseburg, Rep. 92 Althoff B, Nr. 192, vol. 1, Bll. 119–20.

## CHAPTER TEN

## Morphology, Biology, and the Zoological Professoriate

I f anatomy's situation within medicine deflected rewards away from both evolutionary morphology and Roux's *Entwicklungsmechanik* and toward descriptive studies (especially at the cellular level), one might expect the situation to be different in zoology, where both evolution and the proximate causes of form were of more central interest. And so it was. Two particular points of contrast with anatomy should be noted at the outset. Whereas relatively few anatomists took up both phylogenetic questions (considered either descriptive or speculative by most) and causal ones, in zoology this mix was common; after all, systematics, physiology, and the animal's relation to its physical environment were all within the recognized purview of zoologists, as were broader questions about the nature and mechanisms of evolution. Another difference lay in who carried out such research. Among anatomists, the combination of evolutionary and causal research interests surfaced primarily among investigators of Roux's generation, and it appears to have found little support among older institute directors. In the zoology community, however, questions concerning the mechanical causes of form, especially under the rubric of biological causes, had been viewed as a legitimate aspect of scientific zoological research since at least the 1860s, and were frequently tied to evolutionary arguments.[1] Thus although this sort

1. Throughout the rest of this chapter, I will use the term "biological" in the late nineteenth-century sense, meaning "concerning the organism in relation to its environment," and dispense with the quotation marks.

of research was carried out by numerous zoological contemporaries of Roux, they were following an existing course that was already being pursued by some members of the previous generation, rather than charting a new one.

In order to investigate zoologists' attitudes toward evolutionary questions and causal-mechanical ones, this chapter begins by surveying a number of leading zoologists who ran vigorous research institutes between the late 1870s and early 1890s. Mainly members of Haeckel's own cohort, Carl Semper at Würzburg, Rudolf Leuckart at Leipzig, August Weismann at Freiburg, and Ernst Ehlers at Göttingen had established their institutes during the upswing of evolutionary research in the late 1860s and 1870s and continued to produce both research and successful researchers up through the turn of the century. Many of their long-term research associates were students from the cohort born around 1850. A brief tour of these zoologists' institutes can give us an idea of the broader front of zoological investigation in which questions about animal form were being pursued in these decades.

Even as this research was going on, the discipline was undergoing a demographic upheaval. Between 1880 and 1898, fourteen zoologists died or retired; the resulting openings and shifts among professors already within the system, as well as one expansion position, yielded a total of twenty-six changes of position, affecting sixteen of the twenty German universities. Despite this massive turnover, however, the perception was one of scarcity: because of the enormous growth in numbers of advanced researchers during these years, there were always many more aspirants than openings.

The second part of the chapter therefore turns to the question of rewards. If only the lucky and talented few were judged suitable to move up into professorships, what went into the judgment of "suitability" by university faculties and state administrations? More specifically for understanding the direction taken by the study of animal form, we may ask whether a candidate's adherence to Darwinism was a factor in these decisions, whether a causal or experimental program was especially valued, and what the professional profile of the discipline looked like by 1898, when this wave of turnovers was completed. The chapter concludes by considering the consequences of these intellectual and institutional developments for understanding the "revolt from morphology" said to have taken place in biology at the turn of the century.

## EVOLUTION AND THE CAUSES OF FORM: RESEARCH IN ZOOLOGY

For historians conditioned by the writings of Haeckel, Driesch, or later writers to believe that causal and evolutionary research into animal form and life were mutually contradictory, or at least required opposed mental frameworks, the activities of Germany's major zoological research institutes in the 1880s and early 1890s might come as a revelation. Investigators at these institutes clearly saw themselves as trying to discover the mechanisms of form change within an evolutionary framework, but the framework was one in which little was settled about the causal mechanisms of evolution. A fair amount of experimental research was being undertaken in these institutes as well, although without the fanfare that would come later. The opposition later expressed between a rigorous causal, experimental approach that rejected evolutionary history and a morphological approach that was historical, descriptive, and speculative simply did not exist in these institutes. None of these men had ever been satisfied with stopping where Haeckel did, with the statement that adaptation and inheritance were the sufficient mechanical causes of form, but that does not mean that they rejected an evolutionary framework. Far from it: their research was very much within that framework. It just was not within Haeckel's particular rendition of it. Instead, it should be viewed as the evolutionary extension of scientific zoology.

Perhaps the clearest example of this is Carl Semper's institute at Würzburg, which illustrates well the ways that causal biological questions and evolutionary-phylogenetic ones could be pursued side by side. Semper, who in the 1860s had initially been an enthusiast of Haeckel's approach, had become disenchanted with both Haeckel's dogmatic style of argumentation and his phylogenetic conclusions by the mid-1870s (although he shared the latter's penchant for polemics). Starting from comparisons of the urogenital systems of the shark and various segmented worms or annelids, Semper developed a theory that the vertebrate phylum originated from the annelids.[2] Although the theory conflicted both with the phylogenies cast by Haeckel and Gegenbaur and with the way they employed the comparative method, this work fell squarely into the sort of comparative and embryological studies done in the service of phylogeny that epitomized the evolutionary morphology of the period.

Between the mid-1870s and the early 1880s, when he was trying to gain further support for his theory, Semper was also supervising

2. C. Semper, *Die Verwandtschaftsbeziehungen* (1875).

quite a number of advanced students. Some of them, such as Johann Wilhelm Spengel (1852–1921) and Maximilian Braun (1850–1930), investigated the urogenital systems of different lower vertebrates in up-to-date morphological style, applying the new serial section techniques that Haeckel so deplored. These men were clearly pursuing their research within the framework of Semper's phylogenetic theory. At the same time, however, other researchers in the institute made use of similar new techniques to approach the study of animal form in quite a different way. At the annual meeting of German naturalists and physicians in 1879, three young Würzburg researchers presented the results of their investigations on regeneration, arguing over the extent to which regeneration was similar to embryological development. These investigations clearly involved manipulation—hacking off the relevant tissue or structure—but the experimental aspects were not emphasized in the brief reports of the proceedings. Nor did the researchers view their work as a challenge to the legitimacy of phylogenetic studies. The intellectual framework within which these men were working oriented them instead toward the significance of regeneration for illuminating phylogenetic recapitulation and the germ-layer doctrine.[3] Thus although Semper himself did not conduct such research, he certainly was not close-minded about the appropriateness of experimental methods for answering morphological questions, or about supporting researchers with such interests.

Semper also took an interest in the relationship between the organism and its physical environment. As early as 1868, he had argued for the need to investigate through experiment "the influence of light and heat, the degree of moisture, nutrition etc. on the living animal,"

---

3. P. Fraisse, "Über die Regeneration von Organen" (1879); M. Flesch, "Zeichnungen über die Entstehung des Schwanzendes" (1879); J. Carriere, Über die Regeneration" (1879); in part because few records from the University of Würzburg survived the Second World War, little information is available on these researchers. Fraisse and Carriere (1854–93) both went on to pursue their interests at other zoological institutes into the mid-1890s. Fraisse worked as *Extraordinarius* in Leuckart's institute at Leipzig from 1881 to 1896, then disappears from available records; Carriere held a similar position at Strassburg, first under Oscar Schmidt, later under Alexander Goette, between 1885 and his early death from diphtheria in 1893. On Fraisse, see Wunderlich, *Leuckart,* p. 54; and F. Churchill, "Regeneration, 1885–1901" (1991), p. 116–19. Churchill's essay also lays out succinctly the intellectual orientation of the regeneration research of the early 1880s. A notice of Carriere's death from diphtheria appeared in 1893 (*Zoologischer Anzeiger* 16 [14 August 1893]: 324). Flesch (1852–?) was employed as prosector and *Privatdozent* for anatomy under Kölliker when he presented his research; he later became professor of anatomy at the veterinary school in Bern and eventually left academia for medical practice (see Feser, "Das anatomische Institut," p. 66).

and he renewed his call in a lecture series in 1877 on the natural conditions of existence of animals.[4] The lack of appropriate facilities in his institute, however, hampered him from developing this line of research. Only in 1889 was a new institute building completed, with the indoor aquaria, breeding rooms, and outdoor breeding ponds necessary for raising the experimental animals under controlled conditions, but by that time Semper's health was poor. He retired a few years later, leaving the impressive new facilities to the use of his successor Theodor Boveri, who would become famous as an experimenter.[5]

Elsewhere, the sort of biological research Semper advocated was actually being carried out. Probably the largest institute to promote such research between the late 1870s and the early 1890s was that of the intrepid Rudolph Leuckart. An intellectual generation older than Haeckel's cohort and in many ways the spiritual leader of scientific zoology, Leuckart also had a large, varied, and ever growing group of advanced researchers working in his institute at Leipzig up to his death in 1898. His own research in the last twenty years of his life was confined almost exclusively to the life cycles of parasitic worms, but he did not shy away from discussing the major issues of the day with his students and colleagues. At the first official meeting of the German Zoological Society in 1891, for example, Leuckart opened the proceedings with a challenge to his audience: "People have even maintained that the phenomena of inheritance and adaptive variation by themselves suffice as a causal basis for the theory of descent. As if inheritance and adaptation were simple, mechanically operating forces and not the results of processes that themselves require a causal explanation! Only when it has been possible to trace the processes back to their causes, only then will there appear the possibility of a disposition in the sense of causality. Until then we are operating with these factors just as the old physiologists did with their *Lebenskraft.*"[6] Nearly forty years after his first efforts to sidestep the *Lebenskraft* by focusing on vital processes in terms of mechanical causes, Leuckart found himself pressing home the same message, now in an evolutionary context, to a community in which nearly all the members were now his juniors.[7]

---

4. Semper, *Reisen im Archipel der Philippinen,* p. 229; the lecture series, held at the Lowell Institute near Boston, was published as *Die natürlichen Existenzbedingungen der Tiere* (1880) and in English as *The Natural Conditions of Existence as They Affect Animal Life* (1881).

5. A. Schuberg, "Das neue zoologisch-zootomisch Institut" (1895).

6. R. Leuckart, Opening address (1891), p. 10.

7. Bergmann and Leuckart, *Anatomisch-physiologisch Uebersicht.*

Judging from the titles of the dissertations completed under Leuckart during the 1880s and 1890s, he continued to assign the sort of small-scale anatomical and embryological studies that had been typical of zoology dissertations for decades, although it is clear from those published in the *Zeitschrift für wissenschaftliche Zoologie* that Leuckart's students were equipped with the most up-to-date staining and slicing techniques for conducting such research.[8] The advanced researchers connected with the institute show a more varied set of research interests, however, ones that would take them into numerous biological problem areas. Carl Chun, who had completed his dissertation under Leuckart in 1874 and then worked as Leuckart's assistant from 1878 to 1883, produced research in this period that demonstrated the combination of technical skills and biological questions that were Leuckart's trademark. In addition to demonstrating fine microscopic technique, Chun's work on the ctenophores and siphonophores, two groups of jellyfish that occupied most of his attention in the 1880s, showed his concern with their peculiar life cycles, structure, function, and relationship to their marine surroundings, especially in relation to the depths at which they lived. By the end of the decade, his research had led him to important new ideas about how deep-water organisms survived, a project of great biological significance.[9] And only a few years after he received the professorship of zoology at Königsberg in 1883, Chun was requesting funds to support experimental work on symbiosis (another "biological" problem) being conducted by his assistant Karl Brandt (1854–1931).[10] These projects did not prevent Chun from pursuing a more traditional problem area, however: in the 1890s he took on the classification of the coelenterates for the massive project known as *Bronns Klassen und Ordnungen*.[11]

Heinrich Simroth (1851–1917), a *Realschule* teacher who had studied under Leuckart and who lectured at Leipzig as a *Privatdozent* beginning in 1889, pursued a similar combination of biological, natu-

---

8. A list of Leuckart's Leipzig dissertators and their dissertation titles is given in Wunderlich, *Leuckart,* pp. 41–49; for examples of Leipzig dissertations published in the *Zeitschrift,* see K. R. Krieger, "Über das Centralnervensystem" (1880); R. Rössler, "Beiträge zur Anatomie der Phalangiden" (1881–82); P. Schiemenz, "Über das Herkommen des Futtersaftes" (1883).

9. On Chun, see W. Pfeffer, "Carl Chun" (1914); E. L. Mills, "Alexander Agassiz, Carl Chun" (1980).

10. Chun to Kultusministerium, 20 February 1886. SBPK, Lc 1898 (29): C. Chun.

11. C. Chun, *Coelenterata* (1889–1916); it should be noted, however, that Chun was too distracted by other projects to complete the classification project before he died (Pfeffer, "Chun," p. 13).

ral-historical, systematic, and anatomical research on a different group of organisms, the mollusks, becoming one of Germany's leading authorities on land snails. In pursuit of a complete understanding of this group, he, too, employed research techniques ranging from observation of the organisms in their environment to fine microscopic anatomy to experiments. Nor did he neglect evolutionary issues, arguing that the leeches (*Nacktschnecken*) were the product of convergent evolution and not a natural phyletic group. Here again, biological issues about the relationship between organism and environment came into play.[12] Rounding out the group of longtime researchers at Leipzig were Paul Fraisse (1851–1909), who taught as an *Extraordinarius* from 1881 to 1896, and who made his reputation with experiments on regeneration; and William Marshall (1845–1907), best known for his popular natural history writings on insects and marine life, which reflected both biological interests and phylogenetic ones.[13] Virtually all of these long-term colleagues at Leuckart's institute pursued biological lines of research, sometimes connected to evolutionary and classificatory considerations, sometimes divorced from them.

Although it was far smaller and supported few advanced researchers, Freiburg's zoology institute also turned out an important amount of research on biological questions concerned with form, most of it coming from the hand of the institute's director August Weismann, the member of his generation whose evolutionary interests had always been most actively directed toward the biological problems of generation and adaptation. Weismann's own investigations of the early 1870s on seasonal dimorphism in butterflies have already been alluded to (p. 180); simultaneously he was also pursuing a related line of research on seasonal differences in generation in water fleas. Both of these studies were concerned in part with the direct and indirect influences of the environment (especially temperature and food) on the life cycle of the organisms involved, which drew him into a complex knot of problems on sexual and asexual reproduction, adaptation, and development, all pursued from an evolutionary perspective.

Not surprisingly, the researchers coming out of Weismann's insti-

12. P. Ehrmann, "Heinrich Simroth" (1916–17); Ehrmann mentions several other biological problems Simroth wrote about as well.

13. On Fraisse and Marshall at Leipzig, see Wunderlich, *Leuckart*, p. 54; information on Marshall also appears in G. Wagenitz, *Göttinger Biologen 1737–1945* (1988), p. 118; for an example of Marshall's interests, see *Tiefsee und ihr Leben* (1888).

tute tended to be preoccupied with similar biological questions. As early as 1877, his former student (and later avid opponent) Theodor Eimer (1843–98) showed his debt to his teacher both theoretically and methodologically. His presentation at the fiftieth meeting of German naturalists and physicians in Munich, on the color variations in lizards and their correlation to different environments, was the starting point of a long train of research on the direct influence of the environment on organisms, which would eventually lead Eimer to a complete commitment to the heritability of acquired characteristics and his related theory of orthogenetic evolution.[14] Other researchers working in Weismann's institute, such as his brother-in-law and second in command August Gruber (1853–1938), and Marie von Chauvin (?–?), one of the few women zoological investigators in this period, developed their own lines of experiment within this broad program.[15]

Weismann's own interests in the relationship between evolution, adaptation, and generation led him through detailed empirical studies to develop his theory of the continuity of the germ plasm, first presented in nascent form in 1883 and then further elaborated over the next decade. This theory held that each organism possessed two different kinds of cells: body, or somatic, cells—which, during development, divided and differentiated into the body's tissues and organs—and germinal cells, which contained material that was transmitted from parents to their offspring. Weismann argued that these latter cells, which were the sole transmitters of parental characteristics, were

14. T. Eimer, "Neue Beobachtungen über das Variiren" (1877), pp. 179–82; Eimer also gave another talk at the same meeting, "Ueber künstliche Teilbarkeit," concerning experiments he had been conducting over the previous several years on jellyfish, in which he had cut up their bell-shaped bodies in various ways to establish the function of their nervous systems and the relation of the nervous systems to their two germ layers. Although he did view the experiments as controls of his descriptive morphological work, like other experimenters of this time, he did not make much fuss about the methods; his later monograph *Die Medusen* (1878), which expanded on his results with the jellyfish, was dedicated to Weismann.

15. Gruber has been deeply overshadowed by Weismann; beyond what is revealed in his own published work, and various mentions in relation to Weismann, I have been unable to find anything more than the most rudimentary information about him. I have been unable to find any biographical information at all on Marie von Chauvin; she conducted a series of important experiments on the Mexican axolotl beginning in the early 1870s; the Royal Society *Catalogue of Scientific Papers* lists six works by her, the longest of which are "Über das Anpassungsvermögen" (1877) and "Über die Verwandlungsfähigkeit" (1885); see also A. Weismann, *Studien zur Descendenz-Theorie* (1875–76), pp. 231–33, which quotes directly from Chauvin's previously unpublished experiments.

segregated from the somatic cells and were protected from the effects of the environment; thus changes effected on the somatic cells through interaction of the environment could not influence the germinal cells, and the "inheritance of acquired characteristics" was impossible.

This theory was probably the most visible and controversial intellectual hot spot of evolutionary theory of the late 1880s and early 1890s; one might say that it was for that period what Haeckel's gastraea theory had been to the 1870s. Like Haeckel's theory, it made claims and raised questions both at the grand level of evolutionary change and at the level of microscopic observation (and, indeed, at the level of invisible submicroscopic events). Dovetailing with numerous other lines of research on fertilization, cytology, and germ-cell development, it sparked an uproar over the inheritance of acquired characteristics and the "all-sufficiency" of natural selection, while at the same time inviting further empirical research on the nature of the cell, its role in the transmission of characters, and the subsequent development of the organism. Like Haeckel's gastraea theory, it was attacked as speculative, reinforcing Weismann's reputation as a man who sometimes went beyond the appropriate limits of zoological theorizing.[16] And like Haeckel's theory, it helped frame a considerable amount of research among younger investigators in the decade or so following its appearance.

These examples indicate the existence of considerable biological research on the causes of form in German zoology beginning in the late 1870s. However, not all the leading German zoology institutes were as open to the biological questions of adaptation and inheritance as those of Semper, Leuckart, and Weismann. At Göttingen Ernst Ehlers, who had a reasonably wide circle of students, and who, through his editorship of the *Zeitschrift für wissenschaftliche Zoologie,* held considerable power, was always cool toward grand theorizing, evolutionary or otherwise, and not apparently very interested in supporting experimental research. Both his own work and that of his

16. The path of Weismann's own research and its conjunction with other lines of investigation have been studied in exquisite detail by F. Churchill, "August Weismann," "Weismann's Continuity," "Weismann, Hydromedusae," (1986), and "From Heredity Theory to *Vererbung.*" On Weismann's life and career more generally, see E. Gaupp, *August Weismann: Sein Leben* (1917), and K. Sander, ed., August Weismann (1834–1914) (1987–88). One might also note that like Haeckel's theory, which was widely disseminated through his *Anthropogenie,* aspects of Weismann's germ-plasm theory were picked up by people well outside of zoology (see D. Bellomy, "'Social Darwinism' Revisited" [1984], pp. 77–96).

students consisted largely in systematics and descriptive anatomy and embryology. Because these works combined systematics with embryology, it might be tempting to view them as "phylogenetic speculations," but in fact they contain little of a Haeckelian character. These works generally shunned pronouncements about large-scale relationships between phyla and used embryology for classificatory purposes in a manner quite typical of pre-Darwinian researchers, as just one source of comparison among many. Among the many researchers who worked with Ehlers, those who continued on in zoology, such as Johannes Spengel and Hubert Ludwig (1852–1913), tended to share his descriptive and classificatory orientation, sidestepping controversy (and broad-scale theoretical novelty) by becoming authorities on relatively narrow areas of the animal kingdom.

These thumbnail sketches, brief as they are, may be composed into a picture of the research fostered by the leaders of the zoological community in their institutes. This research was oriented around evolutionary questions as they pertained to form, function, inheritance, adaptation, and the systematic organization of animal life. Concern with the causes and evolutionary significance of animal form held a place in this research, but it was not slavishly devoted to a Haeckelian program: neither the construction of large-scale phylogenetic trees nor efforts to verify or extend the biogenetic law were among its central features. At the same time, it was not exclusively oriented toward passive observation of organisms but also incorporated experimental manipulation where that was seen as a fruitful approach.

Thus the fear expressed by Haeckel and Carl Theodor von Siebold around 1880, that zoology was being taken over by men adept at slicing up organisms and tracing their development but lacking interest in the whole organism in relation to its conditions of life, was not borne out. What one finds instead is something more like a stratification of problems according to the researcher's level of advancement. Embryological slice-and-stain problems appear to have been agreed on as eminently appropriate topics for zoological dissertations and even some *Habilitationsschriften* through the 1880s and 1890s; however, more advanced researchers who continued to develop their ideas over that time span often moved on to biological problems that were both more technically and intellectually complex and considered less descriptive, more causal. In historical accounts, this research has disappeared into the large gap created by the rhetorical dichotomy between experimental and evolutionary approaches; but at the time it was clearly where much of the action was.

APPOINTMENTS AND ORIENTATIONS IN ZOOLOGY

Just as the range of zoological research orientations differed from that in anatomy, with attendant differences in the shape and significance of morphological questions in the disciplines, so, too, did the values reflected in hiring decisions. As in anatomy, experience in running an institute was important, although less so at smaller and less prestigious universities. Also as in anatomy, teaching ability was highly valued; here, even more than in anatomy, the breadth or variety of the candidate's research was taken to be a mark of teaching competency. Zoology, however, was not a basic requirement for any large group of students in the way that anatomy was for medical students; zoologists did not have to provide the scaffolding upon which the rest of a physician's education was to be built. Zoologists did have their own constituency, of course—future teachers of natural history, especially at *Realschulen,* might take a doctorate in zoology, and other science teachers might study it as a secondary field—but students' practical concerns were not at the core of the zoology professor's role, as they were for anatomists. In this discipline, then, one might expect research to be more central in appointment decisions.

As well as looking out for any patterns in the broad thrust of recommendations and appointments for professorships, we may ask how the particular research orientation of Haeckelian evolutionary morphology fared in the appointment decisions of the 1880s and 1890s relative to the biological and experimental approaches to animal life and form. Were these addressed explicitly? Were individuals hired or rejected, considered or not considered on the basis of these or other research orientations? Here it is useful to divide our examination of the appointments into two parts, one covering the 1880s, when a cluster of scholars born in the years around 1850 received their first professorships, and the other covering the 1890s, when a second cohort, born around 1860, entered the professoriate.

The turnovers in the 1880s were catalyzed by a cluster of five deaths and retirements of Prussian zoologists born in the 1810s, who had held positions since the 1850s. Within the space of a few years, Halle, Königsberg, Breslau, Bonn, and Berlin—all the traditional Prussian universities except the smallest and least important University of Greifswald—as well as several non-Prussian universities, the most important of which was the Bavarian University of Munich, had an opportunity to invigorate their zoology programs with new blood. When the older Prussian zoologists had been appointed in the 1850s and early 1860s, university and state administrators had been concerned to find men who would devote themselves to the care and expansion of the already securely established museums; as a result, Prussian zoology had remained dominated by systematic zoologists, and the movement for scientific zoology had made virtually no inroads. Now, in the 1880s, both university faculties and higher administrators saw a chance, however belatedly, to change the situation. As the faculty at Halle wrote in its recommendation, zoology was no longer purely systematics, and it was imperative to hire "a man who has carried out sound work through original research in the areas of histology and comparative morphology as well as embryology, and who does not stand distant from physiology."[17] At Bonn, the university's *Curator* did not make such a direct statement, but indirectly displayed the same concern. He noted in his recommendations that the retired professor of zoology had neglected the biological and microscopic side of zoology—he had only a single, older microscope in the museum. But modern teaching and research of a scientific character required laboratory courses in animal dissection and microscopy. The *Curator* thought it would not be possible to hire a man of the younger generation without providing new funds to upgrade zoology in this direction.[18] The Prussian education minister Gossler expressed the same sentiment in discussing the replacement position at the University of Berlin in 1883: modern zoologists, he wrote, "have begun to regard the animal no longer as a purely formal given, but to analyze the structure and development of the individual parts and to connect a creature's performance [*Leistungsfahigkeit*] with its organization. Only in this way did zoology become a *Wissenschaft*. . . . But although the majority of the other German universities took part and collaborated in this most significant transformation, most of the Prussian universities (as a consequence of accidental circumstances) remained almost entirely untouched by it. . . . Scientific zoology in this sense has therefore had no representation in Berlin up until now."[19] Even at Munich, where scientific zoology had long had a home under Carl Theodor von Siebold, one of its pioneers, his retirement was taken as an occasion to strengthen it still further. The faculty search committee suggested that the only way to acquire a first-rate scientific zoologist was to separate the direction of the zoology institute from that of the museum (to be run by a systematist as conservator). In this way, the search committee hoped, the professor of zoology would not be burdened by the

17. Philosophische Fakultät Halle, Vorschläge (Abschrift ohne Datum), GStA Merseburg, Halle Phil. Fak., Bll. 104–5.

18. GStA Merseburg, Bonn Phil. Fak., Bll. 305–9.

19. [Gossler] to Königlichen Staats- und Finanzminister Herrn von Scholz, 20 December 1883, GStA Merseburg, Berlin Phil. Fak., Bll. 92–92RS.

many administrative headaches of running the museum and would be free to develop his more "scientific" efforts in the institute.[20]

The view of scientific zoology represented in these comments is quite broad. For a better sense of how these general ideas were translated into appointments, we must turn to the specific recommendation lists of the early 1880s and their outcomes. What is most striking about these lists is the extent to which the same two names came up over and over again at both Prussian and non-Prussian universities, a fact that suggests considerable consensus about who the best zoologists were. August Weismann was considered at the universities of Berlin, Breslau, Bonn, Strassburg, and Munich, while Franz Eilhard Schulze was considered at Breslau, Bonn, Munich, and Berlin (by the time the Strassburg position came open, Schulze had already accepted at Berlin). What made these men the "best"?

To begin with, they were assumed to be available and personally suitable. At both Berlin and Munich, which were vying for Germany's best zoologists in the same year (1883), the names of Rudolf Leuckart, Carl Claus, and Ernst Haeckel were also raised along with Weismann and Schulze as "the most outstanding representatives of modern zoology."[21] But Leuckart, well situated at Leipzig and already sixty-one years old, was considered too old to pick up and move; Claus was so well paid at Vienna that he seemed out of reach; and Haeckel was viewed by university decision makers in both the Protestant north and the Catholic south as unacceptable because of his extreme anticlerical views.[22] As seasoned professors, Weismann and Schulze had both already demonstrated their abilities as teachers and institute directors (as in anatomy, these were always serious considerations), Weismann at Freiburg and Schulze at the German-speaking Austrian university in Graz. More important, though, within the research tradition of scientific zoology, they had both shown considerable originality, versatility, and technical mastery. As the Munich recommendation noted, Wiesmann had begun his career with outstanding histological studies; he had moved on to elegant and original embryological inves-

20. Despite persistent and long-continuing efforts, this division was not even partially realized until 1898; the definitive split between museum and institute director did not take place until 1917 (see BHStA, file MK 19432, vol. 5; file MK 11750, vol. 6).

21. [Gossler] to Kgl. Staats- und Finanzminister Herrn von Scholz, 20 Dec. 1883, GStA Merseburg, Berlin Phil. Fak., Bl. 94RS.

22. Search committee to philosophical faculty, 30 November 1883, UA München, Senatsakten, "Zoologie und vergleichende Anatomie." At Munich the comment was made of Haeckel that "doubts were raised on account of his extreme natural-philosophical partisan stance [seiner extremen naturphilosophischen Parteistellung]."

tigations; and he had confronted the most important theoretical challenge to scientific zoology by exploring Darwinian evolution, which he interpreted "not as an axiom, but as a theory that is to be subjected to critique by the facts."[23] Letters describing Schulze's assets emphasized his combination of comparative anatomical, histological, and systematic research as especially exemplary.[24]

These recommendations do not merely reflect Weismann's and Schulze's strengths, but highlight more generally the values shared by the leading scientific zoologists and the expectations for zoology held by the broader scientific community.[25] Thus at Munich following Weismann's refusal of the position, the search committee proposed Otto Bütschli, arguing that "no other person among Germany's zoologists unites in himself to the same degree *the* qualities" they required—biological interests as well as comparative anatomical ones and proven skill as a teacher and institute director, including the ability to teach the latest in microscopic and histological techniques.[26] At smaller universities making appointments at the same time, a similar set of values held. Morphological research, insofar as it consisted of microscopical, embryological, and even systematic investigations, certainly received its share of attention, but at the same time, in accordance with the long-established program of scientific zoology, biological research was valued as well. In addition, teaching breadth and administrative experience could tip the balance toward one candidate or another.[27]

23. Ibid.

24. See, e.g., Waldeyer to Althoff, 24 August 1883, GStA Merseburg, Berlin Phil. Fak., Bl. 70.

25. Significantly, this blend of qualifications was emphasized at both the level of individual university faculties and that of higher administrators. Even the Prussian university administrator Althoff, known for not abiding by faculty wishes, appears always to have taken his choices for zoology professors from the faculty recommendation lists, just as ministers in other states did.

26. Search Committee to medical faculty, 26 July 1884, UA München, Senatsakten, "Zoologie und vergleichende Anatomie." The qualities desired by the Munich search committee are inferred from their previous recommendations of 30 November 1883.

27. Although it was considered customary for a faculty to propose three names, sometimes there would be two lists, one of more senior scholars and a second list of more junior ones considered more likely though less desirable candidates. This meant that sometimes as many as six names were sent up on a recommendation list. At the less prestigious universities, where younger scholars came under serious consideration more often, the names of Hermann Grenacher (1843–1923), Otto Bütschli, Richard Hertwig, and Hubert Ludwig repeatedly came up; Grenacher was on the faculty short lists at Halle, Königsberg, and Giessen in 1881, in the latter two as the first-rung candidate; Bütschli received offers from Rostock (1882), Königsberg (1883), and Munich (1884), and was on the list at Halle (1881), Strassburg (1886), and Leipzig (1898);

Glaringly absent from most of these recommendations is any explicit discussion of evolutionary theory. In fact, in the few places where evolution is mentioned, both professors making recommendations and higher administrators expressed their uneasiness. We have already seen an expression of that disquiet in the refusals at Munich and Berlin to consider Haeckel's candidacy, despite the acknowledgment that he was one of the leading figures in zoology. But the worries over Darwinism and Haeckelism spilled over to others as well. At Munich, for example, the recommendation stressed Weismann's empirically based criticisms of Darwinism rather than his evolutionary speculations. After Weis-

---

Hertwig was considered at Königsberg (1881), Breslau (1881), Bonn (1883), Munich (1884–85), and later, as a senior scholar, in Leipzig (1898); he also was under consideration at the Austrian university of Graz in 1884; Ludwig was listed as a secondary or tertiary candidate at Breslau (1881), Giessen (1881)—where he was eventually hired—Munich (1884–85), and the Austrian university of Prague (1885) before finally being named number 1 at Bonn in 1887 (information gathered from diverse correspondence, university archives, and state archives). Recommendations for the younger zoologists described their strengths in similar terms as for the older ones. At Bonn, Richard Hertwig was commended for his "exact and penetrating investigations" that contributed substantially to the "current views of the organization and mode of life of the lower animals"; at Munich he was considered an "excellently schooled microscopist" (Philosophische Fakultät, Bonn to Curator, Bonn, 2 December 1882, GStA Merseburg, Bonn Phil. Fak., Bll. 316RS–317; search committee's suggestions at Munich, 16 November 1884, UA München, Senatsakten, "Zoologie und vergleichende Anatomie"; Hertwig himself, reflecting the same values, regarded Bütschli as the best man to be his successor at Königsberg (R. Hertwig to Althoff, 30 July 1883; GStA Merseburg, Rep. 92 Althoff B, Nr. 68, Bd. 2, Bll. 60–61). Less information is available on Grenacher, who had taught at Rostock since 1873 and was considered for Halle, Königsberg, and Giessen in 1881; his strengths were in the realm of histology and microscopic technique. At Halle, where he was eventually hired, his research was initially considered "one-sidedly histological," but this liability appears eventually to have been outweighed by his experience of a year teaching at a forestry academy, which was seen as fitting in with the recent incorporation of agricultural studies at Halle (Philosophische Fakultät Vorschläge [Abschrift ohne Datum], GStA Merseburg, Halle Phil. Fak., Bll. 104–5); on agricultural interests see ? to Staats- und Finanz-Minister Bitter, 3 November 1881, Halle Phil. Fak., Bl. 112. Ludwig, who was always number 2 or lower on the lists, was also considered somewhat one-sided, since nearly all his work concerned the structure and systematics of echinoderms, but this appears to have been made up for by broader administrative experience, first as director of the city natural history and ethnology museum in Bremen and later as professor at the stepping-stone University of Giessen; Ludwig also undoubtedly benefited from the persistent efforts of his patron Ernst Ehlers at Göttingen (see, e.g., Anton Schneider's recommendations for his replacement at Giessen, 20 January 1881, UA Giessen, "Die Wiederbesetzung der durch den Weggang des Hrn. Professors Dr. Schneider in Erledigung kommenden Professur [1881] durch Dr. Hubert Ludwig, January 1881–April 1887"; recommendations of Munich zoology search committee, 16 November 1884, UA München, Senatsakten: "Zoologie und vergleichende Anatomie"; for further documentation of Ehlers's role as patron to Ludwig, see UB Göttingen, Cod. MS E. Ehlers 1158, "Hubert Ludwig," which contains 217 letters from Ludwig to Ehlers).

mann refused the position there and the faculty turned to consider Otto Bütschli, one professor worried that they might be picking a Haeckelian and had to be reassured that "Bütschli was a student of Pagenstecher's and absolutely not an agitator for an extreme orientation."[28]

The situation in Prussia reveals a similar nervousness, if a somewhat different process. By the time the Bonn position came open in late 1882, Friedrich Althoff had been charged within the ministry of culture with overseeing university appointments, and he took a more aggressive stance than his predecessor in seeking out opinions beyond those of the revelant faculty. Althoff wrote at least two zoology professors asking them to compare Schulze and Weismann, the leading candidates recommended by Bonn's faculty for the zoology position, and their responses appear to have been incorporated into his recommendations, not only for the Bonn position but for Berlin several months later as well.

Karl Möbius, professor of zoology at Kiel, noted that Schulze was probably the more important histologist, while Weismann excelled as a "biological experimenter." The latter was "richer in ideas than Schulze," but in conducting his research he had sometimes trespassed onto the territory of others;[29] here "Schulze stands entirely blameless." Möbius gave the edge to Schulze as a teacher as well. The recommendation from Ernst Ehlers, professor at Göttingen and the influential editor of the *Zeitschrift für wissenschaftliche Zoologie,* could only have deepened the incipient shadows surrounding Weismann. Ehlers, too, put Schulze ahead of Weismann, because the latter's theoretical work, with regard to both his specific research problems and his more general writings on the theory of descent, tended toward generalizations whose foundations were "not always entirely sure enough." Moreover, Weismann had spoken on the subject of descent in such popular fora as the annual meeting of German naturalists and physicians, which typically drew in a large nonprofessional audience; Schulze, by contrast, "has kept his distance from such things."[30]

28. Protokoll über die Fakultätssitzung von 26. Juli 1884, UA München, Dekanatsakten; here it is unfortunate that there is little information on the discussions leading to the ultimate choice of Richard Hertwig, who was, after all, Haeckel's student!

29. Weismann was involved in a number of priority disputes in the 1870s (see L. K. Nyhart, "Writing Zoologically," esp. pp. 55–61).

30. Möbius to Althoff, 30 January 1883, GStA Merseburg, Rep. 92 Althoff B, Nr. 134, Bd. 1, Bl. 6; Ehlers to Althoff, 30 Jan. 1883, GStA Merseburg, Rep. 92 Althoff B, Nr. 32, Bd. 2, Bl. 175; a draft of Ehlers's letter is in UB Göttingen Cod. MS E. Ehlers 25, Beilage 1; we should note that these reservations appeared even before Weismann's germ-plasm theory, which cemented the opinions of many concerning his penchant for unsupported speculation.

Neither Schulze nor Weismann, apparently, was interested in the Bonn job, and it went to the third man on the list, the much more junior Richard Hertwig (who would depart almost immediately for Munich). But just under a year later, Ehlers's concerns about public discussions of evolution were transmitted in an intensified form in a letter sent from Althoff's superior, the culture minister Gossler, to the finance minister in an effort to secure funds to meet Schulze's demands for the zoology professorship at Berlin. In making his pitch for Schulze, the Berlin faculty's sole candidate, Gossler presented a larger case (already quoted above) for Schulze's ability to bring Berlin into the age of scientific zoology. However, scientific zoology was also subject to abuse because of its close association with Darwin's theory, which some had turned into a "new and refined worldview." "And the champions of this view haven't contented themselves with pursuing the discussion within learned circles; they have swamped the popular literature with it. . . . Your Excellency will agree with me that as much as freedom of scientific research and persuasion must be guaranteed, it appears in no way advisable to bring a man to Berlin who preaches scientific hypotheses to the people as irrefutible truths."[31] On the face of it, this consideration applied more to Haeckel than to Weismann, but Ehlers's letter suggests that the latter, too, had been tarred at least a little with the dangerous brush of evolutionary speculation and popularization.

That this was known to be a serious point is suggested by Schulze's own assurances to the Prussian ministry. Schulze later confessed to Haeckel that the latter had "influenced me and my fate, especially my scientific beliefs and entire orientation, more deeply than you yourself know," and he recalled having read through the *Generelle Morphologie* three times when it appeared in the mid-1860s.[32] But he was neither disposed to advertise this intellectual debt nor to follow Haeckel in his more polemical or speculative tendencies. In a note appended to an outline of his teaching plans sent to Althoff during the negotiations, Schulze felt compelled to spell out his stance on Darwinism (something he never did explicitly in published work). "Persuaded of the truth of descent, I hold Darwin's theory of selection as suitable to explain many phenomena but do not assume that natural selection is the sole force that determines the organization of individual animal form. As much as I will confess and represent my conviction firmly,

31. Gossler to Scholz, 20 December 1883, GStA Merseburg, Berlin Phil. Fak., Bll. 93–94.
32. Schulze to Haeckel, 23 July 1894, quoted in G. Tembrock, "Franz Eilhard Schulze" (1966), p. 142.

unconcerned with the prejudices of others, I am just as little inclined by my entire nature to proceed to provoke anyone."[33] If Althoff was seeking a guarantee that Schulze would not speculate publicly in the manner of Haeckel or Weismann, he seems to have found it. This assurance, too, made its way into the final recommendation for Schulze: "He has never tried, through the popularization of immature hypotheses, to gain the approval of the masses. His activity in his academic teaching post is all the more avid and successful."[34]

The tone of the documents surrounding Schulze's appointment to Berlin, with their expressions of nervousness about the popularization of evolutionary theory and their emphasis on Schulze's technical and empirical merits as a histologist, embryologist, and comparative anatomist—in a word, a cautious morphologist—is reflected in other appointments made throughout Germany in the 1880s. Those who successfully moved up into their first professorships during this decade tended to be characterized by theoretical caution and technical proficiency of a kind best appreciated by other professionals. Even Richard Hertwig and Carl Chun, who cast their theoretical nets widely in their attempts to answer broad biological questions, did not seek to stir the world to the depths sought by Haeckel or Weismann. Far more modest still were the aspirations of Hubert Ludwig, Johann Wilhelm Spengel, and Hermann Grenacher, who won appointments at Bonn, Giessen, and Halle, respectively. These men conducted research that was morphological, insofar as it dealt with embryology, histology, comparative anatomy, and systematics, and they made important technical and empirical contributions in these areas, but their work was intellectually bland with regard to broader questions. Although all three believed in evolutionary descent, none was committed to grand theorizing. Like Schulze, they assumed the reality of evolution and pursued classificatory questions within a broadly phylogenetic framework, but generally did not try to make large claims about its mechanisms or the linkages between phyla. Nor did they devote themselves to telling the public about evolution, choosing instead to debate the finer technical points within the confines of professional publications and meeting presentations.

This professional orientation marks all of the successful zoologists of the cohort born around 1850. Reaching out to the public with politically and emotionally charged ideas was not their approach;

33. Schulze to Althoff, 27 August 1883, GStA Merseburg, Rep. 92 Althoff B, Nr. 173, Bd. 1, Bll. 86–87.

34. [Gossler] to Kgl. Staats- und Finanzminister Herrn von Scholz, 20 December 1883, GStA Merseburg, Berlin Phil. Fak., Bl. 94RS.

even those members of the cohort who did develop innovative research areas, most prominently Bütschli and Richard Hertwig, took up problems that had fairly narrow professional resonance. To the extent that zoologists who advanced in this decade did seek to reach a wider, nonacademic public with their research, the connections they made were of a practical nature: at the north German coastal university of Kiel, Karl Brandt followed the example of his predecessor Karl Möbius in conducting research on plankton that was important to aquaculture and the fisheries industry; similarly, at Rostock, Maximilian Braun addressed questions of practical interest for fisheries, and after he moved to Königsberg he was active in the East Prussian Fisheries Society, sitting on its board of directors for fifteen years.[35] In such associations these men acted as scientific specialists offering their expertise to an economically interested public, rather than philosophical thinkers propounding a particular worldview to the masses. The scholars who did put forth more philosophically sweeping ideas accessible to public debate were not primarily those appointed in the 1880s, but those who had established their careers in the more theoretically freewheeling decades of the 1860s and 1870s, namely, Haeckel, Weismann and Eimer.

In the 1890s another cluster of older professors (not mostly the leaders of the field) was replaced by a group of seven younger scholars born between 1858 and 1862.[36] Although one might have expected by this time to see some greater representation of a new, experimental direction in this younger group, in fact, this is not especially prominent; instead, these men reflect the spectrum of interests within the profession. Only two of the seven, Theodor Boveri (1862–1915) and Eugen Korschelt (1858–1946), made reputations as experimentalists, and there is little to suggest that at the time of their appointments it was the experimental aspect of their work that especially favored them. Boveri gained his reputation as an innovative cytologist and

35. J. Reibisch, "Karl Brandt" (1933); O. Koehler, "Maximilian Braun" (1930–31), pp. xiv, xx.

36. Those who died or retired in the 1890s were Anton Schneider (d. 1890), Karl Semper (d. 1893), Richard Greeff (1829–92), Adolf Gerstaecker (1828–95), Emil Selenka (1842–1902), Theodor Eimer (d. 1898), and the Nestor of German zoology, Rudolph Leuckart (d. 1898). The seven new slots opened up were taken by the following first-time professors: Gustav Wilhelm Müller, Oswald Seeliger (1858–1908), Eugen Korschelt, Friedrich Blochmann, Willy Kükenthal, Theodor Boveri, and Albert Fleischmann. (Of course, these are not one-to-one replacements; these men generally moved into the less prestigious positions, while established professors moved up to the more prestigious ones.) It should be noted that this younger group does not comprise one of my morphological generations.

embryologist working under Richard Hertwig at Munich, and his appointment at Würzburg in 1893 may well have rested on his path-breaking, physiologically oriented work on fertilization and cell division, some of which was experimental. However, at that point in his career Boveri had also published two comparative-anatomical studies and two systematic ones, thereby establishing more standard zoological credentials as well. And even before that time, Hertwig had been promoting him as an exceptional teacher.[37] Korschelt was also impeccably credentialed, having studied with Bütschli, Leuckart, and Weismann, and having worked as a *Privatdozent* and assistant under Schulze in Berlin. It is worth noting that he began publishing the experimental research on regeneration that would establish him as a "modern" researcher only well after he arrived at Marburg in 1893. The official considerations that went into his appointment there (where he was the most junior candidate on the list) emphasized the synthetic command of his field he had shown with the textbook in comparative embryology he had coauthored and his experience working at the Berlin zoological institute, which had given him "the best opportunity [for one not already an institute director] to acquaint himself with the direction of such an institution." Despite his later work on regeneration, Korschelt would remain best known for the more traditionally oriented work represented in his textbooks.[38]

The research areas of other members of this age cohort who gained professorships tend to reflect the range of existing orientations rather than dramatic theoretical or technical departures. Gustav Wilhelm Müller (1857–1940), appointed in 1895 at Greifswald, where he had spent nearly all his career until then, was a systematist and natural historian. Friedrich Blochmann (1858–1931), a student of Bütschli who was appointed at Rostock in 1891 and Tübingen in 1898, was known early on for his research in the area of generation, cell division, and early development. Willy Kükenthal (1861–1922) was a Haeckel

37. The state-level appointment records for the University of Würzburg between 1870 and 1900 were destroyed in World War II. However, a letter from Boveri to his sister-in-law Victoire Boveri indicates that the main reviewer for the position was the plant physiologist Julius von Sachs, who would be likely to be well disposed toward the sort of work Boveri was doing (Boveri to Victoire Boveri, 30 January 1893, StaBi München, Boveri Nachlass Ana 389, C 1); on Boveri's early publications, see H. Spemann, "Theodor Boveri" (1917); and F. Baltzer, *Theodor Boveri*; see also R. Hertwig to Althoff, 29 June 1896, quoted in T. J. Horder and P. Weindling, "Hans Spemann" (1986), pp. 204–5.

38. Philosophische Fakultät to Kurator Steinmetz, 29 October 1892, Staatsarchiv Marburg, "Zoologie," H. Querner, "Zoologie in der Aera von Eugen Korschelt" (1980).

protégé who worked primarily on the ontogeny and phylogeny of marine mammals, taking an interest in the convergence of such adaptations to marine life as flippers and their consequences for systematics. Having spent nine years as Ritter Professor of Phylogeny at Jena (a position carrying the rank of salaried *Extraordinarius*), he was appointed *Ordinarius* at Breslau in 1898 in the wake of vigorous efforts by Haeckel—one of Haeckel's rare successful placements of a student in Germany.[39]

Among the men appointed to professorships in this period just one was stridently critical of an existing zoological orientation. This was Albert Fleischmann (1862–1942), who replaced his teacher Emil Selenka at the Bavarian University of Erlangen in 1896. Fleischmann conducted much of his early research within the evolutionary-morphological tradition, taking up such topics as the comparative anatomy and development of the placenta (with consequences for phylogeny). By the time of his appointment, however, his efforts to unite comparative anatomical and developmental research into a broader synthesis had led him to express doubts about what he called the "Darwin-Haeckel" hypothesis of descent. Both the faculty and the ministry at Erlangen emphasized that Fleischmann's skeptical view of Darwinism "cannot not be understood as a disadvantage" to his candidacy; indeed, after Fleischmann's appointment, Haeckel suggested that it fulfilled the "heartfelt wish of the Bavarian provincial government . . . that among the representatives of science no more 'Darwinists' be appointed."[40] While there may be some truth to this, it should also be noted that Fleischmann, who was at the bottom of the list of candidates and the sole nonprofessor on it, was also the least costly candidate and the only Bavarian. This latter consideration appears to have been particularly important, affecting other academic appointments as well during the mid-1890s.[41] In any case, although

39. On Müller, see R. Keilbach, "Chronik des zoologischen Instituts" (1956), pp. 565–66; on Kükenthal, see K. Heider, "Kükenthal" (1924); on Blochmann, see R. Vogel, "Blochmann" (1931). The first choice on the university's list of recommendations was Friedrich Blochmann; on this point and on Haeckel's role in Kükenthal's appointment at Breslau, see Uschmann, *Geschichte der Zoologie*, p. 157.

40. Senat to Ministerium des Innern, 29 February 1896; Ministerium des Innern to Prince Luitpold, 22 March 1896, BHStA München, file MK 11496; Haeckel's comment is quoted in A. Fleischmann, *Descendenztheorie* (1901), p. iv.

41. Johannes Rückert to Max Fürbringer, 21 January 1897 and 21 April 1897, SB, Fürbringer Nachlass, letters 2143, 2144; concerning the possibility of Fürbringer's protégé Richard Semon succeeding Rückert at the veterinary school in Munich, Rückert emphasized that since Semon was neither born nor trained in Bavaria, there was no hope of his receiving the position.

Fleischmann's opposition to Darwin's theory produced two tomes devoted to its refutation, it did not lead him to take up radically new approaches: rather, opposing any theorizing that might be considered speculative, for the rest of his career he pursued a program of doggedly descriptive research in comparative vertebrate embryology.[42]

Among this cohort, then, there is little evidence that one research orientation was winning out over the others, must less that there was any sort of extensive effort to promote radically new directions of research. Among the Prussian appointees of this decade, the ones already developing biological lines of research were not the youngest professors just entering the system (Korschelt, Müller, and Kükenthal), but those from the previous entering cohort (Chun, Brandt, and Braun). Althoff's possible interest in gaining more biologically or even experimentally inclined men might be seen in the advancement of Chun and Braun to more prestigious universities in Prussia, but the events surrounding the appointments of Korschelt at Marburg, Müller at Greifswald, and Kükenthal at Breslau suggest that local circumstances, financial considerations, and pressure from various powerful voices also had their effects.

What do these new zoological appointments of the 1880s and 1890s tell us about the relative fortunes of evolutionary morphology and experimental embryology in this period? Most evidently, they suggest dissipation of support for Haeckel's program; among the new professors of the 1880s and 1890s only Willy Kükenthal could be called a true follower, and even he did not follow his master in all respects.[43] By the end of the turnovers of the 1890s, one would be hard pressed to find anywhere among university zoologists Haeckel's own tightly bound package of commitments to large-scale phylogenetic questions, the public and unyielding claims for the truth of particular phylogenetic schemes, and the use of the biogenetic law as providing a reliable guide in constructing phylogenies, particularly as expressed through the gastraea theory. A number of his precepts remained and were incorporated into the accepted zoological wisdom: the aim of evolutionary classification; the use of comparative embryology and comparative anatomy to adjudicate taxonomic divisions (al-

42. H.-J. Stammer, "Albert Fleischmann" (1952).

43. Kükenthal's years of running the zoological practica as Ritter-Professor at Jena, which culminated in the publication of his widely used Leitfaden für das zoologische Praktikum (Jena, 1898), had helped his professional prospects by making it clear that he was a dedicated laboratory teacher and not just a Haeckelian theorist; see Uschmann (Geschichte der Zoologie p. 157), who notes that this lab manual was continued by Kükenthal's student E. Matthes into a thirteenth edition (1952).

though the methodological problems raised in the *Kompetenzkonflikt* surely gave some researchers pause); and the acknowledgment that ontogeny could, under certain circumstances, reveal clues about the ancestral past. But these softened tenets were hardly unique to Haeckel's program of evolutionary morphology.

Perhaps still more significant than adherence to or rejection of any particular theoretical principles was the change in tone among the group appointed in the 1880s and 1890s. The professional appointments of this period appear to have favored or reinforced a reluctance among younger zoologists to present their debates before a wider public, or even to treat issues in publicly accessible terms. Emblematic of this inward-turning, "professionalizing" trend was the founding in 1890 of the German Zoological Society. Earlier, the annual gathering of German zoologists took place under the aegis of the meetings of German physicians and naturalists. The general sessions of these meetings, which were open to the public and were widely reported in the press, had often been used to raise theoretical issues of popular interest. In these open sessions, for example, Haeckel and Virchow had sparred over the teaching of evolution in the schools in 1877, and Haeckel had renewed his defense of evolutionary monism in the same forum in 1882, while Virchow had continued his criticisms of descent theory at meetings in 1886 and 1887; Weismann had spoken on sexual selection and the continuity of the germ plasm in 1885 and on the inheritance of acquired characters in 1888; in 1886 the botanist Ferdinand Cohn had spoken on the origin of life and the connection between life and soul.[44] The Zoological Society meetings, by contrast, were generally divided among reviews of recent research results, laboratory demonstrations, and discussions of professional matters such as a project to reform zoological nomenclature and an effort to gain funding for a biological station in Helgoland.[45]

In this new professionalized atmosphere, the public and publicity-seeking aspect of Haeckel's approach may well have blended with the existing technical and philosophical criticisms of his theories and methods to make his program of evolutionary morphology seem "unprofessional" in a number of reinforcing ways. The documents surrounding appointments in the 1880s certainly indicate political worries about preaching evolution to the masses, worries that no doubt

44. See *Tageblatt der Versammlung Deutsche Naturforscher und Ärzte* for the years in question; see also H. Querner, "Darwins Deszendenz- und Selektionslehre" (1975).

45. A. Geus and H. Querner, *Deutsche Zoologische Gesellschaft 1890–1990* (1990), esp. pp. 26–53.

stemmed from Haeckel's infamous 1877 Munich speech, in which he advocated replacing religious instruction with evolutionary monism in the grammar schools. Although there was no necessary, logical connection between popularizing and speculative theorizing per se, it does appear that the younger zoologists linked the two and stayed away from the sort of scientific theorizing that could be exalted into a worldview.

If speculative phylogenetics lost out in the orientations advanced through professional appointments, one cannot say that it was replaced by a single dominant new approach to studying animal form. Experimental biology was slightly on the increase, but descriptive morphological research remained a strong component of the field. In fact, the interests of the community of zoology professors viewed as a whole suggest that the dichotomy posed by Haeckel and Driesch has directed our gaze away from the major sites of change in zoology in this period. These were the institutes established in the 1860s, largely by Haeckel's generation, who sought to unify the scientific zoology of the 1850s with evolutionary concepts. The empirical directions of that program were modified to some degree by their students (mainly of the cohort born around 1850), and by their students' students, but a surprising number of the principles by which the community operated remained consonant with those articulated decades before.

## RHETORIC, RESEARCH PROGRAMS, AND THE "REVOLT FROM MORPHOLOGY"

This chapter and the previous two have taken three different slices through the community of researchers concerned with animal form, examining the narrower circle of vertebrate evolutionary morphologists, the anatomists seeking the mechanical causes of form through embryological experiments, and the scientific zoologists of the 1880s and 1890s. An important component of each of these stories is the mode of intellectual change. Within the group of Haeckel and Gegenbaur followers who focused on the large-scale phylogenetic question of the origin of the vertebrates, the debate over Gegenbaur's *Archipterygium* theory evolved by the mid-1890s into a broader set of struggles over the relative value of the two leading morphological methods, comparative anatomy of adult characters and embryology. The second story described both Roux's own presentation of *Entwicklungsmechanik* as an extension of existing morphological concerns and its interpretation by others as marking a sharp break with the past. The third story, in this chapter, showed a number of leading zoological institutes promoting a similar continuity between research often treated as evolutionary and morphological and research considered experimental or causal.

What ties these stories together and makes them part of a single larger story are the considerations of age groupings and chronology. In both anatomical and zoological research and the programmatic discussions, it was the cohort born around 1850 that was most prominent in negotiating between evolutionary questions and those concerning the mechanics of the developing form. This is particularly striking among the anatomists, where the handful of men (Born, Strasser, Roux, O. Hertwig, Rabl) most closely involved in both areas of research in this period were all born between 1849 and 1853. But it is also evident among zoologists: Otto Bütschli and Richard Hertwig; Leuckart's associates Carl Chun, Heinrich Simroth, and Paul Fraisse; Weismann's student, brother-in-law, and colleague August Gruber; Semper's student Justus Carriere; and Chun's associate Karl Brandt were born between the mid-1840s and the early 1850s. They all conducted experimental and/or biological research in a context of furthering evolutionary understanding. Among these zoologists there was clearly greater continuity with the interests of their teachers than was the case among the anatomists—a number of these zoologists' mentors were also conducting such research—but there is no doubt that the sheer amount of such research mushroomed substantially in the 1880s and 1890s, or that this group was responsible for much of the expansion of this orientation.

Up to the early 1890s, mechanical and biological investigations were certainly distinguished from strictly descriptive studies of the sort that was often the subject of anatomical and zoological dissertations, but they do not appear to have been widely viewed as philosophically opposed to either description or evolutionary explanation. Only in the middle and later 1890s did it become more common to construe them as inherently incompatible intellectual orientations. This chronology is significant. Certainly one can find powerful personalities, such as Haeckel and His, who had earlier ordered their views according to this split. Those who later wanted to could reach back even further to the pre-Darwinian rhetoric of opposition between physicalist physiologists and "idealist" morphologists. These dichotomies allowed later writers to project a historical continuity onto the current debate, as if it were an eternal one rather than one growing out of a rapidly changing historical situation that crystallized only in the 1890s. But Haeckel and His's disputes in the mid-1870s do not

appear to have incited many others then to view the world of morphological research in the same way. What happened to reify this division
after the early 1890s?

Obviously Roux's program had something to do with it. His persistence in calling attention to it kept it from fading into the background,
and the excitement and controversy surrounding the specific embryological experiments he and others conducted contributed much to the
sense that this was a truly novel approach that might be productive
in a way that evolutionary morphology no longer seemed to be. Once
in the public domain, *Entwicklungsmechanik* afforded a new jump-
ing-off point for people with more extreme views than his own, who
did see the situation as more dichotomous than he himself did. This
view of a natural division between the two approaches then gathered
momentum as a new framework within which people could align
themselves (or more likely, place their opponents), even if earlier they
might not have viewed their work as being on one side or the other.

It seems likely that the *Kompetenzkonflikt* contributed to the divide
as well. To the extent that the morphological method was understood
to depend on the biogenetic law, the conflict damaged it seriously. By
the time Fürbringer and Rabl came to verbal blows in the mid-1890s,
neither was seeking to save recapitulation as a methodological device.
Fürbringer was touting the virtues of adult comparative anatomy,
while Rabl had developed a way of interpreting embryological evidence that did not depend on recapitulation at all. Even beyond the
biogenetic law, the controversy cast into the limelight the weaknesses
of the morphological enterprise as pursued by Haeckel and Gegenbaur, at least to anyone with a mind to view it that way. And here it
seems likely that the appearance of Roux's program and its strident
interpretations by the young Driesch and Dreyer in the early 1890s,
coupled with the *Kompetenzkonflikt,* would invite neophyte researchers casting around for a problem area in the mid-1890s to perceive the
Haeckel-Gegenbaur program of evolutionary morphology as having
outlived its fruitfulness. Certainly the timing of the attacks on evolutionary morphology must have given an especially sharp edge to the
*Kompetenzkonflikt* itself as Fürbringer, for one, saw the enterprise so
carefully erected by his revered teacher Gegenbaur threatened with
oblivion.

Driesch, Dreyer, and others entering the field in the early 1890s
contributed substantially to this shift in construal. Only with the
emergence of their generation, born too late to have experienced the
1880s as researchers themselves, did the opposition between a causal-
mechanical-experimental approach and a descriptive-historical-specu-

lative approach to morphology become widely considered a given. Although there would still be individuals who tried to knit the two approaches together, younger members of the community viewed the issue as overcoming an opposition rather than as simply shifting between two different but not necessarily incompatible approaches to understanding form.[46]

We may speculate on one last possible contribution to the new perception that evolutionary thought and causal analysis were incommensurable. The philosophical tone and orientation of Driesch and Dreyer's writings suggest that there were new outside authorities to draw on for this conception of a deep split between the causal-mechanical and evolutionary-morphological approaches. A broader reconsideration of scientific and historical epistemology was taking place in the mid-1890s and early 1900s, developed mainly by such neo-Kantian philosophers as Otto Liebmann, Wilhelm Windelband, and Heinrich Rickert, whose ideas gained the attention of biologists. The distinction these philosophers drew between historical knowledge and natural-scientific knowledge clearly presented a challenge to the epistemological status of evolutionary theory as both a kind of history and a legitimate scientific theory. It appears that younger scholars, who perhaps like Driesch were introduced to them during their university studies, may have picked up on them first. By the early twentieth century these philosophical discussions drew quite a number of more established German biologists onto unfamiliarly philosophical terrain.[47] These broader philosophical debates, which partly concerned whether evolutionary arguments could count as natural science or not, certainly reinforced the sense that the evolutionary-morphological and causal-mechanical approaches to understanding the organism were incommensurable.

The interrelated stories sketched above suggest a solution to the

46. A good example of this is Hermann Braus, one of Gegenbaur's last students and son-in-law of Gegenbaur's lead protégé Max Fürbringer. Braus, a contemporary of Driesch, Dreyer, and Hans Spemann, hoped to create an "experimental morphology" that would blend the phylogenetic concerns of the Gegenbaur school with the techniques of *Entwicklungsmechanik* (see Braus, "Versuch einer experimentellen Morphologie" (1903), pp. 2076–77, and esp. "Vorwort" (1906–9), pp. 1–37). This pattern of a gradual dichotomization reified by a younger generation, followed by attempts to overcome it, has striking parallels in other fields; one needs only to think of the internalist-externalist dichotomy in the history of science.

47. For example, T. Boveri, *Die Organismen als historische Wesen* (1906); O. Bütschli, "Kants Lehre von der Kausalität" (1905); S. Becher, *Erkenntnistheoretische Untersuchungen zu Stuart Mills Theorie der Kausalität* (1906).

problem of the intellectual "revolt from morphology" that has raised so much discussion among historians of biology. It is, as Garland Allen supposed, a generational issue, but it has a slightly different cast. The generation of Roux, the Hertwigs, Rabl, and others born in the three or four years on each side of 1850 experienced the development of new methods and questions about the causes of form as continuous with older ones.[48] If these new problems rapidly expanded into autonomous research areas, there seems to have been little perception among this generation that they were fundamentally irreconcilable with the older questions. For the slightly younger cohort of zoologists born in the years around 1860, too, the perceptions seem to have been similar, although more work would need to be done on them to be sure.

People embarking on their first morphological research in the early to mid-1890s, however—the generation born after 1864—tended to be much more impatient with the evolutionary approach. Although Driesch and Dreyer appear to have led the way here, they were not alone. Curt Herbst (1866–1946), who had joined his close friend Driesch and Dreyer in their Jena anti-Haeckelian reading circle as a student, was not given to polemics, but his work on the effects of different chemical media on development was wholeheartedly outside of the evolutionary framework. Hans Spemann (1868–1941) recalled his frustration as a young researcher in the mid-1890s that his supervisor, the experimentalist Theodor Boveri, would not give him an experimental topic for either his doctoral dissertation or his *Habilitationsschrift*, but insisted that he prove his skills on a descriptive embryological project and a comparative anatomical one, respectively. Spemann also authored a powerful 1915 essay damning the concept of homology, an idea central to the evolutionary morphologi-

48. There were differences among these researchers, to be sure; most prominent are the contrasts between anatomists and zoologists. Whereas relatively few anatomists exhibited interest in both the mechanical causes of form and evolution, many zoologists did so; moreover, anatomists and zoologists had quite different orientations within their respective disciplines to draw on for their construal of mechanical explanations; mechanical anatomy, which usually looked to physical forces acting on or in the body as sources of explanation, was a long-standing orientation within anatomy, from which Roux, Strasser, Barfurth, and His borrowed, and to which the new *Entwicklungsmechanik* was often understood to be connected. By contrast, zoologists were more apt to interpret the mechanical causes of form in terms of the natural environment, and especially in terms of the ways that differences in that environment produced evolutionarily significant differences in organic forms and functions. Despite these disciplinary differences, however, the theme of continuity with the past remains prominent.

cal enterprise.[49] Even Otto Maas (1867–1916), whose introductory textbook in experimental embryology sought to be fair to the older morphological approach, argued that "at least for the discovery [*Ermittelung*] of the conditions of causal dependency, experiment has a greater value, whereas comparative observation . . . contributes more as a helper [*mehr helfend dazutritt*], whether by preparing for or by subsequently securing" the knowledge gained from experiment.[50] As a causal explanation of form, evolution is given no weight in Maas's book. These examples suggest that although the relationship between evolutionary morphology and causal-analytical embryology was viewed as continuous by the teachers of this younger group, they themselves tended to dismiss the former as at best descriptive and nonexplanatory, at worst hopelessly speculative and methodologically bankrupt.

This rendition of events appears to provide a neat solution to the question of evolutionary versus discontinuous change in biological mentalities around the turn of the century: we might best consider it a two-generation process in which an older group, who brought about many of the new possibilities, nevertheless did not cast off completely the goals and values they had learned as students in the heyday of grand evolutionary theorizing; they were followed by a younger generation not tied by personal experience to this past, which appeared to them tiredly old-fashioned, boring, and unfruitful.

There are two problems with this story, however. First, although Driesch, Herbst, and Spemann certainly exhibited a new interest in experimental embryology and a concomitant lack of enthusiasm for evolutionary morphology, it may be skewing the picture to view them as representative of their entire generation. They may have been revolting from the morphology of their teachers—Driesch and Herbst from Haeckel and Spemann from his one-time teacher Gegenbaur— but for most zoologists, the choices were not limited to accepting or rejecting the Haeckel-Gegenbaur orientation. Let us consider Richard Hesse (1868–1944), who would eventually become professor of zoology at Berlin (1926–35): drawing from the biological orientation of the scientific zoologists, he coauthored a prominent work entitled *Tierbau und Tierleben in ihrem Zusammenhang betrachtet* (The connections between animal form and animal life), wrote important es-

49. J. M. Oppenheimer, "Curt Herbst's Contributions" (1991); H. Spemann, *Forschung und Leben* (1943), pp. 171–74; "Zur Geschichte und Kritik des Begriffs der Homologie" (1915).

50. O. Maas, *Einführung in die experimentelle Entwickelungsgeschichte (Entwickelungsmechanik)* (1903), p. 7.

says on ecology and animal distribution, and studied the microscopic structure of visual cells from a phylogenetic perspective. That Hesse by no means rejected evolution or the morphological evidence for it may be seen in his *Abstammungslehre und Darwinismus* (Descent theory and Darwinism), an introductory text that went through multiple editions.[51] Why should we consider Driesch and the other experimentalists to be more representative of the modern interests of the new generation than Hesse? Other active ecologists and biogeographers abounded in their generation, although few of them ever gained appointments at the universities under consideration here.[52] These considerations suggest that however much the story of a revolt from morphology might hold for a few men of this generation, the larger picture indicates a greater continuity of interests within the broadly biological orientation of scientific zoology. It might even be worth exploring the idea that others in Driesch's generation found the biological, nonmorphological lines of research as fruitful in part *because* they sidestepped the dichotomy between experimental and historical analyses that had emerged within the morphological orientation.

There is another problem with the "revolt from morphology" story as well. Concentrating purely on intellectual change provides an incomplete picture of the fate of morphological research at the end of the century. If one stands back to look at the larger institutional and disciplinary contexts of anatomy and zoology, these generational differences acquire a somewhat different significance than one might expect. We might think that the younger generation naturally takes over from the older one, all in due course, and ideas that once seemed radical eventually come to be viewed as normal. In the case of these disciplines, however, the selection of professors militated against this process. In the 1880s and 1890s, the university professorships were going not to the neophytes agitating for self-conscious experimentalism, but to men in Roux's generation and their slightly younger contemporaries. Although the research of these more established scientists often led them into new intellectual territory, most would continue as teachers to defend the legitimacy and importance of the evolutionary

51. *Neue Deutsche Biographie*, s.v. "Hesse, Richard"; Hesse, *Abstammungslehre und Darwinismus* (1912).

52. Prominent among these were Robert Lauterborn (1869–1952), a leading hydrobiologist who studied animal communities in rivers; the famous ecological theorist Jakob von Uexküll (1864–1944); and Hans Lohmann (1863–1934), a plankton researcher who pushed physiological research at his lab in Hamburg; on German ecology see Jahn, Löther, and Senglaub, *Geschichte der Biologie*, pp. 612–16; for brief references to each of these men, see Geus and Querner, *Deutsche Zoologische Gesellschaft*, passim.

framework.[53] By the time Driesch's cohort was of an age to be considered eligible for professorships in the early 1900s, however, there were none to be had.

Between 1898 and 1908 not a single full professorship in zoology came open in the traditional German universities. Spemann, Herbst, Hesse, and the heredity researcher Valentin Haecker (1864–1927) were the only members of their cohort ever to attain such positions, which started opening up again at the end of the first decade of the century. Driesch, as is well known, moved into philosophy—in any case, as an independently wealthy man, he did not need to look for a position in zoology. Dreyer had long since gone into business. Otto Maas died in 1916 without ever gaining a full professorship. It was only in the mid- to late 1910s that the trickle of positions expanded to a wave, as the last hangers-on from Haeckel's generation, as well as a few younger men, retired or died; most of these posts went to men of the newest intellectual generation, born in the 1870s and 1880s. Thus with regard to full professorships and directorships of zoology institutes, the generation of Driesch and Spemann was almost wholly skipped.[54]

In anatomy, where six professorships opened up between 1897 and 1911, only one went to a member of the cohort of the mid-1860s. The rest went to older scholars who had been waiting patiently in prosectorships or other subordinate positions—a reinforcement of the values of the previous generation. Again, it was not until the mid- to late 1910s that more chairs came available, and a few members of the generation of the latter 1860s finally arrived at the heads of their own institutes. Although the situation was not quite as severe as in zoology, here, too, the same generation was pinched for leadership positions.

The significance of this is not that these men were without an impact on their profession—after all, some of them taught students and continued to publish—but that the job squeeze inhibited the rapid development of certain new areas of research and of new ways of thinking about animal form and biology more generally. Men hoping

53. A notable exception to this attitude was Oscar Hertwig, who eventually came to repudiate evolutionary research altogether.

54. This is true only within the confines of the twenty traditional universities under consideration here. The new universities of Hamburg, Münster, and Frankfurt, founded in this period, did provide some openings for zoology professors, and with the raising of the *Technische Hochschulen* to parity with the universities of Ph.D.-granting institutions (1900), the institutional landscape changed considerably, although the status hierarchy remained.

for professorships had to present themselves as well-rounded stewards of the traditional anatomical and zoological subjects, not as specialists devoted to a single problem. Under these circumstances, two of the biggest growth areas in America—genetics and experimental embryology, which were perceived in Germany as specialties—did not flourish in Germany in these early years.[55]

In retrospect, this has made Germany appear something of a traditionalist backwater in the era of the new biology. There seems little doubt that the emphasis on teaching coverage in the 1880s and 1890s and job scarcity thereafter maintained the prominence of older traditions in German zoology and anatomy well into the twentieth century. We must be careful, however, about what traditions we point to. It would be hard to claim that the evolutionary program initiated by Gegenbaur and Haeckel persisted as an important part of university-level biology. Rather, the traditional morphology of the early twentieth century was that of the scientific zoologists. Nor did this tradition stand still—it evolved as well, into new research programs in cytology, protozoology, deep-water ecology, symbiosis, regeneration, and even genetics. These areas beckon us to broaden our outlook beyond the simple dichotomy of evolutionary morphology and experimental biology, to study further the manifold directions taken by Germans seeking to understand animal form and life in the early twentieth century.

55. On early German genetics, see the welcome new book by Jonathan Harwood, *Styles of Scientific Thought*.

## CHAPTER ELEVEN

# Morphology and Disciplinary Development

## Observations and Reflections

The proper evaluation of the significance of a scholar requires consideration of the teachers to whom he owes his training and the state of the science at the time when he began his research.
*Robert Bonnet, "Moritz Nußbaum"*

The years between 1900 and 1914 marked the gradual close of an era in morphology. It was not that Haeckelian evolutionary morphology was overtaken by *Entwicklungsmechanik*—as we have seen, neither orientation was marked by particular success in this period. Rather, what closed was a particular kind of memory. Kölliker, the last of his generation still active, died in 1905. With the deaths of Gegenbaur in 1903 and His in 1905, and the retirements of Haeckel in 1909 and Weismann in 1912, the next generation, too, bowed out of the morphological community. Only Ernst Ehlers, indefatigable as professor of zoology at Göttingen and editor of the *Zeitschrift für wissenschaftliche Zoologie,* continued to carry his generation's torch of memory to the end of the 1910s. All these men had been educated in the 1850s, when morphology emerged out of an earlier anatomico-physiological approach to research on life and gained its own identity. None of them actively continued to champion the views they had learned during their early university years; they had all long since developed their own, independent research programs. Nevertheless, their thinking was shaped fundamentally by an approach to form not yet informed by Darwinian concerns, one in which the problems of individual development, generation, and adaptation were viewed as the keys to understanding the unity and diversity exhibited by animal forms. The direct link that these men provided to the past was severed by their deaths and retirements. Those who sought to revive a consciously idealistic morphology in the early twentieth century (and there were a few)[1] did so without benefit of those memories—for them early nineteenth-century idealistic morphology provided a philosophical exemplar, largely unattached to the flesh-and-blood people who had produced the original model.

This feeling for the passage of time, with its dying away of particular memories and their re-creation by others as history is, to my mind, an essential part of understanding intellectual change, especially change in science, notorious for its forward-lookingness and attendant short memory. If one is concerned to understand the shape and values of a scientific community as part of understanding what is going on inside it intellectually, then the timing and significance of its changing membership must be critically attended to. One of the advantages of a study that traverses a century is that one can observe the past turning from individual memories (however reconstructed they themselves might be as time passed) into history. Studies of shorter time periods, as valuable as they can be in their greater attention to broader contexts, must necessarily miss this aspect of change.

This theme, and others concerning the nature of change within a scientific community, have more often than not remained implicit in my narrative and analysis of morphology's history. In what follows, I hope to draw them out. First, a review of the reconstructions developed here leads to some suggestions for new directions in which we might take the history of nineteenth- and early twentieth-century German morphology and biology. The next section, on the interactions of disciplinary, institutional, and generational structures, seeks to recapitulate and evaluate the utility of these analytical vantage points. The book ends with some reflections on the dialectics of power and values in mediating scientific change.

## THE HISTORIES OF MORPHOLOGY AND BIOLOGY

This study has suggested several revisions that need to be made in our picture of the intellectual history of German morphology. Most important, the powerful tripartite chronological narrative set up in E. S. Russell's *Form and Function,* dividing morphology into idealistic, evolutionary, and causal phases, needs to be refined, at least in the German case, to make room for the research orientation known as "scientific zoology." Not only did this orientation, which began in the late 1840s, span all three of Russell's eras, but among German zoologists its prominence blurred the sharp boundaries historians

1. R. Trienes, "Type Concept Revisited" (1989).

have drawn among them. Scientific zoologists had little difficulty in negotiating between the typological thinking associated with idealism and the idea of evolutionary development. Of course, many of them retained certain idealistic elements in their evolutionary thought, such as a conviction that somewhere, under it all, the course of evolution must not have been wrought purely by accident, but they were nonetheless open not only to descent but also to natural selection, despite its potentially disturbing philosophical consequences. The shift in thought was mediated by a conviction, widespread among the scientific zoologists, of the need to confine their science entirely to natural laws and causes and an accompanying value placed on ideas that were heuristically fruitful, even if their ultimate metaphysical implications might be disquieting.

Central to the smoothness of this transition was the concept of adaptation. Although some older scientists such as von Baer could not relinquish the idea that adaptation meant an unchangeable state associated with fixity of form, the promoters of scientific zoology and their students were able to transfer their interests in the relationship of the organism to its surrounding environment from an idealistic framework to an evolutionary one. Leuckart's emphasis on the functional purpose of morphological structures, which he retained throughout his long teaching career, could easily be reinterpreted as a consequence of natural selection, especially by scientists schooled to shun outdated notions of teleology. Leuckart continued to defend the language of purpose as a useful heuristic, and his students did the same.[2] But this was not the same as retaining a commitment to idealism. It meant instead a more flexible approach to thinking about form, one that could be fruitful in both idealistic and evolutionary frameworks.

Even more important than its role in mediating between idealistic and evolutionary morphology is scientific zoology's bridging of Russell's evolutionary and causal phases of morphology. Although Haeckel, Driesch, and many later commentators claimed that evolutionary and causal morphology were two distinct and incompatible approaches to form, I have argued that their position was not the predominant one. In fact, most of their community before 1900 would have viewed this dichotomy as false. The orientation of scientific zool-

2. See, e.g., the brief discussion of the teaching philosophy of Leuckart's student Karl Kraepelin in G. A. Erdmann, *Geschichte der Entwicklung und Methodik* (1887), p. 135.

ogy filled that putative gap with its concentration on research problems such as generation, inheritance, and the effect on the organism of its conditions of existence, in which both immediate mechanical causes and more mediate evolutionary concerns were relevant.

In this revised historical picture, the evolutionary program of Haeckel and Gegenbaur holds a considerably smaller place. Neither man was able to make it appeal to large segments of his discipline: Gegenbaur fought a losing battle for elbowroom in an anatomy community that was largely turned away from evolutionary questions and toward ones that had a greater claim to medical relevance, while Haeckel was unable to equip his students with the technical skills they needed to compete in the academic job market. As judged by their ability to gain followers who achieved university professorships, Haeckel and Gegenbaur had limited success.

If this picture of the professional communities is accurate, it leaves us with a nagging question. How and why did the German evolutionary morphology taught by Haeckel and Gegenbaur gain the historical reputation of having so dominated the late nineteenth-century life sciences? Several answers seem possible. The traditional intellectual historians' focus on great men surely has been a contributor, but one cannot attribute either Gegenbaur's reputation or (especially) Haeckel's entirely to the prejudices of intellectual historians. It may be more to the point that biologists of Driesch's generation and later, especially experimental biologists, used Haeckel—the extreme case—as an exemplar of all that was wrong with "old-fashioned" morphology in seeking to build the case for their own approach. By taking the extreme as representative, however, they cast into deep shadow the broader, less polemically useful tradition of scientific zoology.

Two other sources, deriving more directly from the experiences of people at the time, seem to have been even more important in the identification of late nineteenth-century German morphology with Haeckel and Gegenbaur. One stems from Haeckel's role as a public figure. Haeckel aimed at a broad audience in his writings, not only in his early *Natürliche Schöpfungsgeschichte* (Natural history of creation [1st ed., 1868]), but more especially later in his career with such accessible and comprehensive works as the *Welträtzel* (Riddle of the universe [Bonn: Strauss 1899]), and his efforts in connection with the German Monist League, which he helped to found in 1906. Moreover, as Alfred Kelly has shown, some of his ideas reached an even wider audience through other popularizers, including the best-selling writer

Wilhelm Bölsche, whose writings could be found from middle-class parlors to workingmen's libraries.[3]

The issue of "Haeckelism," especially in its anti-Christian aspects, was also prominent in discussions concerning school reform. Beginning in 1877, Haeckel publicly advocated the replacement of religious education in the schools with evolutionary monism; the perceived dangers of this view were heightened by the fact that in some schools, evolutionists were already active in teaching their natural history classes from an evolutionary and developmental perspective. The most famous case was of the Prussian *Oberrealschule* teacher Hermann Müller (brother of the Darwinian naturalist Fritz Müller), who was accused of teaching irreligion to his students in 1876. Although he was cleared of the charges, the case was taken by conservatives as representative of the dangers of evolution to impressionable minds. Thereafter, the teaching of descent in the schools was officially discouraged, and the impression rapidly spread, though apparently falsely, that it was forbidden.[4] It is possible, however, that in practice, teachers other than Hermann Müller may also have exposed their students to more Haeckelism and Darwinism than appears in formal curricula.[5]

Information about high-school teachers and their classroom practices is scant, but available hints and scraps do allow us to speculate that Darwin's and Haeckel's ideas did reach high-school students. In Prussia, to be sure, the biology-poor curriculum provided little opportunity for teachers at either humanistic *Gymnasien* or the more technically oriented *Realgymnasien* to teach natural history at the upper levels (where exposure to Darwinian ideas would be most likely). In more liberal states, the situation for teachers at the pre-university levels appears to have been different. To begin with, in

3. E. Krause, *Ernst Haeckel* (1984), pp. 112–19; the most readily available English-language source on Haeckel and the Monist League is D. Gasman's notoriously polemical *Scientific Origins of National Socialism: Social Darwinism in Ernst Haeckel and the German Monist League* (1971); unfortunately, Gasman tells us virtually nothing about the organization or composition of the League, or about Haeckel's role in it; a source closer to the action (which has been unavailable to me), is W. Breitenbach, *Die Gründung und erste Entwicklung des Deutschen Monistenbundes* (Brackwede, 1913); Kelly, *Descent of Darwin* (1981).

4. Kelly, *Descent of Darwin*, pp. 60–64| P. Depdolla, "Hermann Müller-Lippstadt" (1941).

5. On natural history teaching, see I. Scheele, *Von Lüben bis Schmeil;* one Prussian teaching reformer and enthusiast of Haeckel was Gustav Adolf Erdmann, who appears to have taught at a preparatory school for teachers (Erdmann, *Geschichte der Entwicklung und Methodik,* esp. pp. 118–24, 129–30); Erdmann lists several other teachers whom he identifies as proponents of evolution, p. 131.

some states more time in the curriculum was devoted to natural history. In Saxony, for example, the 1878 revisions gave at least an hour a week to natural history in the *Realschulen* in each of its eight grades.[6] Heinrich Simroth, who taught such courses at the *Realschule* in Leipzig while he worked in Leuckart's laboratory, was a Darwinian who may well have conveyed his ideas to his students.[7] In Haeckel's home Thuringian States, the *Oberlehrer* at the teachers' seminar in Gotha taught "Darwin's theory" to his class 4 and human descent to class 3. Between Haeckel and educators such as this one, who taught descent to future teachers, these states may well have many high-school teachers spreading the evolutionary gospel in their classes.[8]

Through popular dissemination and possibly through the education system as well, the broad evolutionary program associated with Haeckel made its way into many different niches of German society, justifying his reputation as a dominating cultural figure in late nineteenth-century Germany, even if he dominated his own discipline somewhat less. Nevertheless, we should not attribute the standard characterization of late nineteenth-century German morphology—especially in internalist histories of science—entirely to Haeckel's popular influence. For this we should probably look to an additional source, namely, the foreign researchers who worked in their laboratories and carried their orientation back home. E. Ray Lankester, for example, studied with Haeckel in 1871. He was responsible for spreading Haeckel's approach in England, both by overseeing the English translation of the *Natürliche Schöpfungsgeschichte* and in his own research and teaching at the universities of London and Oxford; he sent a number of students to work with Haeckel. Less well known are Achille Quadri, Conrad Keller, Berthold Hatschek, August Langhoffer, and Nicolaus Leon, who studied with Haeckel and presumably carried his ideas as representative of German morphology to their home universities in (respectively) Siena, Zürich, Prague, Zagreb, and Bucharest.[9] Similarly, Gegenbaur students became professors (mainly of zoology) at the universities of Utrecht, Helsinki, Stockholm, Copenhagen, and Cambridge.[10] In all of these universities, it seems likely

6. The hours devoted to natural history in the new Saxon curriculum are listed in "Mathematische und naturwissenschafliche Lehrfächer" (1877), p. 471.

7. Ehrmann, "Simroth."

8. Erdmann, *Geschichte der Entwicklung und Methodik*, pp. 139–40. Uschmann (*Geschichte der Zoologie*, pp. 192–96) lists a number of *Gymnasium* and *Realschule* teachers among the doctoral students who worked under Haeckel between 1883 and 1909.

9. Uschmann, *Geschichte der Zoologie*, pp. 127–29, 132, 193.

10. On Gegenbaur's foreign students, see pp. 210, 212.

that these professors, and subsequently their students and colleagues, would have identified German morphology with the approach of Haeckel and Gegenbaur. It is thus plausible that the tradition within the international scientific community, that German animal biology was predominantly influenced by Haeckel and Gegenbaur, results in some measure from an international reputation derived from foreigners' limited experiences.[11]

However we may account for the disjunction between the picture of German morphology presented by standard histories and the story I have told here, my story redirects our attention toward various new issues within and outside of Germany, four of which seem especially worthwhile. To begin with, historians need to pay greater attention to what Russell and later writers have called "functional morphology" within the German intellectual context. My story suggests that a fairer account of German biological activity, in zoology at any rate, will have to look beyond Haeckel to such lesser-known figures as Carl Theodor von Siebold and Rudolf Leuckart, and beyond the stereotype of evolutionary-tree construction toward other activities surrounding such problems as adaptation and generation. Historians working on the history of ideas of inheritance and generation, most notably Fred Churchill and his students, have already begun to flesh out this picture, but much remains to be done.[12] If we see scientific zoology rather than Haeckelian evolutionary morphology as the predominating intellectual agenda of the mid- to late nineteenth-century German zoology,

11. A notable exception to this experience seems to be the United States, where few zoologists went to study with Haeckel and Gegenbaur; the paleontologist William Berryman Scott is the only American of any note who did so. A sizable fraction of American zoologists studying in Germany in the late nineteenth century, including most prominently C. O. Whitman (Chicago), E. L. Mark (Harvard), and John Henry Comstock (Cornell), took their most lasting impressions not from Haeckel or Gegenbaur, but from Leuckart's lab in Leipzig. They certainly would have come away with a very different conception of what Germans were up to than those who studied with Haeckel or Gegenbaur, although they might have found it expedient to bash Haeckelian science just as other scientific zoologists did. The names of nearly twenty American students of Leuckart are interspersed in a list of Leuckart students in Wunderlich, *Leuckart*, pp. 41–57; although this does include some students who did not receive Ph.D.'s under him, the list is incomplete; Comstock, for example, is not on the list; on Comstock, see P. Henson, "Evolution and Taxonomy" (1990); on American impressions of German zoology see P. J. Pauly, "American Biologists in Wilhelmian Germany." Of course, other factors besides their raw experiences would have affected which views Americans took home with them as quintessentially German (cf. Warner, "Remembering Paris").

12. See esp. Churchill, "From Heredity Theory to *Vererbung*," among other works; Jacobs, "From Unit to Unity"; M. Richmond, "Richard Goldschmidt" (1986); see also M. S. Saha, "Carl Correns" (1984); and G. Robinson, *Prelude to Genetics* (1979).

the lines of intellectual change might look quite different. To take just one example, a considerable amount of agitation for high-school natural history teaching reform in the 1880s and 1890s was focused around increasing the biological content of natural history education, that is, examining the organism in relation to its environment.[13] The source of this reform movement is mysterious if one believes that everyone in the biological community was preoccupied with inventing hypothetical ancestors and speculative phylogenetic trees, but it becomes more explicable in light of scientific zoology. Not only was this problem area a well-established one that many teaching candidates would have been likely to learn, but it also had the advantage of being neutral to Darwinism: a curriculum devoted to biological concepts could serve to open a discussion to evolutionary issues or not, at the discretion of the teacher.

A second new direction for research would be to begin investigating more deeply what was going on in German biology in the first two decades of the twentieth century. There can be little doubt that for Driesch and a few other members of his generation, the "revolt from morphology" was an overriding theme, one which still deserves further exploration. (Just why did Driesch and some of his contemporaries so vehemently reject evolutionary biology?) If we shift our attention away from Haeckelian morphology toward scientific zoology as the dominant tradition from which Driesch's generation was drawing, it appears as though there would be less to revolt against. Indeed, it seems that historians' emphasis on experimental embryology and genetics as representative of modern approaches has deflected us from numerous lines of zoological research considered modern and promising in early twentieth-century Germany, including ecological problems, symbiosis, protozoology, and animal behavior, as well as new approaches to traditional problem areas such as biogeography and classification.

If we are to embark on serious explorations of these research areas, we must simultaneously head in a third new direction as well and move away from an exclusive focus on the universities. The idea that the traditional universities were the fountainhead of innovative and important research has structured most accounts of nineteenth-century German science, including this one. But at least for zoology, this assumption has come about in the absence of serious historical research into other sites of activity. Beginning in the late 1870s, the traditional universities were producing more professionally aspiring

---

13. Scheele, *Von Lüben bis Schmeil,* chap. 3.

zoologists than they could employ as professors, and it is worth investigating where they went.[14] *Technische Hochschulen* and forestry and veterinary schools provided a few positions. Secondary-school teaching (more likely at a *Realschule* than at a *Gymnasium*) was a career that might allow time for some research, and additionally in the 1860s and early 1870s, there was a demand for teachers qualified to teach natural history. (By the early 1880s, however, there were already complaints about the overproduction of prospective math and science teachers.)[15] Public and private natural history museums offered some opportunities, although it was difficult to return to academia after such a career track. From the mid-1870s on, the most exciting nonacademic institutions to work at were probably the new marine stations such as those at Naples and Trieste, the Prussian marine station founded at Helgoland on the North Sea in 1892, and the private inland limnological station at Plön. Research fellowships, assistantships, and staff positions at these places also provided new sources of support for underemployed zoologists trying to stay in the profession.[16] The culmination of the drive for institutional alternatives (not just for zoologists) appeared with the Kaiser-Wilhelm-Institut für Biologie, founded in 1914 in good measure to accommodate researchers having trouble getting university professorships to support their novel biological orientations.[17] By investigating these sites of research and employment in the decades around the turn of the twentieth century, we would gain a more balanced picture of the zoological community and its activities.

14. For a preliminary survey, see L. K. Nyhart, "Beyond the Institutes" (1993).

15. On the earlier situation, see Ackermann, "Verschiedene Notizen" (1875), which quotes from L. Wiese, *Das höhere Schulwesen in Preußen;* on the later situation, see J. C. V. Hoffmann, "Zur Verständigung" (1881), p. 4; the concern about overproduction was repeated often in the pages of the *Zeitschrift für mathematischen und naturwissenschaftlichen Unterricht* in the early 1880s.

16. J. W. Spengel, for example, spent a year working as the librarian at Naples, then seven years as director of the museum of natural history and folk history at Bremen before finally gaining a university professorship; there is some indication that his nonacademic positions made it increasingly difficult for him to be considered for professorships; for a good sense of the vicissitudes and anxieties of a struggling zoologist, see Spengel to Ehlers, UB Göttingen, Cod. MS E. Ehlers 1851, esp. letters 2–38, 1876–89.

17. On the founding of the Kaiser-Wilhelm-Institute für Biologie, see L. Burchardt, *Wissenschaftspolitik* (1975), pp. 108–15; on the Kaiser-Wilhelm-Gesellschaft more generally, see, in addition to Burchardt, R. Vierhaus and B. vom Brocke, eds., *Forschung im Spannungsfeld von Politik und Gesellschaft* (1990); the lack of university positions was a major complaint expressed in the suggestions submitted by biologists for the structure and coverage of the KWI-Biologie (see, e.g., W. Roux, *Gutachten* [1912]).

Finally, a richer understanding of German morphology and biology more generally in the nineteenth and early twentieth centuries ultimately requires that we look beyond disciplinary and university issues toward broader cultural and political ones. As I have hinted, the increased reticence of zoologists over Darwinism and their more professionalized outlook in the 1890s may be connected to political and religious concerns, but to understand this better, we would need to know much more about the political and religious views of these scientists, as well as about their concept of their role as bearers of culture in their society. This is true for all of the scientists in my story, not just those active in the 1890s. For example, one might seek a connection between the scientific issues taken up by the founders of scientific zoology in the 1840s and 1850s and the political events surrounding the Revolution of 1848 and the reaction thereafter; one might also investigate the reactions of non-Prussian zoologists and anatomists to the unification of Germany, as well as the effects of the *Kulturkampf* on morphologists' views concerning the roles of science and religion in their society. Again, by moving beyond the universities to such locations as natural history museums and zoos, one might be able to draw interesting and important connections between zoology and the cultural and political history of Germany during the period of empire building around the turn of the century.[18] New historical research in any of these directions would enrich and broaden the picture I have drawn of the history of morphology, though I like to think it would not contradict it.

## DISCIPLINARY, INSTITUTIONAL, AND GENERATIONAL CHANGE

The implications of this study for the history of science go beyond the intellectual history of morphology; I hope I have raised some different ways of thinking about certain structural dynamics of intellectual change. Although, as I have indicated above, there is certainly more to be said about the cultural history of morphology, my chief

---

18. For works that have linked other German life scientists to their political and cultural context, see e.g., F. Gregory, *Scientific Materialism* (1977); Lenoir, "Social Interests"; and Weindling, *Darwinism and Social Darwinism*, chaps. 8, 9. The literature on German zoos is scanty and generally unscholarly; for an exception, see I. Jahn, "Zoologische Gärten" (1992); there is even less of value on natural history museums; a model for such research might be S. Marchand, "Archaeology and Cultural Politics" (1992); Cittadino (*Nature as the Laboratory*) offers provocative insights into the relationship of botanists to German imperial ambitions.

concern here has been to examine the intellectual changes in relation to the development of disciplines and disciplinary communities within the German universities. To do this, I have focused the study primarily around the question of how practitioners of a given intellectual orientation—morphology—made and preserved a place for themselves in the university setting.

As any reader knows who has made it this far, the story I have told is detailed and complex. To account for morphology's fortunes, much of the analysis has revolved around two sorts of social structure. One kind is bureaucratic; such structures include the disciplines as recognized through professorships, institutes, and state funding; their internal hierarchies of *Assistenten, Privatdozenten,* and *Extraordinarien;* and the larger bureaucracies of which the disciplines were a part—the university faculties, each university as a whole, and the university system managed by the German states. The other sort of structure is the age structure of the disciplinary community.

Before we dive into discussing the roles and relationships between these two sorts of structures, two points of clarification are in order. First, these structures alone cannot account for intellectual change; they are merely necessary elements of the story, and not sufficient ones. As I argue in the next section, the intellectual and social values of the bureaucratic and scholarly communities, and the distribution of power within each that enforced those values, are indispensible ingredients of a satisfactory accounting of an orientation's fortunes. Second, even if we add power and values to our list of elements to take into consideration, we must be careful about the sort of account these elements yield. If we think of these structures as independently acting causal factors that explain the history of morphology (or any other orientation), then that history will seem hopelessly overdetermined. As I hope my narrative has suggested, and the following analysis will reinforce, this is not exactly what I intend. Each of these features interacted with others in particular ways, and through these interactions morphology was constituted as an orientation. Rather than seeking a causal *explanation* for the history of morphology, then, I am hoping for a kind of *understanding* of it that is more fluid but no less analytical. That said, let us turn to the particular features and their interactions.

The important role of the community's age structure is perhaps most clearly indicated by the fact that one can summarize the story of nineteenth-century German morphology as a generational narrative (see table 1.1). As a heuristic, the generational approach has both strengths and limitations. By focusing our attention on horizontal, age-related groupings instead of the more vertical approach underlying many histories of science organized around research traditions, programs, schools, or the genealogy of ideas, the generational viewpoint helps us re-see familiar stories in new ways. For example Wilhelm His, Wilhelm Roux, and Hans Driesch are usually clumped as causal-mechanical thinkers opposed to Haeckelian morphology. If Haeckelism is to be the main point of departure, then this is a reasonable grouping (at least Haeckel thought so). But it might also be useful to see what separates them and joins them to other thinkers instead. A perspective that attends to generations can help us do this, for in fact these three men hailed from three different age cohorts for which the overriding intellectual and professional concerns differed. His (1831–1904), a close contemporary of Haeckel, was, like numerous other anatomists of his generation, preoccupied with the implications of the cell theory as it had developed up to the late 1850s, and was concerned to develop it in mechanical directions over the next several decades. For him Haeckel's orientation was a direct rival. Furthermore, His came into his discipline when it was expanding in a direction that favored microscopic anatomy, and he does not appear to have been overly concerned with fighting for elbowroom within his discipline in the 1860s and 1870s. Roux (1850–1924), a member of the next intellectual generation, was more inclined to try to deal with both mechanical and evolutionary levels of causation. In comparison to His, he seems to have been much more professionally self-aware, trying to situate his program strategically to make it appeal in a more institutionally competitive climate. Driesch (1867–1941), still a generation younger, entered the debates in the early 1890s, by which time His and Haeckel had been going at each other for over twenty years. His perspective was shaped by what was by now a longstanding feud between the two, and by the fact that Roux's program, already under way (at least on paper), could be interpreted in a more extreme fashion than Roux himself was inclined to do. By helping us to focus on timing and different experiences, attending to generations helps us understand some differences in tone between Driesch and Roux.

However, this same example also points up the limitations of generational analysis, for if we drop Driesch into his generation, he suddenly appears much less representative than one might have assumed. Experimental embryology and the extreme anti-Darwinian approach developed by Driesch do not appear to have been predominating concerns of his generation as a whole. Nor did Driesch share in another leading concern for his generation—the lack of university professorships—for he was independently wealthy and does not appear to have been concerned to find a position teaching zoology. Thus here the generational analysis seems to fail us, for it does not really help us understand the ways in which Driesch fit into broader trends of his age group. It appears that the key issues for this generation, as a generation, are yet to be found.

My generational story is limited in another way as well, for it does not pretend to substitute for a history of German biology in general. As appendix 1 shows, the generational lineup I have produced for morphology encompasses only a portion of all German professors of anatomy and zoology: if we were to focus on those other anatomists and zoologists, we might ask what they had in common that sets them apart from the morphologists. The answer seems to be cell theory and histology. Jakob Henle and Karl Reichert (who worked together in the mid- to late 1830s as students of Johannes Müller), born around 1810, were among the most active early theorists on cells and tissues as was Theodor Schwann (1810–82) (who is excluded from app. 1 because his professorship was outside of Germany). The famous cytologist and histologist Max Schultze, born in 1825, stands with Gegenbaur (b. 1826) as clearly in between professorial cohorts. A third small cluster of microscopists who focused on histology and were born in the early 1840s, includes the zoologists Franz Eilhard Schulze, Alexander Goette, and Hermann Grenacher. Indeed, based upon these examples, one might profitably set up an alternative generational schema of tissue and cell biologists, offset by about seven or eight years from the morphological generations.

The difference between these two generational stories helps to highlight the highly contingent relationship between institutional and intellectual developments. There were certainly fewer histologically inclined animal biologists than morphologically inclined ones. We may speculate that if more chairs had come open in anatomy in the late 1830s or the 1860s, the cohort of Henle and one of Schultze's contemporaries, along with their particular microscopical interests, might have been much more fully represented in the professoriate, with consequences for the intellectual directions taken by the discipline. Had no chairs in zoology opened up in the late 1860s, when Haeckel and Weismann's cohort launched their evolutionary research programs, the course of German zoology, too, might have looked quite different. Imagine if Haeckel and Weismann had given up hopes of a professorial career and become secondary school teachers, and the more histologically inclined contemporaries of Schulze and Goette had predominated in the 1870s and after![19]

Indeed, one of the broader points raised by the story of morphology is the crucial role played by institutional opportunity in enhancing or diminishing the fortunes of a given orientation. As readers familiar with the sociological history of German science will have recognized, this is much the same point made over thirty years ago by Awraham Zloczower in his extraordinarily influential Master's thesis on career opportunities and physiological research in late nineteenth-century Germany.[20] The ghost of Zloczower's thesis has hovered over these pages, at times shimmering into near visibility as I have discussed the fate of morphology in relation to career opportunities in anatomy and zoology. Certainly career opportunities played an important role in the fortunes of morphology. Sometimes they had to be seized, as the scientific zoologists seized them in the 1850s and 1860s when they turned a rejection from physiology into a pioneering program for zoology. But sometimes such opportunities seem to have fallen into a generation's collective lap, as in the case of the first generation of Darwinian zoologists. Conversely, lack of institutional opportunity, as faced by Driesch and Spemann's generation, may have helped to slow down the growth of *Entwicklungsmechanik*.

We must be cautious, however, in attributing too much to career opportunities alone, as represented by professorships, in explaining the history of a particular orientation. For one thing, as Zloczower pointed out in his thesis, members of a community with a dearth of professorships might respond in several ways that maintained career opportunities.[21] Discouraged prospective scholars might, of course,

19. This scenario may not be as far-fetched as it sounds; the evidence from German botany, scarce as it currently is, suggests the possibility of just such a pattern, in which very few botanists born in the early 1830s (the group that in zoology took up Darwinism with the greatest enthusiasm) gained professorships, while a group born in the early to mid-1840s, more preoccupied with cellular anatomy and physiology—including the cell physiologist Wilhelm Pfeffer (b. 1845) and the cellular anatomist Edouard Strasburger (b. 1844)—did obtain professorships. Of course, a fuller analysis would have to take into account the already-existing orientations within that discipline; for a start, see Cittadino, *Nature as the Laboratory;* Junker, *Darwinismus und Botanik.*

20. A. Zloczower, *Career Opportunities;* Zloczower aimed to link the production of new ideas in physiology with the institutional conditions facing young physiologists, arguing that during an expansive period, when university careers in physiology were possible, more new ideas were generated, whereas in the 1890s when no more positions were available, the number of important new ideas declined, presumably because bright young men siphoned themselves off into other realms that looked as though they had more career potential.

21. Ibid., esp. pp. 5, 100–125.

move out of the discipline into others that seemed to have more potential for a professorship. But the discipline might develop "waiting positions" through vertical stratification. This is what happened with the position of prosector, or head laboratory instructor, in anatomy.[22] Such intermediate academic positions were not available to zoologists facing a similar job crunch, for with their much smaller student constituency, they had a correspondingly greater difficulty in persuading their states of the need for longer-term salaried positions with the title of *Extraordinarius*. Instead, as I have indicated earlier in this chapter, they had to look to institutions outside the universities for support.

In the late nineteenth and early twentieth centuries, therefore, when the immediate possibilities for full professorships were extremely slender in both anatomy and zoology, persistent and lucky members of both disciplines could find ways to continue their research, despite more limited professional possibilities than had existed for their elders. A lack of full professorships did not by itself entirely close out either career opportunities or the possibility for developing a particular intellectual orientation. Indeed, it may have aided orientations such as marine ecology, which benefited from its connection to practical concerns among fisheries.

Noting that younger scholars could find ways to continue to pursue research is hardly the same as saying that a lack of available professorships had no effect on a discipline. I would agree with Zloczower that during times of expansion, when money was poured into the university system and it was possible to establish new professorships, intellectual novelty was more apt to be accommodated institutionally. (It is worth emphasizing that in the case of the fragmenting of physiology at midcentury, not only novelty, in the form of experimental physiology, but also the older tradition it displaced—namely, morphology—

22. Prosectorial positions were an old tradition in German anatomy; although they had originated as posts most often given to men of lesser education, already by the eighteenth century they were sometimes used as stepping stones to an *Ordinarius* and institute directorship. Usually younger prosectors held the title and salary of an *Extraordinarius,* but it was not uncommon for an older prosector to gain the title of *Ordinarius,* especially if he had served a long time and was unlikely to be named the head of the anatomy institute; on the history of the prosectorship, see Schierhorn, "Prosektor." At times when hopes for professorships were slim, as in the 1890s and 1900s, a prospective anatomist devoted to his calling might well view a prosectorship as an acceptable place to pass a good portion of his career, as did Gegenbaur's disciples Ruge and Maurer. In the early 1900s, Max Fürbringer brought up the age problem explicitly, noting that many of Germany's best anatomists were stuck in subordinate positions (Fürbringer to Schultze [draft, n.d., but evidently November 1906], and Fürbringer to Pflüger, 19 November 1906, SB, Fürbringer Nachlass, D-2-k, "Bonner Berufungsangelegenheit," items 206, 207.

was accommodated as well. Both can happen if the pie is getting bigger.) Conversely, in the absence of expansion, the university structure itself acted more often as a brake on radical innovation, largely because of the length of time individuals held professorships. And here again, the role of generational structure demands attention. Generational turnovers constitute an important part of the structural story because a wave of retirements could open up opportunities to a new generational cohort even in the absence of actual expansion. This is what happened with Prussian zoology in the early 1880s, when within a few years, nearly all the old professors retired or died. Nor need we confine the story to the sciences concerned with animal form: Klaus Köhnke, in his history of nineteenth-century neo-Kantianism, noted that the neo-Kantian generation born in the 1840s profited enormously from the opening up of philosophy professorships in the 1870s.[23] Other examples could doubtless be found if they were looked for.

If one is trying to account for the gross level of intellectual energy and novelty in a discipline, as Zloczower was, I am inclined to think that attention to demography and institutional structures can provide an important element of an explanation. Indeed, I would argue that without taking them (or their equivalents in other communities) into account, we will not understand the dynamics of intellectual change very well at all. However, once the question is shifted from Zloczower's concern with the quantity of 'important discoveries' to the success of particular intellectual orientations relative to others, other elements must necessarily be brought into play. To understand the status of morphology or any other orientation within a discipline, one must attend more directly to considerations of power and values that such institutional/structural explanations leave implicit, at best. Thus two other fundamental questions for this study concern where the power of selection among orientations lay and what values were sustained by those holding it.

## POWER AND VALUES IN THE GERMAN UNIVERSITIES

In the German university system, the power to influence disciplinary development was shared by university faculties and state educational administrators. The balance between them is most visible in the pattern of emergence of new (nonindustrial) disciplines: typically, the initiative came from the faculties, but the ministry, through its power

23. K. C. Köhnke, *Entstehung und Aufstieg* (1986), pp. 306–7.

to create new positions and its control of funding levels, opened or closed the door on formal institutional support for a prospective discipline.[24] The balance of power was also evident in the traditional right of the faculty to name a list of candidates for existing professorships and the state education minister's right to choose among them. Occasionally the minister could impose on the faculty a person from outside their list, but that so threatened the traditional balance of power—and even possibly the harmonious internal functioning of the faculty itself—that administrators rarely made use of their right to do so. Even the late nineteenth-century Prussian cultural minister Friedrich Althoff, who was notorious for using (or abusing) his power of appointment to circumvent faculty wishes, made little use of it as a way of directing intellectual developments in anatomy and zoology. It is possible that he simply did not care enough about these disciplines to alienate faculty over them, but that itself suggests the need for him to be prudent in exercising his power—it was not unlimited, and its limits were to some degree shaped by the professoriate.

Within individual disciplines the bureaucratic power structure was reinforced by, and in turn reinforced, the authority of age. Older members of the community held considerable direct power over the fortunes of younger men and their ideas. Not only could their recommendations further or diminish prospects for a career, but their longevity itself, which in German zoology and anatomy could mean a professorship maintained over as many as fifty years, meant the persistence of older perspectives and the fending off of younger men's access to positions of power. As Jonathan Harwood has recently emphasized for the case of German genetics, younger scholars were heavily dependent on a sponsoring *Ordinarius* for support of their research. Younger researchers could try to develop new lines of inquiry, but they had little say in the channeling of funds, space, or other resources. Occasionally a younger researcher would appeal directly to a state official for support, but this was frowned on as undermining faculty authority and autonomy.[25]

24. Of course, state policy considerations could, in some cases, be a driving force in supporting a new orientation, but these tended to come out more in the level of funding for it and the selection of professors to fill existing slots than in the creation of new chairs; for examples of the former, see Tuchman, "From the Lecture to the Laboratory"; Lenoir, "Science for the Clinic." The initiative in founding and developing disciplines with industrial links, such as chemistry and certain areas of physics, undoubtedly had other sources.

25. Harwood, *Styles of Scientific Thought*, pp. 166–72; this is part of a more extensive and very valuable discussion of power relationships in the German universities, pp. 156–77; Roux's appeals to Althoff, discussed in chap. 9 above, exemplify the

Attention to the interaction of the bureaucratic structure and age structure of the community draws renewed attention to what Kuhn reminded us of long ago: that when a new generation emerges with new ideas, the older ones do not disappear and their perspectives do not vanish immediately.[26] This feeling for the power of older members of the community is, I think, important, for it helps us take more seriously positions that we might take as merely rhetorical. For instance, it might be easy to dismiss as rhetorical bluster or dead-horse-beating attacks characterizing ideas as "naturphilosophisch" in the 1830s, but in fact, some proponents of *Naturphilosophie* were still alive, kicking, and making themselves heard then. Burdach's *naturphilosophisch* inclinations are still evident in 1839, in the concluding chapter he wrote to the six-volume collaborative work *Die Physiologie als Erfahrungswissenschaft* (Physiology as a science of experience).[27] Lorenz Oken (1779–1851), Georg August Goldfuss (1782–1848), and C. G. Carus (1789–1869) continued writing and publishing into the 1840s. To be sure, their moment of ascendance had long since passed, but that did not mean they had become entirely inconsequential. Similarly, the cohort of Kölliker (1817–1905) and Leuckart (1822–98), who pioneered scientific zoology in the late 1840s and early 1850s, continued to inculcate the values of intellectual rigor and epistemological modesty in their students into the 1890s, teaching them to eye the intervening Darwinian and Haeckelian theories with skepticism. They also controlled access to two of the largest and most productive research laboratories in their disciplines, with their attendant postdoctoral opportunities for research and employment. If their students did not always listen, they nevertheless had to come to grips with their teachers' attitudes.

If the power relationship between the ministry and the professoriate and that within the disciplinary communities have been the most evident in shaping the fortunes of particular orientations, another power relationship within the professoriate itself was also significant. R. Steven Turner pointed out long ago that professors carry a dual loyalty: to their institution and to their discipline.[28] Although Turner did not

---

boorishly direct approach. Of course, seniority carries authority in other systems, too, but the German case appears to have been extreme. By contrasting the German institute structure with the less hierarchical American department system, Harwood nicely shows how this age dependency could be less severe, and what its effect was on the growth of the new orientation of genetics.

26. T. Kuhn, *Structure of Scientific Revolutions* (1970), pp. 150–52.

27. Burdach, ed., *Die Physiologie als Erfahrungswissenschaft*, vol. 6 (1840), pp. 588–614.

28. Turner, "Growth of Professorial Research" p. 160.

put it this way, this dual loyalty reflects two different nodes of power, one held by members of the discipline, the other by the members of the local community. Being considered a good scholar by other members of one's discipline was by midcentury necessary to achieving a professorship, but it was not the only consideration. Although the *Ordinarius* has most often been represented as the ruler of his disciplinary kingdom at a university, he was not, in fact, autonomous. All requests for funding and lists of recommendation for appointments had to be approved by the faculty senate, made up of all the *Ordinarien* in the faculty. While there was undoubtedly considerable mutual logrolling in terms of requests for funding, appointment lists were much more often contested, negotiated, and made to accommodate the interests of the other professors. Thus if we consider just the level of the professoriate, decisions about the direction a discipline would take were not decided exclusively by its own membership: the power to shape the discipline was shared by its members with those in surrounding disciplines, university by university.

A disciplinary orientation gained support, then, through a combination of negotiations involving the membership of the discipline itself (where greater authority was accorded to senior members), the local cast of characters in related disciplines, and the state administrator in charge of universities. What they were negotiating over, at a basic level, always involved values. In any given decision, the balance of power might be different, and the values sustained by the decision might differ accordingly. However, the orientations that emerged as dominant, I would argue, were those that tended to reflect values agreed upon by these different groups. Two values in particular have been central to the history of morphology that I have told here, and their frequent conflicts, with different outcomes at different times, contributed much to the turns of the story.

Fundamental to the idea of *Wissenschaft,* and common to anatomists, zoologists, their colleagues in the medical and philosophical faculties, and state administrators concerned with education, was the value of intellectual innovation. Research did not mean simply accumulating ever more empirical data, although that was certainly an important part of it. It also meant developing new methods and new intellectual frameworks. Although there were always objections to particular innovations, without the idea that innovation in general was good, research itself would be a hollow enterprise. Now, of course, there were limits to the acceptability of novel ideas—truly iconoclastic ideas were unlikely to get a hearing, and those perceived to threaten the social order were likely to be quashed at the state level, if not before. In addition, any innovation is likely to raise some negative response from those whose work it challenges. It is neverthe-less worth mentioning this basic value, because if it did not exist, knowledge would not change at all. This is not the only value shared by the scientific community, but it is surely one of the most impor-tant.[29] It is also especially relevant to the story of morphology because of its relationship to another basic value within both anatomy and zoology: intellectual breadth.

Elaborating on a theme suggested by the work of Fritz Ringer, Jonathan Harwood has recently argued that the commitment to breadth among early twentieth-century biologists derived in part from the self-perception of the professoriate as the society's culture bearers and partly from the "one-professor-one-discipline" structure of the university that demanded broad teaching skills.[30] As the history of morphology shows, the commitment to breadth among life scientists has roots stretching back at least as far as the early nineteenth century. Not only were individual scholars praised for their breadth of knowl-edge, but objections to new orientations were often expressed in terms of their perceived narrowness. Such arguments were made by the morphologically oriented physiologist Rudolph Wagner against the physicalist physiologists in the 1840s and 1850s, and by Haeckel against the "one-sidedly" microscopical morphologists of the 1870s. Proponents of new orientations took two characteristic stances: some-times (but by no means always) they acknowledged their "nar-rowness" but claimed that the quantity of knowledge in the discipline had grown too large for a single person to master it. Whether or not this argument was used, innovators virtually always presented their orientation as more *wissenschaftlich* than existing ones, holding the prospect of penetrating more deeply, more fully into nature. To the extent that such claims also presented the possibility of finding gen-eral, fundamental laws of nature, they were also claims to breadth.

As the preceding discussion suggests, the sometimes conflicting val-ues of innovation and breadth mapped readily onto the realms of research and teaching: teaching was supposed to be founded upon a broad research base, but a new research orientation, it was under-

29. So much of the science studies literature emphasizes barriers to innovation thrown up by self-protective scientists, one begins to wonder how any intellectual change ever takes place; the answer lies, I think, in this underlying commitment to innovation.

30. Harwood, *Styles of Scientific Thought*, chaps. 4, 5.

stood, might require intensive digging into one area to the exclusion
of others. Innovators often had to struggle against being considered
too narrow to teach well, and differences in the value placed on innovation and breadth often turned into conflicts over teaching and research. Indeed, despite the nineteenth-century German university's
mythological status as the fountainhead of the modern research university, research in the university never supplanted teaching—it was
added to it. Teaching always remained a fundamental mission of the
university. The centrality of this point was underscored at the turn of
the twentieth century by Althoff in advocating the need to establish
institutions devoted solely to research: "In many cases it is not possible to fill chairs with persons equally talented for teaching and research. The highest gifts on the latter side seem to exclude the former,
since for the researcher it is necessary to concentrate on a particular
area, whereas for the university instructor the most many-sided education and handling of broad areas must be assumed."[31] In expressing
regret that there was no appropriate institutional location for the
talented researcher not equally gifted for teaching, Althoff implicitly
acknowledged that universities could not become institutions devoted
purely to research.

Considerations of innovation/research and breadth/teaching were
played out in different ways in the two disciplines I have focused on
here, as they interacted with each other and with other values particular to each discipline. In anatomy, the primary constellation of values
throughout the century revolved around the relationship between utility to medical practice and *Wissenschaft* for its own sake. Although
this relationship was an issue in most other areas of medicine as well,
in anatomy it took on a unique character because of anatomy's pedagogical role as the fundament of all medical education. Medical faculties seeking strong enrollment in the preclinical years (which included
especially those located away from large clinics) were particularly
concerned with prospective professors' teaching skills, including the
teaching of both gross anatomy (important for diagnostic and surgical
skills) and microscopic anatomy (important for pathological understanding). Within the discipline, however, gross anatomy did not have
the research cachet that microscopic anatomy did—it was often seen
to be a closed field of research, whereas microscopic anatomy remained wide open. Thus a microscopic anatomist who was competent
in gross anatomy was generally preferable to the reverse.

31. F. Schmidt-Ott, citing "Althoff's rich experiences," quoted in J. Lemmerich,
ed. *Dokumente zur Gründung* (1981), p. 43.

For much of the century, morphologists could successfully claim to bridge the two in their teaching, by placing both macroscopic and microscopic structures into a larger comparative framework that would cement the myriad facts of anatomy into a unified whole that students could grasp and keep referring back to. What they were less successful at was linking that picture to practical medicine. During the early and middle decades of the century, this appears not to have been an overriding problem. But by the 1870s and 1880s, medical faculties were much more concerned that anatomists present their subject in a way more directly conducive to practical medical learning. Furthermore, by then the comparative perspective was doubly tarnished by its recent association with Darwinism: on the one hand, it was tied to a potentially socially dangerous doctrine, and on the other hand, that doctrine itself, along with those who exercised it, was perceived by most medical men to be more suitably located in the philosophical faculties.

In zoology, where university teaching for practical ends meant something quite different (namely, the education of high-school teachers), the concerns over breadth and innovation interacted with interestingly different results. Breadth continued to be a trait valued among zoologists from the early nineteenth century all the way through the early twentieth century, but it could be construed in different ways. For most of the nineteenth century, when zoologists referred to "breadth," they seem generally to have meant research-based knowledge of many different taxa, preferably from different classes. But this is not the only way to construe breadth. Theoretical breadth also counted, at least after midcentury, and could be achieved either by working in numerous different problem areas or by developing a comprehensive framework that unified them. Haeckel's gastraea theory embodied both kinds of breadth as well as being innovative—this may help account for the felt need to take it seriously by those who might otherwise have discounted it as dogmatic puffery.

One of morphology's features that probably contributed to its survival and success over the course of the century was that it was fruitful for people who valued breadth. With comparison as its central method, it invited breadth of a taxonomic kind; to the extent that the organization of the whole animal was viewed as the locus through which to study development, taxonomy, reproduction, and adaptation, it also invited problem-oriented breadth. Where it began to look less fruitful by the latter part of the century was in terms of innovation. Thus those who valued innovation more than breadth (usually younger scholars) would be more apt to sacrifice breadth or try to reinterpret it to accommodate innovative approaches, as did those who sought to breathe new life into morphology by introducing experimental methods.

Although it has been useful to set up innovation and breadth as distinct and opposing values, they were not actually as separate as all that. In fact, their interaction produced one of the most interesting and significant changes in the nature of zoology. Over the century, I would argue, intellectual innovations very gradually wrought a change in the kind of breadth zoologists sought to achieve. Although systematic breadth appears to have remained the predominant way of understanding it among zoologists, by the end of the period covered here, such areas as genetics and cell biology could compete (though not especially successfully in the universities) for claims to breadth. A claim that the cell, for example, was the fundamental site of organic activity ranging from reproduction through development to physiological functions, was a claim to breadth quite different from one based on knowledge of numerous different taxonomic groups.[32] By 1930 one obituarist regarded as common a division within the zoological community between "systematic experts on fauna" [*systematische Faunisten*] and "problem zoologists" [*Problemzoologen*], a contrast that suggests a gradual refashioning of what counted as "broad" in biology. Harwood's discussion of breadth among geneticists implicitly suggests such a change, in remarking on the number of geneticists working in "*both* developmental *and* evolutionary problems"—not, we might note, "genetics in both vertebrates and invertebrates."[33]

This gradual reshaping of the construction of breadth, which was by no means complete at the end of the period covered here, points us to one final observation about breadth and innovation. Much has been made of the scholarly community's complaints and fears about fragmentation and specialization in the early twentieth century, which were many and vocal.[34] Without denying the reality of specialization,

32. O. Hertwig made this claim in the foreword to the 2d ed. of his textbook *Allgemeine Biologie* (formerly titled *Die Zelle und die Gewebe* [Cells and tissues]), when he wrote, "I designate as 'general biology' that science that, taking a comprehensive viewpoint, concerns the morphology and physiology of the cell and the related large questions concerning life: the elementary structure and basic characteristics of the living substance, the problems of reproduction, inheritance, development, the nature of the species, and so forth" (O. Hertwig, *Allgemeine Biologie* [1912], p. vii).

33. O. Koehler, "Maximilian Braun" (1930), p. v; Harwood, *Styles of Scientific Thought*, p. 183.

34. F. Ringer, *Decline of the German Mandarins* (1969); Harwood, *Styles of Scientific Thought*; such complaints were heard in the American community as well at the

I would note that it was a classic response in this setting to react to innovation by calling it narrow or specialized. This should not blind us to the much deeper transformation in zoological research taking place in this period, from a predominant way of looking at nature in terms of taxonomic entities—one which persisted long past the era of early nineteenth-century systematists and into the evolutionary framework—to one organized primarily around problems and processes such as development, heredity, cellular function, and the effects of the physical environment on the individual. Although this transformation remained incompletely institutionalized in Germany because of the priority accorded to the traditional organization of teaching matter, it was nonetheless visible. As I hope to have shown in this book, its seeds had already begun to sprout some fifty years earlier, when the approach to form and function that came to be known as scientific zoology was transplanted from medical faculties to the fertile new soil of zoology.

---

same time (Harwood, *Styles of Scientific Thought,* pp. 23–24; Maienschein, *Transforming Traditions,* pp. 296–300).

# APPENDIXES

APPENDIX 1 Anatomists and Zoologists Holding German University Professorships 1810–1918, by Year of Birth.

| GENERATION* | ZOOLOGISTS | ANATOMISTS |
|---|---|---|
| | | DÖLLINGER, 1770 |
| | | Rudolphi, Rosenmüller, 1771 |
| | | Autenrieth, 1772 |
| | | Fuchs, 1774 |
| | | BURDACH, Langenbeck, 1776 |
| | Gravenhorst, 1777 | G. Fleischmann, 1777 |
| | | Bartels, 1778 |
| 1. 1776–1782 | OKEN, 1779 | L. F. Froriep, WILBRAND, 1779 |
| | Lichtenstein, 1780 | Rosenthal, 1780 |
| | | MECKEL, Tiedemann, 1781 |
| | GOLDFUSS, Nitzsch, 1782 | Bünger, 1782 |
| | | Otto, 1786 |
| | | A. F. Mayer, 1787 |
| | Herold, 1790 | |
| | | VON BAER, 1792 |
| | | RATHKE, Quittenbaum, 1793 |
| | | Perleb, F. S. Leuckart, Buchegger, Rapp, 1794 |
| | | M. I. Weber, C. A. S. SCHULTZE, E. H. Weber, E. Schneider, Schlemm, 1795 |
| | | Huschke, 1797 |
| | | Barkow, 1798 |
| 2. 1795–1805 | Pöppig, 1799 | |
| | BRONN, 1800 | J. MÜLLER, Volkmann, 1801 |
| | | ARNOLD, d'Alton, 1803 |
| | Berthold, C. T. E. SIEBOLD, 1804 | Kobelt, 1804 |
| | R. WAGNER, 1805 | |
| | Burmeister, 1807 | Bischoff, 1807 |
| | Stannius, 1808 | Behn, 1808 |
| | Troschel, 1809 | Henle, 1809 |
| | | A. Müller, 1810 |
| | Grube, 1811 | REICHERT, Budge, 1811 |
| | | F. L. Fick, 1813 |
| | | Bergmann, 1814 |
| | F. WILL, Peters, 1815 | |
| | | ECKER, 1816 |
| | Zaddach, 1817 | KÖLLIKER, 1817 |
| 3. 1815–1823 | | Bruch, 1819 |
| | Giebel, 1820 | J. Gerlach, Luschka, 1820 |
| | LEYDIG, 1821 | Lieberkühn, Helmholtz, 1821 |
| | LEUCKART, 1822 | Welcker, Claudius, Eckhard, 1822 |
| | O. SCHMIDT, 1823 | |
| | Pagenstecher, Möbius, Rütimeyer, 1824 | |
| | | M. Schultze, 1825 |
| | | GEGENBAUR, 1826 |
| | Gerstaecker, Greeff, 1827 | |
| | Meissner, 1828 | |

# APPENDIX 1 *continued*

| GENERATION* | ZOOLOGISTS | ANATOMISTS |
|---|---|---|
| | | Sommer, KUPFFER, 1829 |
| **4. 1829–1835** | Schneider, 1830 | |
| | SEMPER, 1831 | His, Braune, La Valette, 1831 |
| | Keferstein, 1832 | Rüdinger, 1832 |
| | WEISMANN, HAECKEL, 1834 | Henke, 1834 |
| | CLAUS, EHLERS, 1835 | Aeby, Eberth, 1835 |
| | | Waldeyer, 1836 |
| | | Stieda, 1837 |
| | F. E. SCHULZE, GOETTE, 1840 | |
| | | HASSE, 1841 |
| | SELENKA, 1842 | |
| | EIMER, Grenacher, 1843 | Flemming, 1843 |
| | | Schwalbe, 1844 |
| | | Merkel, 1845 |
| | | FÜRBRINGER, 1846 |
| | | Gasser, 1847 |
| | Bütschli, 1848 | WIEDERSHEIM, 1848 |
| | | Barfurth, Brunn, Stoehr, |
| **5. 1846–1854** | | O. HERTWIG, A. Froriep, |
| | | 1849 |
| | R. HERTWIG, Braun, 1850 | ROUX, 1850 |
| | H. Ludwig, Chun, Spengel, | L. Gerlach, Bonnet, 1851 |
| | 1852 | C. RABL, 1853 |
| | Brandt, 1854 | Rückert, 1854 |
| | HEIDER, 1856 | Spee, 1855 |
| | G. W. Müller, 1857 | Strahl, 1857 |
| | Seeliger, Blochmann, | |
| | KORSCHELT, 1858 | O. Schultze, MAURER, 1859 |
| | Kükenthal, 1861 | KEIBEL, 1861 |
| | Boveri, Plate, Fleischmann, | |
| | 1862 | |
| | Brauer, 1863 | |
| | Haecker, 1864 | M. Heidenhain, 1864 |
| | | Gaupp, 1865 |
| | | MOLLIER, Held, 1866 |
| **6. 1864–1870** | | KALLIUS, 1867 |
| | | BRAUS, 1868 |
| | SPEMANN, Hesse, 1869 | Sobotta, 1869 |
| | | C. Peter, 1870 |
| | Meisenheimer, 1873 | |
| | Schleip, 1879 | |
| | Becher, 1884 | |
| | Kühn, 1885 | |

*Note:* Names in capitals indicate scholars primarily concerned with animal form. The division into "anatomists" and "zoologists" is somewhat artificial for those men born before about 1825 (see chaps. 2 and 3).

*"Generation" refers to generations of morphologists, as represented in table 1.1.

APPENDIX 2   Professorships in Zoology, 1850–1918

| UNIVERSITIES | PROFESSOR IN 1850 | 1850–1859 | 1860–1869 | 1870–1879 | 1880–1889 | 1890–1899 | 1900–1909 | 1910–1918 | |
|---|---|---|---|---|---|---|---|---|---|
| Berlin | ←1813 Lichtenstein | 1857 Peters | | | 1884 F. E. Schulze | | | 1917 Heider | →1923 |
| | | | | | 1887 Möbius | | 1906 Brauer ao | 1914 Brauer oP | |
| | | | | | | | | 1918 Kükenthal | →1922 |
| Bonn | ←1849 Troschel ao | | | | 1883 R. Hertwig [1885 Leydig] 1887 H. Ludwig | | | 1914 R. Hesse | →1925 |
| Breslau | ←1811 Gravenhorst [+NH] | 1857 Grube | | | 1881 Schneider | 1891 Chun 1898 Kükenthal | | | →1918 |
| Greifswald | [1852 Münter B, NH] | | 1864 Buchholz ao | 1876 Gerstaecker | | 1895 G. W. Müller | | | →1923 |
| Halle | ←1842 Burmeister | | 1861 Giebel | | 1881 Grenacher | | 1909 Haecker | | →1927 |
| Königsberg | ←1834 Rathke [+A, vA in med.] | | 1860 Zaddach | | 1881 R. Hertwig 1883 Chun | 1891 M. Braun | | | →1921 |
| Kiel | 1852 Behn [+A in med.] | | 1868 Möbius | | 1888 Brandt | | | | →1922 |
| Marburg | ←1824 Herold [+P, NH] | | | 1871 Greeff ao 1872 oP | | 1893 Korschelt | | | →1928 |
| Göttingen | ←1836 Berthold [+vA, P in med.] | | 1861 R. Wagner [+vA in med.] 1864 Keferstein ao 1868 oP | 1870 Claus 1874 Ehlers | | | | | →1919 |

| | | | | | | |
|---|---|---|---|---|---|---|
| **Strassburg** (founded 1872) | 1872 O. Schmidt | | 1886 Goette | | | →1918 |
| **Munich** | 1861 Siebold | | 1885 Hertwig | | | →1925 |
| **Würzburg** | ←1830 Leiblein | 1869 Semper | | 1893 Boveri | 1916 Schleip | →1948 |
| **Erlangen** | ←1848 Will | 1869 Ehlers | 1874 Selenka | 1898 Fleischmann | | →1933 |
| **Leipzig** | ←1846 Pöppig | 1869 Leuckart | | 1898 Chun | 1914 Meisenheimer | →1933 |
| **Jena** | 1855 Gegenbaur [+A in med.] | 1862 Haeckel ao 1865 oP | | | 1909 Plate | →1934/5 |
| **Giessen** | 1850 Leuckart ao 1855 oP | 1869 Schneider | | 1881 Ludwig 1887 Spengel | | →1921 |
| **Heidelberg** | ←1837 Bronn | 1863 Pagenstecher ao 1866 oP | 1878 Bütschli | | | →1918 |
| **Freiburg** | 1863 Weismann ao 1867 oP | | | | 1912 Kühn | →1918 |
| **Tübingen** | 1857 Leydig | 1875 Eimer | | 1898 Blochmann | | →1925 |
| **Rostock** | 1864 F. E. Schulze ao | 1871 Schulze oP 1873 Grenacher | 1882 Goette 1887 Braun | 1891 Blochmann 1898 Seeliger | 1908 Spemann 1914 Becher | →1921 |

*Note:* All professorships are in philosophical faculty unless otherwise noted. A = anatomy; B = botany; ao = *ausserordentliche* professor; med. = medical faculty; NH = natural history; oP = *ordentliche* (full) professor; P = physiology; vA = *vergleichende* (comparative) anatomy; brackets indicate that zoology teaching is one of a number of several teaching duties associated with the professorship; it does not have an independent professorship devoted to it; left-pointing arrow ← and date in first chronological entry indicate date when professor teaching zoology in 1850 gained his professorship; right-pointing arrow → in last chronological entry indicates when 1918 professor gave it up.

APPENDIX 3   Professorships in Anatomy, 1850–1918

| UNIVERSITIES | PROFESSOR IN 1850 | 1850–1859 | 1860–1869 | 1870–1879 | 1880–1889 | 1890–1899 | 1900–1909 | 1910–1918 |
|---|---|---|---|---|---|---|---|---|
| Berlin | ←1833 J. Müller [+P] | 1858 Reichert | | | 1883 Waldeyer | | | 1917 Fick →1935 |
| | | | | | 1888 O. Hertwig | | | →1922 |
| Bonn | ←1819 Mayer [+P, vA, pA] | 1855 Helmholtz [+P] 1859 M. Schultze | | 1875–1907 La Valette St. George 1875–87 Leydig [+vA] | | | 1907 Bonnet →1918 | |
| Breslau | ←1835 Barkow | | | 1873 Hasse | | | | 1915 Gaupp 1917 Kallius →1918 |
| Greifswald | ←1831 C. A. S. Schultze [+P, pA] | 1856 Budge [+P to 1872] | | | 1888 Sommer | 1895 Bonnet | | 1917 Peter →1935 |
| Halle | ←1834 d'Alton [+P] | 1854 Volkmann ao [+P] | 1867 Volkmann oP [+P to 1870] 1866–94 Welcker | 1876 Steudener | 1881–94 Eberth | 1895 Roux | | →1921 |
| Königsberg | ←1834 Rathke [+vA] | | 1860 A. Müller | 1876 Kupffer | 1880 Schwalbe 1883 Merkel 1885 Stieda | | | 1912 Gaupp 1916 Sobotta →1919 |
| Kiel | ←1848 Behn [+P to 1853, Z] | | 1867 Kupffer | 1876 Flemming | | | | 1912 Graf Spee →1923 |
| Marburg | ←1843 F. L. Fick [+pA] | 1859 Claudius | 1867 Lieberkühn | | 1887 Gasser | | | →1918 |
| Göttingen | ←1814 Langenbeck [+surg.] | 1852 Henle | | | 1885 Merkel | | | →1919 |

| University | | | | | | | | Terminal |
|---|---|---|---|---|---|---|---|---|
| Strasburg (founded 1872) | | | 1872 Waldeyer | 1883 Schwalbe | | | 1914 Keibel | →1918 |
| Munich | ←1832 E. Schneider | 1853 Siebold [+P, vA] | 1854 Bischoff [+P to 1856] | | 1880 Kupffer; 1880 Rüdinger | 1897 Rückert | 1902 Mollier | | →1935; →1922 |
| Würzburg | ←1848 Kölliker [+P, vA to 1865] | | | | | 1897 Stöhr | | 1911 O. Schultze | →1920 |
| Erlangen | | 1850 J. Gerlach [+P to 1872] | | | | 1891 L. Gerlach | | | →1918 |
| Leipzig | ←1821 E. H. Weber [+P to 1865] | | | 1872–1904 His; 1872–1892 Braune | | | 1904 Rabl | 1917 Held | →1934 |
| Jena | ←1838 Huschke [+P] | 1855 Gegenbaur ao 1858 oF | | 1873 Schwalbe | 1881 O. Hertwig; 1888 Fürbringer | | 1901 Maurer | | →1932 |
| Giessen | ←1845 Bischoff [+P] | 1855 Bruch | 1860 Eckhard [+P] | | | 1891 Bonnet; 1895 Strahl | | | →1920 |
| Heidelberg | ←1844 Henle [+P, vA] | 1852 Arnold [+P to 1858, vA] | | 1873 Gegenbaur | | | 1901 Fürbringer | 1912 Braus | →1921 |
| Freiburg | ←1847 Kobelt | 1857 Ecker | | | 1883 Wiedersheim | | | | →1918 |
| Tübingen | ←1845 Arnold [+P, Path.] | 1852 Luschka [+Path. to 1867] | | 1875 Henke | | 1895 Froriep | | 1917 Heidenhain | →1933 |
| Rostock | ←1831 Quittenbaum [+surg.] | 1852 Bergmann [+P, vA] | 1866 Henke | 1872 Merkel | 1883 Brunn | 1896 Barfurth | | | →1920 |

*Note:* ao = *ausserordentliche* professor; oP = *ordentliche* (full) professor; P = physiology; pA = pathological anatomy; Path. = pathology; surg. = surgery; vA = *vergleichende* (comparative) anatomy; Z = zoology; left-pointing arrow ← and date in first chronological entry indicate date when professor teaching zoology in 1850 gained his professorship; right-pointing arrow → in last chronological entry indicates when 1918 professor gave it up. For simplicity, only the year in which a professor began his appointment is listed, occasionally a year or two might pass between the death or retirement of one professor and the appointment of his successor.

# ARCHIVAL SOURCES

BGLA = Badische Generallandesarchiv, Karlsruhe: file 235/19905, "Lehr-kanzel der Zoologie an der Universität Heidelberg, 1862–1948"; file 235/29858, "Universität Heidelberg Dienste–Lehrstuhl für Anatomie 18731952."

BHStA München = Bayerische Hauptstaatsarchiv, Munich: file MK 11072, "Hohe Schulen in genere; das Studium der Medizin in Genere, Bd. 6, 1871–72"; file MK 11415, "Hohe Schule Würzburg; Anatomisches Institut, Bd. 1, 1823–76"; file MK 11496, Universität Erlangen, Philo-sophische Fakultät, Lehrstellen, Bd. 2, 1895–1910"; file MK 11750, "Akademie der Wissenschaften; Zoologisches Institut und Zoologisch-zootomische Sammlung, Bd. 6, 1916–"; file MK 19432, "Akademie der Wissenschaften; Zoologisches Institut und Zoologisch-zootomische Sammlung, Bd. 5, Wissenschaftliche Personal, 1885–1915"; file M. Inn. 61291, "Studium der Medizin, Facultäts-Prüfungen, Bildung der Senate-Prüfungsfragen, 1847–1859, 1859–1869."

GStA Merseburg = Geheimer Staatsarchiv Merseburg

Berlin Med. Fak. = Rep. 76, Vᵃ, Sekt. 2, Tit. IV, Nr. 46, Bd. 2, "Die Anstellung und Besoldung der ausserordentlichen und ordentlichen Pro-fessoren in der medizinischen Fakultät der Universität zu Berlin" (January 1858–June 1862).

Berlin Phil. Fak. = Rep. 76, Vᵃ, Sekt. 2, Tit. IV, Nr. 47, Bd. 19, "Die Anstellung und Besoldung der ausserordentlichen und ordentlichen Pro-fessoren in der philosophischen Fakultät an der Universität zu Berlin" (August 1883–September 1884).

Berlin Zool. Sammlung = Rep. 76, Vᵃ, Sekt. 2, Tit. X, Nr. 63, Bd. 3: "Die zoologische und zootomische Sammlung der Universität zu Berlin, December 1856–December 1860."

Bonn Med. Fak. = Rep. 76, Vᵃ, Sekt. 3, Tit. IV, Nr. 39, Bd. 6, "Die Anstel-lung und Besoldung der ausserordentlichen und ordentlichen Professoren in der medizinischen Fakultät der Universität zu Bonn" (January 1872–March 1878).

Bonn Phil. Fak. = Rep. 76, Vᵃ, Sekt. 3, Tit. IV, Nr. 10, Bd. 14, "Die Anstellung und Besoldung der ausserordentlichen und ordentlichen Professoren der philosophischen Fakultät der Universität zu Bonn" (October 1880–April 1883).

Bonn Zool. Prof. = Rep. 76, Vᵃ, Sekt. 3, Tit. IV, Nr. 28, "Acta betr. die Wiederbesetzung der durch den Tod des Professor Dr. Goldfuss in Bonn erledigten ordentlichen Professur der Zoologie bei der dortigen königlichen Universität."

Breslau Med. Fak. = Rep. 76, Vᵃ, Sekt. 4, Tit. IV, Nr. 35, Bd. 4, "Die Anstellung und Besoldung der ausserordentlichen and ordentlichen Professoren in der medizinischen Fakultät der Universität zu Breslau" (January 1869–September 1874).

Halle Med. Fak. = Rep. 76, Vᵃ, Sekt. 8, Tit. IV, Nr. 33, "Die Anstellung und Besoldung der ausserordentlichen und ordentlichen Professoren in der medizinischen Fakultät der Universität zu Halle" (Bd. 4, January 1872–September 1881; Bd. 7, April 1894–September 1900).

Halle Phil. Fak. = Rep. 76, Vᵃ, Sekt. 8, Tit. IV, Nr. 34, Bd. 13, "Die Anstellung und Besoldung der ausserordentlichen und ordentlichen Professoren der philosophischen Fakultät der Universität zu Halle" (January 1880–December 1882).

Königsberg Professoren = Rep. 76, Vᵃ, Sekt. 11, Tit. IV, Nr. 1, Bd. 4, "Die Anstellung und Besoldung der ausserordentlichen und ordentlichen Professoren bei der Universität zu Königsberg" (September 1811–December 1814).

Königsberg Anat. Theater = Rep. 76, Vᵃ, Sekt. 11, Tit. X, Nr. 10, Bd. 1, "Das anatomische Theater der Universität zu Königsberg" (November 1809–May 1828).

Rep. 92 Althoff = Althoff Nachlass.

Medhist. Inst. Zürich = Medizinhistorisches Institut der Universität Zürich: Johannes Brunner, lecture notes on Blumenbach's course on physiology, winter semester 1804–5; Franz Sidler, lecture notes from Heinrich Müller's lectures on *vergleichende Anatomie* and Albert von Kölliker's lectures on *Physiologie der Menschen,* both dated Würzburg, 1861; Strasser Nachlass.

SB = Senckenbergisches Bibliothek, Frankfurt am Main: Fürbringer Nachlass.

SBPK = Staatsbibliothek Preussischer Kulturbesitz, Berlin: 3k 1855 (5): C. T. von Siebold; Lc 1851: R. Leuckart; Lc 1860 (9): E. Ehlers; Lc 1875 (13): E. Haeckel; Lc 1889 (23): A. Weismann; Lc 1898 (29): Chun; Lc 1898 (29): R. Hertwig.

Staatsarchiv Marburg

Zoologisches Institut = Best. 16 Rep. VI, Kl. 14, Nr. 2, "Königliche Universitäts-Kuratorium Marburg; Zoologisches Institut zu Marburg, Bd. 1, 1817–1892."

Zoologie = file 307d, Acc. 1966/10, Nr. 17, "Zoologie, Entwicklungsgeschichte 1892–1934."

StaBi München = Staatsbibliothek München, Handschriftenabteilung: Boveri Nachlass.

UA Giessen = Universitätsarchiv Giessen: Akten der Landes-Universität zu Giessen.

UA Göttingen = Universitätsarchiv Göttingen: Personalakte Keferstein, 4)IVb, Nr. 143; Personalakte Merkel, 4)IVb, Nr. 124.

UA Halle = Universitätsarchiv Halle: Rep 29, Med. Fak. I: Medizinische Fakultät Dekanatsbücher.

UA Heidelberg = Universitätsarchiv Heidelberg: Akten der philosophischen Facultät 1877–78.

UA Leipzig = Universitätsarchiv Leipzig: Personalakte Rudolf Leuckart.

UA München = Archiv der Ludwig-Maximilians-Universität München.
  Senatsakten: "Anatomie als Disziplin, 1835–78"; "Zoologie und vergleichende Anatomie, Bd. I, 1832–1929."
  Dekanatsakten: OCI 10, Dekanat der philosophischen Fakultät, Sekt. 2, 1883–84.

UB Göttingen = Universitätsbibliothek Göttingen, Handschriftenabteilung: Cod. MS Ernst Ehlers; Cod. MS Rudolph Wagner.

UB Heidelberg = Universitätsbibliothek Heidelberg, Handschriftenabteilung: Braus Nachlass; Fürbringer Nachlass.

Weismann, August. "Erinnerungen," typescript autobiography, 1914, copy in personal collection of Professor K. Sander, Zoologisches Institut, Universität Freiburg.

# BIBLIOGRAPHY

## REFERENCE WORKS

*ADB* = *Allgemeine Deutsche Biographie.* 53 vols. 1875–1907. Reprint, Berlin: Duncker & Humblot, 1967–71.
*Biogr. Lex.* I = *Biographisches Lexikon der hervorragenden Ärzte aller Zeiten and Völker.* Edited by W. Haberling, F. Hubotter, and H. Vierordt. 2d ed. 5 vols. Berlin: Urban & Schwarzenberg, 1929–35.
*Biogr. Lex.* II = *Biographisches Lexikon der hervorragenden Ärzte der letzten fünfzig Jahre.* Edited by I. Fischer, 2 vols. Berlin: Urban & Schwarzenberg, 1932.
*DSB* = *Dictionary of Scientific Biography.* Edited by Charles C. Gillispie. 16 vols. New York: Scribner's, 1970–80.

## BOOKS AND ARTICLES

Ackermann, ——. "Verschiedene Notizen." *Zeitschrift für mathematischen und naturwissenschaftlichen Unterricht* 6 (1875): 342–43.
Adams, Mark. "Severtsov and Schmalhausen: Russian Morphology and the Evolutionary Synthesis." In *The Evolutionary Synthesis: Perspectives on the Unification of Biology,* edited by Ernst Mayr and William Provine, pp. 193–225. Cambridge: Harvard University Press, 1980.
Allen, Garland. *Life Science in the Twentieth Century.* New York: John Wiley, 1975.
Amsterdamska, Olga. "Stabilizing Instability: The Controversy over Cyclogenic Theories of Bacterial Variation during the Interwar Period." *Journal of the History of Biology* 24 (1991): 191–222.
Appel, Toby A. *The Cuvier-Geoffroy Debate: French Science in the Decades before Darwin.* New York: Oxford University Press, 1987.
Arnold, Friedrich. *Handbuch der Anatomie des Menschen, mit besonderer*

**375**

*Rücksicht auf Physiologie und praktische Medicin.* Freiburg im Breisgau: A. Emmerling, 1844.

"Aus dem preußischen Abgeordnetenhause 2." *Deutsche Medizinische Wochenschrift* 3 (1877): 117.

Baer, Karl Ernst von. "Über das äussere und innere Skelet: Ein Sendschreiben an Herrn Prof. Heusinger von Prof. Baer." *Archiv für Anatomie und Physiologie* (1826): 327–75.

———. *Über Entwickelungsgeschichte der Thiere: Beobachtung und Reflexion. Erster Theil.* Königsberg: Bornträger, 1828.

———. *Nachrichten über Leben and Schriften.* St. Petersburg: Schmitzdorff, 1866.

———. "Über Darwins Lehre." In *Reden gehalten in wissenschaftlichen Versammlungen und kleinere Aufsätze vermischten Inhalts. Zweiter Theil, Studien aus dem Gebiete der Naturwissenschaften,* pp. 235–480. St. Petersburg, 1876.

———. *Autobiography.* Edited by Jane M. Oppenheimer and translated by H. Schneider. Canton, MA: Science History Publications, 1986.

Balfour, F. M. *Monograph on the Development of Elasmobranch Fishes.* London: Macmillan, 1878.

———. "On the Development of the Skeleton of the Paired Fins of Elasmobranchii Considered in Relation to Its Bearings on the Nature of the Limbs of the Vertebrates." *Proceedings of the Zoological Society of London,* 1881. London: Royal Society, 1881.

Baltzer, F. *Theodor Boveri: Life and Work of a Great Biologist, 1862–1915.* Translated by Dorothea Rudnick. Berkeley: University of California Press, 1967.

Bardeleben, Karl. "Über den Bau der Venenwandung und deren Klappen." *Amtlicher Bericht der 50. Versammlung deutscher Naturforscher und Ärzte,* pp. 228–30. Munich: F. Straub, 1877.

———. "Über den Bau der Arterienwand." *Sitzungsbericht der Jenaischen Gesellschaft für Medizin und Naturwissenschaft,* Sitzung vom 10. Mai 1878. Jena, 1878.

———. "Wilhelm Braune." *Anatomischer Anzeiger* 7 (1892): 440–45.

———. "Georg Hermann von Meyer." *Anatomischer Anzeiger* 7 (1892): 650–54.

Bargmann, W. "K. F. Burdachs Rede über die Geschichte der Anatomie in Königsberg i. Pr." *Zeitschrift für Anatomie und Entwicklungsgeschichte* 116 (1951): 105–11.

Becher, Siegfried. *Erkenntnistheoretische Untersuchungen zu Stuart Mills Theorie der Kausalität.* Abhandlungen zur Philosophie und ihrer Geschichte, ed. Benno Erdmann, Heft 25. Halle: Verlag von Max Niemeyer, 1906.

Bellomy, Donald. "'Social Darwinism' Revisited." *Perspectives in American History,* n.s., 1 (1984): 1–129.

Ben-David, Joseph. "Scientific Productivity and Academic Organization." *American Sociological Review* 25 (1960): 828–43.

———. *The Scientist's Role in Society.* Englewood Cliffs, NJ: Prentice-Hall, 1971.

Ben-David, Joseph, and Randall Collins. "Social Factors in the Origins of a New Science: The Case of Psychology." *American Sociological Review* 31, no. 4 (August 1966): 451–72.

Ben-David, Joseph, and A. Zloczower. "Universities and Academic Systems in Modern Societies." *European Journal of Sociology* 3 (1962): 45–84.

Beneke, Rudolf. *Johann Friedrich Meckel der Jüngere.* Halle: Max Niemeyer, 1934.

Bergmann, Carl, and Rudolf Leuckart. *Anatomisch-physiologische Übersicht des Thierreichs: Vergleichende Anatomie und Physiologie; Ein Lehrbuch für den Unterricht und zum Selbststudium.* Stuttgart: J. B. Müller, 1852.

Bezel, Rudolph. *Der Physiologe Adolf Fick (1829–1901): Seine Zürcher Jahre 1852–1868.* Zürcher Medizinhistorische Abhandlungen, n. s., no. 126 (1979). Zurich: Juris-Verlag, 1979.

Bischoff, Theodor. *Bemerkungen zu dem Reglement für die Prüfung der Ärzte vom 25. September 1869 im früheren norddeutschen Bunde.* Munich: Literarisch-artistische Anstalt, 1871.

Bluntschli, Hans. "Max Fürbringer (1846–1920)." *Anatomischer Anzeiger* 55 (1922): 244–55.

Bonnet, Robert. "Entwickelungsgeschichte. 1. Allgemeines, Lehrbücher, Atlanten, etc." *Ergebnisse der Anatomie und Entwickelungsgeschichte* 1 (1891): 359–85.

———. "Moritz Nußbaum." *Anatomischer Anzeiger* 48 (1915–16): 489–95.

Born, Gustav. "Die Nasenhöhlen und der Thränennasengang der amnioten Wirbelthiere." *Morphologisches Jahrbuch* 5 (1879): 401–29.

———. "Experimentelle Untersuchungen über die Entstehung der Geschlechtsunterschiede." *Jahres-Bericht der Schlesischen Gesellschaft für väterländischen Cultur,* 1881 [printed 1882], Sitzung vom 21. Jan. 1881, pp. 2–23.

Borscheid, Peter. *Naturwissenschaft, Staat und Industrie in Baden, 1848–1914.* Stuttgart: Ernst Kett, 1976.

Boveri, Theodor. *Die Organismen als historische Wesen.* Würzburg: Universitätsdruckerei von H. Stürtz, 1906.

Bowler, Peter J. "The Changing Meaning of 'Evolution.'" *Journal of the History of Ideas* 36 (1975): 95–114.

———. *The Non-Darwinian Revolution: Reinterpreting a Historical Myth.* Baltimore: Johns Hopkins University Press, 1988.

———. *The Mendelian Revolution: The Emergence of Hereditarian Concepts in Modern Science and Society.* Baltimore: Johns Hopkins University Press, 1989.

Bracegirdle, Brian. *A History of Microtechnique.* 2d ed. Ithaca, NY: Cornell University Press, 1978.

Braune, Christian Wilhelm. *Topographisch-anatomischer Atlas: Nach Durchschnitten an gefrornen Cadavern.* Leipzig: Veit & Comp, 1872.

Braus, Hermann. "Versuch einer experimentellen Morphologie." *Münchener Medizinische Wochenschrift* 50 (1903): 2076–77.

———. "Vorwort. Die Morphologie als historische Wissenschaft." *Experimentelle Beiträge zur Morphologie.* Leipzig: Engelmann, 1906–9.

Broman, Thomas. "Herbert Spencer and the Problem of Adaptation." Typescript in possession of the author.

————. "The Transformation of Academic Medicine in Germany, 1780–1820." Ph.D. diss., Princeton University, 1987.

————. "University Reform in Medical Thought at the End of the Eighteenth Century." *Osiris*, 2d ser., 5 (1989): 36–53.

Bronn, Heinrich Georg. *Morphologische Studien über die Gestaltungs-Gesetze der Naturkörper überhaupt und der organischen insbesondere.* Leipzig and Heidelberg: C. F. Winter, 1858.

————. *Untersuchungen über die Entwickelungs-Gesetze der organischen Welt während der Bildungs-Zeit unserer Erd-Oberfläche.* Stuttgart: Schweizerbart, 1858.

————. "Über die Entwicklung der organischen Schöpfung." *Amtlicher Bericht über die 34. Versammlung deutscher Naturforscher und Ärzte.* Carlsruhe: Christian Friedrich Müller, 1859.

Brühl. "Zur Statistik der Vorlesungen über Chemie, Zoologie und Zootomie an allen deutschen (ausserösterreichischen) Universitäten." *Wiener Medizinische Wochenschrift* 20, no. 35 (1870): 827–28.

Budge, Julius. "Die Aufgaben der anatomischen Wissenschaft." *Deutsche Revue* 7 (October-December 1882): 54–65.

Burchardt, Lothar. *Wissenschaftspolitik im Wilhelminischen Deutschland: Vorgeschichte, Gründung und Aufbau der Kaiser-Wilhelm-Gesellschaft zur Förderung der Wissenschaften.* Studien zu Naturwissenschaft, Technik und Wirtschaft im Neunzehnten Jahrhundert, Bd. 1. Göttingen: Vandenhoeck & Ruprecht, 1975.

Burdach, Karl Friedrich. *Die Physiologie.* Leipzig: Weidmann, 1810.

————. *Über die Aufgabe der Morphologie.* Leipzig: Dyk'sche Buchhandlung, 1817.

————. ed. *Die Physiologie als Erfahrungswissenschaft.* Leipzig: Voss, 1830–40.

Burkhardt, Richard W., Jr. *The Spirit of System: Lamarck and Evolutionary Biology.* Cambridge: Harvard University Press, 1977.

Burmeister, Hermann. *Geschichte der Schöpfung. Eine Darstellung des Entwicklungsganges der Erde und ihrer Bewohner.* 3d ed. Leipzig: O. Wigand, 1848.

Bütschli, Otto. "Betrachtungen über Hypothese und Beobachtung." *Verhandlungen der Deutschen Zoologischen Gesellschaft,* 1896, pp. 7–16. Leipzig: W. Engelmann, 1896.

————. "Bemerkungen über die Anwendbarkeit des Experiments in der Entwicklungsmechanik." *Archiv für Entwickelungsmechanik* 5 (1897): 591–93.

————. "Kants Lehre von der Kausalität." *Annalen der Naturphilosophie* 4 (1905): 339–85.

Caneva, Kenneth L. *Robert Mayer and the Conservation of Energy.* Princeton, NJ: Princeton University Press, 1993.

Cannon, Susan Faye. *Science in Culture: The Early Victorian Period.* New York: Dawson, Science History Publications, 1978.

Carriere, Justus. "Über die Regeneration bei den Landpulmonaten." *Tageblatt der Versammlung Deutscher Naturforscher und Ärzte* 52 (1879): 225–26.

Carus, J. Victor. *System der thierischen Morphologie.* Leipzig: Engelmann, 1853.

Chauvin, Marie von. "Über das Anpassungsvermögen der Larven von Salamandra atra." *Zeitschrift für wissenschaftliche Zoologie* 29 (1877): 324–51.

———. "Über die Verwandlungsfähigkeit des mexikanischen Axolotl." *Zeitschrift für wissenschaftliche Zoologie* 41 (1885): 365–89.

Chun, Carl. *Coelenterata (Hohlthiere).* 2 vols. in 1. Dr. H. G. Bronn's Klassen und Ordnungen des Thierreiches, wissenschaftliche dargestellt in Wort und Bild. 2. Bd., 2. Abtheilung. Leipzig: C. F. Winter, 1889–1916.

Churchill, Frederick. "Wilhelm Roux and a Program for Embryology." Ph.D. diss., Harvard University, 1966.

———. "August Weismann and a Break from Tradition." *Journal of the History of Biology* 1 (1968): 91–112.

———. "Roux, Wilhelm." *Dictionary of Scientific Biography.*

———. "Weismann's Continuity of the Germ-plasm in Historical Perspective." In "August Weismann (1834–1914) und die theoretische Biologie des 19. Jahrhunderts," edited by Klaus Sander, *Freiburger Universitätsblätter* 24, no. 87–88 (July 1985):107–24.

———. "Weismann, Hydromedusae, and the Biogenetic Imperative: A Reconsideration." In *A History of Embryology,* edited by T. J. Horder, J. A. Witkowski, and C. C. Wylie, pp. 7–33. Cambridge, New York: Cambridge University Press, 1986.

———. "From Heredity Theory to *Vererbung*: The Transmission Problem, 1850–1915," *Isis* 78 (1987): 337–64.

———. "Regeneration, 1885–1901." In *A History of Regeneration Research: Milestones in the Evolution of a Science,* edited by Charles E. Dinsmore. Cambridge and New York: Cambridge University Press, 1991.

Cittadino, Eugene. *Nature as the Laboratory: Darwinian Plant Ecology in the German Empire, 1880–1900.* Cambridge and New York: Cambridge University Press, 1990.

Claus, Carl. *Die freilebenden Copepoden, mit besonderer Berücksichtigung der Fauna Deutschlands, der Nordsee und des Mittelmeers.* Leipzig: W. Engelmann, 1863.

———. *Die Typenlehre und E. Haeckel's Gastraea-Theorie.* Vienna: Manz, 1874.

———. *Grundzüge der Zoologie.* 4th ed. Marburg: N. G. Elwert, 1880.

Coleman, William. *Georges Cuvier, Zoologist: A Study in the History of Evolution Theory.* Cambridge: Harvard University Press, 1964.

———. "Cell, Nucleus, and Inheritance: An Historical Study." *Proceedings of the American Philosophical Society* 109 (1965): 124–58.

———. *Biology in the Nineteenth Century.* New York: John Wiley, 1971.

———. "Morphology between Type Concept and Descent Theory," *Journal of the History of Medicine* 31 (1976): 149–75.

———. "Gegenbaur, Carl." *Dictionary of Scientific Biography.*

Darwin, Charles. *Über die Entstehung der Arten.* 2d ed. Translated by H. G. Bronn. Stuttgart: Schweizerbart, 1860.

———. *On the Origin of Species*. London: J. Murray, 1859. Facsimile reprint, Cambridge: Harvard University Press, 1964.

Darwin, Francis, and A. C. Seward, eds. *More Letters of Charles Darwin: A Record of His Work in a Series of Hitherto Unpublished Letters*. 2 vols. London: Murray, 1903.

DeJager, Timothy F. "G. R. Treviranus (1776–1837) and the Biology of a World in Transition." Ph. D. diss., University of Toronto, 1991.

Depdolla, Philipp. "Hermann Müller-Lippstadt (1829–1883) und die Entwicklung des biologischen Unterrichts." *Sudhoffs Archiv* 34 (1941): 261–344.

Desmond, Adrian. *The Politics of Evolution: Morphology, Medicine, and Reform in Radical London*. Chicago: University of Chicago Press, 1989.

Dieterici, Karl Friedrich Wilhelm. *Geschichtliche und statistische Nachrichten über die Universitäten im Preußischen Staate*. Berlin: Duncker & Humblot, 1836. Reprint, Aalen: Scientia Verlag, 1982.

Dohrn, Anton. "Studien zur Urgeschichte des Wirbeltierkörpers. 6. Die paarigen und unpaarigen Flossen der Selachier." *Mittheilungen aus der Zoologischen Station zu Neapel* 5 (1884): 161–95.

———. "Studien zur Urgeschichte des Wirbeltierkörpers. 21. Theoretisches über Occipitalsimite und Vagus. Competenzconflict zwischen Ontogenie und vergleichender Anatomie." *Mitteilungen aus der Zoologischen Station zu Neapel* 15, pts. 1, 2 (9 July 1901): 186–279.

Dreyer, Friedrich. *Ziele und Wege biologischer Forschung, beleuchtet an der Hand einer Gerüstbildungsmechanik*. Jena: G. Fischer, 1892.

Driesch, Hans. *Die mathematisch-mechanische Betrachtung morphologischer Probleme der Biologie: Eine kritische Studie*. Jena: G. Fischer, 1891.

———. *Die Biologie als selbständige Grundwissenschaft: Eine kritische Studie*. Leipzig: W. Engelmann, 1893.

———. *The Science and Philosophy of the Organism*. London: Adam & Charles Black, 1908.

———. *The Science and Philosophy of the Organism*. 2d ed., London: A. & C. Black, 1929.

———. *Lebenserinnerungen: Aufzeichnungen eines Forschers und Denkers in entscheidender Zeit*. Basel: E. Reinhardt, 1951.

Du Bois-Reymond, Emil. "Gedächtnisrede auf Johannes Müller." In E. Du Bois-Reymond, *Gesammelte Reden,* edited by Estelle Du Bois-Reymond. 2d ed. Vol. 1, pp. 135–317. Leipzig: Veit, 1912.

———. "Der physiologische Unterricht einst und jetzt: Rede zur Eröffnung des neuen Physiologischen Instituts zu Berlin (16 Nov. 1877)." In E. Du Bois-Reymond, *Gesammelte Reden,* edited by Estelle Du Bois-Reymond. 2d ed. Vol. 1, pp. 630–53. Leipzig: Veit, 1912.

Duchesneau, François. *Genèse de la Théorie Cellulaire*. Montreal: Bellarmin; Paris: J. Vrin, 1987.

Ebbecke, Ulrich. "Das Physiologische Institut." In *Geschichte der Rheinischen Friedrich-Wilhelm-Universität zu Bonn am Rhein: Institute und Seminare,* pp. 72–75. Bonn; Friedrich Cohen Verlag, 1933.

Ecker, Alexander. *Hundert Jahre einer Freiburger Professorenfamilie*. Freiburg, 1886.

Ehlers, Ernst. *Die Borstenwürmer (Annelida chaetopoda) nach systematischen und anatomischen Untersuchungen.* Leipzig: W. Engelmann, 1864–68.

———. "Carl Theodor Ernst von Siebold: Eine biographische Skizze." *Zeitschrift für wissenschaftliche Zoologie* 42 (1885): i–xxiii.

———. "Göttinger Zoologen." In *Festschrift zur Feier des hundertfünfzigjährigen Bestehens der Königlichen Gesellschaft der Wissenschaften zu Göttingen: Beiträge zur Gelehrtengeschichte Göttingens,* pp. 391–494. Berlin: Weidmann, 1901.

Ehrmann, Paul. "Heinrich Simroth: Ein Nachruf." *Sitzungsberichte der naturforschenden Gesellschaft zu Leipzig* 43/44 (1916–17): 47–81 (printed 1918).

Eimer, Theodor. "Neue Beobachtungen über das Variiren der Mauereidechse." *Amtlicher Bericht der 50. Versammlung deutscher Naturforscher und Ärzte,* pp. 179–82. Munich: F. Straub, 1877.

———. *Die Medusen, physiologisch und anatomisch auf ihr Nervensystem untersucht.* Tübingen: Laupp, 1878.

Engard, Charles J. "Introduction." *Goethe's Botanical Writings.* Translated by Bertha Mueller. Pp. 3–18. University Press of Hawaii, 1952. Reprint, Woodbridge, CT: Ox Bow Press, 1989.

Engelmann, Wilhelm. *Bibliotheca Historico-naturalis.* Leipzig: W. Engelmann, 1846.

Erdmann, G. A. *Geschichte der Entwicklung und Methodik der biologischen Naturwissenschaften (Zoologie und Botanik).* Cassel and Berlin: Theodor Fischer, 1887.

Esler, Anthony. "The Generation Gap in Society and History: A Select Bibliography." *Public Administration Series* Bibliography P 1418. Monticello, IL: Vance Bibliographies, 1984.

Eulner, Hans-Heinz. *Die Entwicklung der medizinischen Spezialfächer an den Universitäten des deutschen Sprachgebietes.* Stuttgart: Ferdinand Enke, 1970.

Eulner, Hans-Heinz, and Hermann Hoepke, eds. *Der Briefwechsel zwischen Rudolph Wagner und Jacob Henle, 1838–1862.* Göttingen: Vandenhoeck & Ruprecht, 1979.

Feser, Georg. "Das anatomische Institut in Würzburg, 1847–1903." Inaugural diss., University of Würzburg, 1977.

*Festschrift zum siebenzigsten Geburtstage von Carl Gegenbaur am 21. August 1896.* Leipzig: W. Engelmann, 1896.

Feuer, Lewis. *Einstein and the Generations of Science.* New York: Basic Books, 1984.

Figlio, Karl. "The Metaphor of Organization: An Historiographical Perspective on the Bio-Medical Sciences of the Early Nineteenth Century." *History of Science* 14 (1976): 17–53.

Fischel, A. "Carl Rabl." *Anatomischer Anzeiger* 51 (1918–19): 54–79.

Fischer, Eugen. "Ernst Gaupp." *Anatomischer Anzeiger* 49 (1916–17): 584–91.

Fleischmann, Albert. *Die Descendenztheorie: Gemeinverständliche Vorlesungen über den Auf- und Niedergang einer naturwissenschaftlichen Hypothese.* Leipzig: A. Georgi, 1901.

Flesch, [Max]. "Eine Anzahl Zeichnungen über die Enstehung des Schwanz-endes der Wirbelsäule bei Urodelen." *Tageblatt der Versammlung Deutscher Naturforscher und Ärzte* 52 (1879): 225.

Florkin, Marcel. "Schwann, Theodor." *Dictionary of Scientific Biography.*

Fraisse, Paul. "Über die Regeneration von Organen und Geweben bei Amphibien und Reptilien." *Tageblatt der Versammlung Deutscher Naturforscher und Ärzte* 52 (1879): 223–25.

Frank, Morton H., and Joyce J. Weiss. "The 'Introduction' to Carl Ludwig's *Textbook of Human Physiology (1858)*." *Medical History* 10 (1966): 76–86.

Froriep, August. "Wilhelm Henke." *Anatomischer Anzeiger* 12 (1896): 475–95.

Fruton, Joseph. *Contrasts in Scientific Style: Research Groups in the Chemical and Biochemical Sciences.* Memoirs of the American Philosophical Society 191. Philadelphia: American Philosophical Society, 1990.

Fürbringer, Max. *Untersuchungen zur Morphologie und Systematik der Vögel: Zugleich ein Beitrag zur Anatomie der Stütz- und Bewegungsorgane.* 2 vols. Amsterdam: T. J. van Holkema, 1888.

———. "Über die spino-occipitalen Nerven der Selachier und Holocephalen und ihre vergleichende Morphologie." *Festschrift für Carl Gegenbaur.* Vol. 3, pp. 351–788. Leipzig: Engelmann, 1897.

———. "Morphologische Streitfragen." *Morphologisches Jahrbuch* 30 (1902): 85–274.

———. "Carl Gegenbaur." In *Heidelberger Professoren aus dem 19. Jahrhundert.* Vol. 2, p. 400–466. Heidelberg: Carl Winter, 1903.

———. "Erwiderung." *Anatomischer Anzeiger,* supp. 23 (1903): 190–97.

———. "Wie ich Ernst Haeckel kennen lernte und mit ihm verkehrte und wie er mein Führer in den grössten Stunden meines Lebens wurde." In *Was Wir Ernst Haeckel Verdanken,* edited by Heinrich Schmidt. Vol. 2, pp. 335–50. Leipzig: Unesma, 1914.

Gasman, Daniel. *Scientific Origins of National Socialism: Social Darwinism in Ernst Haeckel and the German Monist League.* London: MacDonald; New York: American Elsevier, 1971.

Gaupp, Ernst. *August Weismann: Sein Leben und sein Werke.* Jena: G. Fischer, 1917.

Gebhard, Walther. "Gustav Born." *Anatomischer Anzeiger* 18 (1900): 139–43.

Gegenbaur, Carl. *Untersuchungen zur vergleichenden Anatomie der Wirbelthiere.* 3 vols. Leipzig: Engelmann, 1864–72.

———. "Über primäre und secundäre Knochenbildung mit besonderer Beziehung auf die Lehre vom Primordialcranium." *Jenaische Zeitschrift für Medizin und Naturwissenschaft* 3 (1866): 54–73. Reprinted in *Gesammelte Abhandlungen von Carl Gegenbaur,* edited by M. Fürbringer and H. Bluntschli, vol. 2, pp. 186–202. Leipzig: Engelmann, 1912.

———. *Grundzüge der vergleichenden Anatomie.* 2d ed. Leipzig: W. Engelmann, 1870.

———. "Über das Skelet der Gliedmassen der Wirbelthiere im Allgemeinen und der Hintergliedmassen der Selachier insbesondere." *Jenaische Zeitschrift für Medizin und Naturwissenschaft* 5 (1870): 397–447. Reprinted in *Gesammelte Abhandlungen von Carl Gegenbaur,* edited by M. Fürbringer and H. Bluntschli. Leipzig: Engelmann, 1912.

———. "Über das Archipterygium." *Jenaische Zeitschrift für Medizin und Naturwissenschaft* 7 (1873): 131–41. Reprinted in *Gesammelte Abhandlungen von Carl Gegenbaur,* edited by M. Fürbringer and H. Bluntschli. Leipzig: Engelmann, 1912.

———. "Die Stellung und Bedeutung der Morphologie." *Morphologisches Jahrbuch* 1 (1875): 1–19.

———. "Zur Morphologie der Gliedmassen der Wirbelthiere," *Morphologisches Jahrbuch* 2 (1876): 396–420. Reprinted in *Gesammelte Abhandlungen von Carl Gegenbaur,* edited by M. Fürbringer and H. Bluntschli, vol. 3, pp. 97–116. Leipzig: Engelmann, 1912.

———. *Elements of Comparative Anatomy.* Translated by F. Jeffrey Bell and E. Ray Lankester. London: Macmillan, 1878.

———. *Grundriß der vergleichenden Anatomie,* 2d ed. Leipzig: Engelmann, 1878.

———. "Zur Gliedmassenfrage: An die Untersuchungen v. Davidoff's angeknüpfte Bemerkungen." *Morphologisches Jahrbuch* 5 (1879): 521–25. Reprinted in *Gesammelte Abhandlungen von Carl Gegenbaur,* edited by M. Fürbringer and H. Bluntschli, vol. 3, pp. 176–81. Leipzig: Engelmann, 1912.

———. Review of E. Rautenfeld, "Morphologische Untersuchung über das Skelet der hinteren Gliedmassen von Ganoiden and Teleostiern." *Morphologisches Jahrbuch* 9 (1884): 325–26. Reprinted in *Gesammelte Abhandlungen von Carl Gegenbaur,* edited by M. Fürbringer and H. Bluntschli, vol. 3, pp. 218–19. Leipzig: Engelmann, 1912.

———. "Ontogenie und Anatomie, in ihrer Wechselbeziehungen betrachtet." *Morphologisches Jahrbuch* 15 (1889): 1–9. Reprinted in *Gesammelte Abhandlungen von Carl Gegenbaur,* edited by M. Fürbringer and H. Bluntschli, vol. 3, pp. 452–59. Leipzig: Engelmann, 1912.

———. "Über Cänogenese." *Verhandlungen der Anatomischen Gesellschaft* 2 (1888): 3–9. In *Anatomischer Anzeiger* 3 (1889). Reprinted in *Gesammelte Abhandlungen von Carl Gegenbaur,* edited M. Fürbringer and H. Bluntschli, vol. 3, pp. 446–51. Leipzig: Engelmann, 1912.

———. "Das Flossenskelett der Crossopterygier und das Archipterygium der Fische." *Morphologisches Jahrbuch* 22 (pt. 1, November 1894, 1895): 119–60. Reprinted in *Gesammelte Abhandlungen von Carl Gegenbaur,* edited by M. Fürbringer and H. Bluntschli, vol. 3, pp. 497–531. Leipzig: Engelmann, 1912.

———. *Vergleichende Anatomie der Wirbelthiere.* Leipzig: Engelmann, 1898.

———. *Erlebtes and Erstrebtes.* Leipzig: Engelmann, 1901.

———. *Gesammelte Abhandlungen.* 3 vols. Edited by Max Fürbringer and Hans Bluntschli. Leipzig: Engelmann, 1912.

Geison, Gerald. *Michael Foster and the Cambridge School of Physiology.* Princeton, NJ: Princeton University Press, 1978.

———. "Scientific Change, Emerging Specialties, and Research Schools." *History of Science* 19 (1981): 20–40.

Geison, Gerald, and Frederic L. Holmes, eds. *Research Schools: Historical Appraisals. Osiris* 8 (1993).

*Gesetz- und Verordnungsblatt für das Großherzogthum Baden.* Karlsruhe: Malsch & Vogel: 1869–1906.

Geus, Armin. "Zoologie." In *Die Naturwissenschaften an der Philipps- Universität Marburg, 1527–1977,* edited by Rudolf Schmitz, pp. 167–73. Marburg: N. G. Elwert, 1978.

Geus, Armin, and Hans Querner, *Deutsche Zoologische Gesellschaft, 1890– 1990: Dokumentation und Geschichte.* Stuttgart and New York: Gustav Fischer, 1990.

Giese, Ernst, and Benno von Hagen. *Geschichte der Medizinischen Fakultät der Friedrich-Schiller-Universität Jena.* Jena: G. Fischer, 1958.

Glees, Paul. "Leydig, Franz." *Dictionary of Scientific Biography.*

Goethe, Johann Wolfgang von. *Morphologische Hefte.* Edited by Dorothea Kuhn. *Die Schriften zur Naturwissenschaft,* Abt. 1, Bd. 9. Weimar: Hermann Böhlaus Nachfolger, 1954.

Goette, Alexander. *Entwicklungsgeschichte der Unke.* Leipzig: Voss, 1875.

Göppert, Ernst. "Emil Gasser." *Anatomischer Anzeiger* 54 (1921): 150–57.

———. "Carl Gegenbaur. Rede zum Gedächtnis seines 100. Geburtsjahres." *Jenaische Zeitschrift für Naturwissenschaft* 62 (1926): 500–518.

———. "Friedrich Maurer und der Kreis um Carl Gegenbaur." *Anatomischer Anzeiger* 85 (1936–37): 313–31.

Gould, Stephen Jay. *Ontogeny and Phylogeny.* Cambridge: Harvard University Press, Belknap Press, 1977.

Gräper, Ludwig. "Carl Hasse." *Anatomischer Anzeiger* 56 (1922): 209–21.

Gregory, Frederick. *Scientific Materialism in Nineteenth Century Germany.* Boston: Reidel, 1977.

———. "Kant, Schelling, and the Administration of Science in the Romantic Era." *Osiris,* 2d ser., 5 (1989): 17–35.

Grobben, Karl. "Carl Claus." *Arbeiten aus dem Zoologischen Institute der Universität Wien* 11, no. 2 (1899): i–xiv.

Gursch, Reinhard. *Die Illustrationen Ernst Haeckels zur Abstammungs- und Entwicklungsgeschichte: Diskussionen im wissenschaftlichen und nichtwissenschaftlichen Schrifttum.* Marburger Schriften zur Medizingeschichte, vol. 1. Frankfurt am Main: Peter D. Lang.

Haberling, W. *Johnnes Müller: Das Leben des rheinischen Naturforschers.* Leipzig, 1924.

Haeckel, Ernst, *Die Radiolarien (Rhizopoda radiaria): Eine Monographie.* Berlin: G. Reimer, 1862.

———. "Die Familie der Rüsselquallen (Medusae Geryonidae)." *Jenaische Zeitschrift für Medizin und Naturwissenschaft* 1 (1864): 435–69; 2 (1866): 93–120, 129–202, 263–322.

———. Über die Entwickelungstheorie Darwin's." *Amtlicher Bericht über die 39. Versammlung Deutscher Naturforscher und Ärzte,* pp. 17–30. Stettin: F. Hessenland, 1864.

———. *Generelle Morphologie der Organismen: Allgemeine Grundzüge der organischen Formen-Wissenschaft, mechanisch begründet durch die von*

*Charles Darwin reformirte Descendenz-Theorie.* 2 vols. Berlin: G. Reimer, 1866.

——. *Anthropogenie oder Entwicklungsgeschichte des Menschen. Keimes-und Stammesgeschichte.* Leipzig: Engelmann, 1874.

——. "Die Gastraea-Theorie, die phylogenetische Classification des Thier-reichs und die Homologie der Keimblätter." *Jenaische Zeitschrift für Naturwissenschaft* 8 (1874): 1–55.

——. *Natürliche Schöpfungsgeschichte.* 5th ed. Berlin: G. Reimer, 1874.

——. "Die Gastrula und die Eifurchung der Thiere." *Jenaische Zeitschrift für Naturwissenshaft* 9 (1875): 402–508.

——. *Ziele und Wege der heutigen Entwicklungsgeschichte.* Jena: H. Dufft, 1875. Also *Jenaische Zeitschrift für Naturwissenschaft,* supp. 10 (1876).

——. *Anthropogenie oder Entwicklungsgeschichte des Menschen: Keimes-und Stammesgeschichte.* 3d ed. Leipzig: Engelmann, 1877.

——. "Nachträge zur Gastraea-Theorie." *Biologische Studien 2: Studien zur Gastraea-Theorie,* pp. 227–70. Jena: Hermann Dufft, 1877.

——. "Über die heutige Entwicklungslehre im Verhältnisse zur Gesamtwissenschaft." *Amtlicher Bericht der 50. Versammlung deutscher Naturforscher and Ärzte,* pp. 14–20. Munich: F. Straub, 1877.

——. *Evolution of Man.* Translation of *Anthropogenie,* 3d ed. New York: D. Appleton, 1879.

——. "Ursprung und Entwickelung der thierischen Gewebe: Ein histogenetischer Beitrag zur Gastraea-Theorie." *Jenaische Zeitschrift für Naturwissenschaft* 18 (1885): 206–76.

——. *Anthropogenie oder Entwicklungsgeschichte des Menschen. Keimes-und Stammesgeschichte.* 4th ed. rev. Leipzig: Engelmann, 1891.

——. *Story of the Development of a Youth: Letters to His Parents 1852–1856.* Translated by G. Barry Gifford. New York: Harper & Brothers, 1923.

Hagen, Joel. "Ecologists and Taxonomists: Divergent Traditions in Twentieth-Century Plant Geography." *Journal of the History of Biology* 19 (1986): 197–214.

Harwood, Jonathan. *Styles of Scientific Thought: The German Genetics Community, 1900–1933.* Chicago: University of Chicago Press, 1993.

Hasselwander, A. "Dem Andenken Emil Rosenbergs." *Anatomischer Anzeiger* 70 (1930): 81–107.

Hatschek, Berthold. "Beiträge zur Entwicklungsgeschichte der Lepidopteren." *Jenaische Zeitschrift für Naturwissenschaft* 11 (1877): 115–48.

——. "Über Entwicklungsgeschichte von Teredo." *Arbeiten aus dem zoologischen Institut der Universität Wien* 3 (1881): 1–42.

——. *Lehrbuch der Zoologie: Eine morphologische Uebersicht des Thierreiches zur Einführung in das Studium dieser Wissenschaft.* Jena: G. Fischer, 1888.

Heidenhain, M. "August Froriep." *Anatomischer Anzeiger* 50 (1917–18): 410–24.

Heider, Karl. "Gedächtnisrede auf Franz Eilhard Schulze." *Sitzungsberichte der Preußischen Akademie der Wissenschaften, Philosophisch-Historische*

*Klasse,* 1922, pp. lxxxvii–xcv. Berlin: Verlag der Akademie der Wissenschaften.

———. "Gedächtnisrede auf Willy Georg Kükenthal." *Sitzungsberichte der Preußischen Akademie der Wissenschaften, Philosophisch-Historische Klasse,* 1924, pp. xcix–ciii. Berlin: Verlag der Akademie der Wissenschaften.

Held, Hans. "Nekrolog auf Carl Rabl, gesprochen am 14. November 1918." *Berichte über die Verhandlungen der sächsischen Gesellschaft der Wissenschaften zu Leipzig. Mathematisch-physische Klasse* 70 (1918): 365–80.

Henle, Jakob. *Allgemeine Anatomie: Lehre von den Mischungs- und Formbestandtheilen des menschlichen Körpers.* Leipzig: Voss, 1841.

———. *Grundriß der Anatomie des Menschen.* Braunschweig: Vieweg, 1880.

———. "Anatomie." In *Meyers Konversations-Lexikon,* 5th ed. Leipzig: Bibliographisches Institut, 1893.

Henson, Pamela M. "Evolution and Taxonomy: J. H. Comstock's Research School in Evolutionary Entomology at Cornell University, 1874–1930." Ph.D. diss., University of Maryland, College Park, 1990.

Hertwig, Oscar. *Der anatomische Unterricht: Vortrag beim Antritt der anatomischen Professur an der Universität Jena am 28. Mai 1881.* Jena: Gustav Fischer, 1881.

———. *Zeit- und Streitfragen der Biologie. Heft 2. Mechanik und Biologie. Mit einem Anhang: Kritische Bemerkungen zu den entwicklungsmechanischen Naturgesetzen von Roux.* Jena: Fischer, 1897.

———. *Allgemeine Biologie.* 4th ed. Jena: G. Fischer, 1912.

Hertwig, Oscar, and Richard Hertwig. *Studien zur Blättertheorie.* Jena, 1879–80.

———. "Die Coelomtheorie: Versuch einer Erklärung des mittleren Keimblattes." *Jenaische Zeitschrift für Wissenschaft* 15 (1882): 1–150.

Hertwig, Richard. "Zoologie und vergleichende Anatomic." In *Die deutschen Universitäten,* edited by Wilhelm Lexis. Vol. 2, chap. 9. Berlin: A. Asher, 1893.

———. "Wie ich Ernst Haeckels Schüler wurde." In *Was Wir Ernst Haeckel Verdanken,* edited by Heinrich Schmidt. Vol. 2, pp. 165–170. Leipzig: Unesma, 1914.

Hesse, Richard. *Abstammungslehre und Darwinismus.* 4th ed. Leipzig: B. G. Teubner, 1912.

Heuss, Theodor. *Anton Dohrn.* Tübingen: Rainer Wunderlich, 1940.

His, Wilhelm. "Beschreibung eines Mikrotoms." *Archiv für mikroskopische Anatomie* 6 (1870): 229–32.

———. Review of *Entwicklungsgeschichte der Unke,* by Alexander Goette. *Zeitschrift für Anatomie und Entwicklungsgeschichte* 1 (1876): 298–306, 465–72.

———. "Über Entwicklungsverhältnisse des academischen Unterrichts." *Rektoratswechsel an der Universität Leipzig am 31. October 1882.* Leipzig: Edelmann, 1882.

Hodge, M. J. S. "Darwin as a Lifelong Generation Theorist." In *The Darwin-*

*ian Heritage,* edited by David Kohn, pp. 207–43. Princeton, NJ: Princeton University Press, 1985.

Hoepke, Hermann. "Jakob Henles Gutachten zur Besetzung des Lehrstuhls für Anatomie an der Universität Berlin, 1883." *Anatomischer Anzeiger* 120 (1967): 221–32.

Hoffmann, J. C. V. "Zur Verständigung." *Zeitschrift für mathematischen und naturwissenschaftlichen Unterricht* 12 (1881): 1–7.

Holmes, Frederic Lawrence. *Between Biology and Medicine: The Formation of Intermediary Metabolism.* Berkeley, CA: Office for the History of Science and Technology, University of California, Berkeley, 1992.

Hoppe, Brigitte. "Die Entwicklung der biologischen Fächer an der Universität München im 19. Jahrhundert unter Berücksichtigung des Unterrichts." In *Die Ludwig-Maximilians-Universität in ihren Fakultäten,* edited by Laetitia Boehm and Johannes Spörl, pp. 354–89. Berlin: Duncker & Humblot, 1972.

Horder, T. J., and Paul J. Weindling. "Hans Spemann and the Organizer." In *A History of Embryology,* edited by T. J. Horder, J. A. Witkowski, and C. C. Wylie, pp. 183–242. Cambridge and New York: Cambridge University Press, 1986.

Horn, Wilhelm. *Das preußische Medicinalwesen: Supplement zur ersten Auflage.* Berlin: Hirschwald, 1863.

"Hox Genes, Fin Folds and Symmetry." *Nature* 364 (15 July 1993): 195–97.

Hull, David. *Darwin and His Critics: The Reception of Darwin's Theory of Evolution by the Scientific Community.* Chicago: University of Chicago Press, 1983.

———. *Science as a Process: An Evolutionary Account of the Social and Conceptual Development of Science.* Chicago: University of Chicago Press, 1988.

Ihering, Hermann von. "Über die Ontogenie von Cyclas und die Homologie der Keimblätter bei den Mollusken." *Zeitschrift für wissenschaftliche Zoologie* 26 (1876): 414–33.

———. *Vergleichende Anatomie des Nervensystemes und Phylogenie der Mollusken.* Leipzig: W. Engelmann, 1877.

Jacobs, Natasha X. "From Unit to Unity: Protozoology, Cell Theory, and the New Concept of Life." *Journal of the History of Biology* 22 (1989): 215–42.

Jaensch, Paul. "Beiträge zur Geschichte des anatomischen Unterrichts an der Universität Marburg." *Zeitschrift für die gesamte Anatomie. 3. Abteilung* 25 (1924): 772–823.

Jahn, Ilse. "Geschichte der Botanik in Jena von der Gründung der Universität bis zur Berufung Pringsheims (1558–1864)." Ph.D. diss., University of Jena, 1963. Typescript.

———. "Ernst Haeckel und die Berliner Zoologen." *Acta historica Leopoldina* 16 (1985): 65–109.

———. "Zoologische Gärten in Stadtkultur und Wissenschaft im 19. Jahrhundert." *Berichte zur Wissenschaftsgeschichte* 15 (1992): 213–25.

Jahn, Ilse, Rolf Löther, and Konrad Senglaub, eds. *Geschichte der Biologie:*

*Theorien, Methoden, Institutionen, Kurzbiographien.* Jena: VEB Gustav Fischer Verlag, 1982.

Jungnickel, Christa, and Russell McCormmach. *Intellectual Mastery of Nature.* 2 vols. Chicago: University of Chicago Press, 1986.

Junker, Thomas. *Darwinismus und Botanik: Rezeption, Kritik, und theoretische Alternativen in Deutschland des 19. Jahrhunderts.* Stuttgart: Deutscher Apotheker Verlag, 1989.

———. "Heinrich Georg Bronn und die *Entstehung der Arten.*" *Sudhoffs Archiv* 75 (1991): 180–208.

Kallius, Ernst. "Friedrich Merkel." *Anatomischer Anzeiger* 54 (1921): 40–54.

Kaufmann, Georg, ed. *Festschrift zur Feier des hundertjährigen Bestehens der Universität Breslau.* Breslau: F. Hirt, 1911.

Keibel, Franz. "Gustav Schwalbe." *Anatomischer Anzeiger* 49 (1916–17): 210–21.

Keilbach, Rolf. "Chronik des zoologischen Instituts und Museums der Ernst-Moritz-Arndt-Universität Greifswald." In *Festschrift zur 500-Jahrfeier der Universität Greifswald.* Vol. 2, pp. 561–71. Greifswald, 1956.

Kelly, Alfred. *The Descent of Darwin: The Popularization of Darwinism in Germany, 1860–1914.* Chapel Hill: University of North Carolina Press, 1981.

Kent, George C. *Comparative Anatomy of the Vertebrates.* 5th ed. St. Louis, MO: C. V. Mosby, 1983.

Kilian, R. *Die Universitäten Deutschlands in medicinisch-naturwissenschaftlicher Hinsicht betrachtet.* Heidelberg and Leipzig, 1828. Reprint, Amsterdam: B. M. Israel, 1966.

Klappenbach, Ruth, and Wolfgang Steinitz. *Wörterbuch der deutschen Gegenwartssprache.* Berlin: Adademie-Verlag, 1967.

Koch, Franz. *Der Anatom Georg Hermann von Meyer, 1815–1892.* Zürcher Medizingeschichtliche Abhandlungen, n.s., no. 131. Zurich: Juris Druck u. Verlag, 1979.

Koch, Herbert. *Geschichte der Stadt Jena.* Stuttgart: Gustav Fischer, 1966.

Koehler, Otto. "Maximilian Braun zum Gedächtnis." *Zoologische Jahrbuch. Abteilung für allgemeine Zoologie und Physiologie der Tiere* 48 (1930/31): i–xxxv.

———. "Die Zoologie an der Universität Freiburg im Breisgau." In *Aus der Geschichte der Naturwissenschaften an der Universität Freiburg,* edited by Eduard Zentgraf, pp. 129–44. Freiburg: Verlag Eberhard Albert Universitätsbuchhandlung, 1957.

Koenigsberger, Leo. *Hermann von Helmholtz.* 3 vols. Braunschweig: Vieweg, 1902.

Köhnke, Klaus Christian. *Entstehung und Aufstieg des Neukantianismus: Die deutsche Universitätsphilosophie zwischen Idealismus und Positivismus.* Frankfurt am Main: Suhrkamp, 1986.

Kölliker, Albert von. *Handbuch der Gewebelehre des Menschen.* Leipzig: Wilhelm Engelmann, 1852.

———. "Über die Darwin'sche Schöpfungstheorie." *Zeitschrift für wissenschaftliche Zoologie* 14 (1864): 174–86.

———. "Anatomisch-systematische Beschreibung der Alcyonarien: Erste Abtheilung: Die Pennatuliden." *Abhandlungen der Senckenbergischen Naturforschenden Gesellschaft* 7 (1869–70): 109–256; 8 (1872): 85–275.

———. *Die Aufgaben der Anatomischen Institute: Eine Rede gehalten bei der Eröffnung der neuen Anatomie in Würzburg am 3. November 1883.* Separat-Abdruck aus den *Verhandlungen der physikalisch-medizinischen Gesellschaft zu Würzburg*, n. s., 18. Würzburg: Stahel'schen Universitäts-Buch- und Kunsthandlung, 1884.

———. *Erinnerungen aus meinem Leben.* Leipzig: W. Engelmann, 1899.

Köpke, Rudolf. *Die Gründung der königlichen Friedrich-Wilhelms-Universität zu Berlin.* Berlin: Gustave Schade, 1860.

Körner, Hans. *Die Würzburger Siebold: Eine Gelehrtenfamilie des 18. und 19. Jahrhunderts.* Neustadt an der Aisch: Verlag Degener & Co., 1967.

Kohler, Robert E. *From Medical Chemistry to Biochemistry.* New York: Cambridge University Press, 1982.

Koller, Gottfried. *Das Leben des Biologen Johannes Müller 1801–1858.* Stuttgart: Wissenschaftliche Verlagsgesellschaft, 1958.

Korschelt, Eugen. "Carl Claus." In *Lebensbilder aus Kurhessen und Waldeck 1830–1930*, edited by Ingeborg Schnack, vol. 1, pp. 61–66. Marburg: N. G. Elwert, 1939.

Korschelt, Eugen, and Karl Heider. *Lehrbuch der vergleichenden Entwicklungsgeschichte der wirbellosen Thiere.* Specieller Theil. Erstes Heft. Jena: G. Fischer, 1890.

Krantz, D. L., and L. Wiggins. "Personal and Impersonal Channels of Recruitment in the Growth of Theory." *Human Development* 16 (1973): 133–56.

Krause, Gerhard. "Aus der Geschichte der Zoologie in Würzburg." *Verhandlungen der Deutschen Zoologischen Gesellschaft vom 26. bis 31. Mai 1969 in Würzburg. Zoologischer Anzeiger*, supp. 33 (1970): 1–6.

Krausse, Erika. *Ernst Haeckel.* Leipzig: B. G. Teubner, 1984.

Kremer, Richard L. "Between *Wissenschaft* and Praxis: Experimental Medicine and the Prussian State, 1807–1848." In *"Einsamkeit und Freiheit" neu Besichtigt: Universitätsreformen und Disziplinenbildung in Preußen als Modell für Wissenschaftspolitik im Europa des 19. Jahrhunderts*, edited by Gert Schubring, pp. 155–70. Stuttgart: Franz Steiner Verlag, 1991.

———. "Building Institutes for Physiology in Prussia, 1836–1846: Contexts, Interests, and Rhetoric." In *The Laboratory Revolution in Medicine*, edited by Andrew Cunningham and Perry Williams, pp. 72–109. Cambridge and New York: Cambridge University Press, 1992.

Krieger, Karl Richard. "Über das Centralnervensystem des Flusskrebses." *Zeitschrift für wissenschaftliche Zoologie* 33 (1880): 527–94.

Kruta, Vladislav. "Weber, Ernst Heinrich." *Dictionary of Scientific Biography.*

Kühn, Alfred. *Anton Dohrn und die Zoologie seiner Zeit.* Pubblicazione della Stazione Zoologica di Napoli. Supplemento 1950.

Kühne, W., and C. Voit. "An unsere Leser!" *Zeitschrift für Biologie* 19, n. s. 1 (1883): 1–4.

Kuhn, Thomas S. *The Structure of Scientific Revolutions.* 2d ed. Chicago: University of Chicago Press, 1970.

Kupka, Horst Werner. "Die Ausgaben der Süddeutschen Länder für die medizinischen und naturwissenschaftlichen Hochschul-Einrichtungen 1848–1914." Inaugural diss., University of Bonn, 1970.

Kupffer, Karl von. "Nikolaus Rüdinger." *Anatomischer Anzeiger* 13 (1897): 219–32.

Larson, James L. "Vital Forces: Regulative Principles or Constitutive Agents? A Strategy in German Physiology, 1786–1802." *Isis* 70 (1979): 235–49.

Lee, Arthur Bolles. *The Microtomist's Vade-mecum: A Handbook of the Methods of Microscopic Anatomy.* 1st ed. London: J. & A. Churchill, 1885.

Lemmerich, J., ed. *Dokumente zur Gründung der Kaiser-Wilhelm-Gesellschaft und der Max-Planck-Gesellschaft zur Förderung der Wissenschaften.* Munich: Max-Planck-Gesellschaft zur Förderung der Wissenschaften, 1981.

Lenoir, Timothy. "Generational Factors in the Origin of *Romantische Naturphilosophie.*" *Journal of the History of Biology* 11 (1978): 57–100.

———. *The Strategy of Life: Teleology and Mechanics in Nineteenth-Century German Biology.* Boston: Reidel, 1982.

———. "Science for the Clinic: Science Policy and the Formation of Carl Ludwig's Institute in Leipzig." In *The Investigative Enterprise: Experimental Physiology in Nineteenth-Century Medicine,* edited by William Coleman and Frederic L. Holmes, pp. 139–78. Berkeley and Los Angeles: University of California Press, 1988.

———. "Social Interests and the Organic Physics of 1847." In *Science in Reflection,* edited by Edna Ullmann-Margalit, pp. 169–91. Dordrecht: Kluwer Academic, 1988.

———. "Laboratories, Medicine, and Public Life in Germany 1830–1849: Ideological Roots of the Institutional Revolution." In *The Laboratory Revolution in Medicine,* edited by Andrew Cunningham and Perry Williams, pp. 14–71. Cambridge and New York: Cambridge University Press, 1992.

Leuckart, Rudolf. *Über die Morphologie und Verwandtschaftsverhältnisse der wirbellosen Thiere: Ein Beitrag zur Charakteristik und Classification der thierischen Formen.* Braunschweig: Vieweg, 1848.

———. "Ist die Morphologie denn wirklich so ganz unberechtigt?" *Zeitschrift für wissenschaftliche Zoologie* 2 (1850): 271–73.

———. "Der Bau der Insekten in seine Beziehungen zu den Leistungen und Lebensverhältnissen dieser Thiere." *Archiv für Naturgeschichte* 17 (1851): 1–25.

———. *Über den Polymorphismus der Individuen oder die Erscheinungen der Arbeitstheilung in der Natur: Ein Beitrag zur Lehre vom Generationswechsel.* Giessen: J. Ricker, 1851.

———. "Bericht über die Leistungen in der Naturgeschichte der niederen Thiere während der Jahre 1848–1853." *Archiv für Naturgeschichte* 20, pt. 2 (1854): 289–473.

———. "Bericht über die wissenschaftlichen Leistungen in der Naturgeschichte der niederen Thiere während der Jahre 1866–1867," *Archiv für Naturgeschichte* 33, pt. 2 (1867): 163–304.

———. "Bericht über die wissenschaftliche Leistungen in der Naturgeschichte der niederen Thiere während der Jahre 1872–1875," *Archiv für Natur-*

*geschichte* 39, pt. 2 (1873): 413–567; 40, pt. 2 (1874): 401–505; 41, pt. 2 (1875): 313–517; 42, pt. 2 (1876): 462–605.

———. Opening address. *Verhandlungen der Deutschen Zoologischen Gesellschaft* 1 (1891): 3–10.

Leydig, Franz. *Vom Baue des thierischen Körpers: Handbuch der vergleichenden anatomie.* Tübingen: H. Laupp, 1864.

Liebau, Gustav. *Das Medizinal-Prüfungswesen im Deutschen Reich: Die Vorschriften über Prüfung der Ärzte, Zahnärzte, Thierärzte, Apotheker und Apothekergehülfen.* Leipzig: Duncker & Humblot, 1890.

Lohff, Brigitte. *Der Suche nach der Wissenschaftlichkeit der Physiologie in der Zeit der Romantik: Ein Beitrag zur Erkenntnisphilosophie in der Medizin.* Stuttgart and New York: G. Fischer, 1990.

Ludwig, Carl. Review of *Über die Morphologie,* by Rudolph Leuckart. *Jahrbücher der in- und ausländischen gesammten Medizin* 61 (1849): 341–43.

———. *Lehrbuch der Physiologie des Menschen.* 2 vols. Leipzig and Heidelberg: C. F. Winter, 1852–56.

———. "Zur Ablehnung der Anmuthungen des Herrn R. Wagner in Göttingen." *Zeitschrift für rationelle Medizin,* n. s. 5 (1854): 269–74.

Maas, Otto. *Einführung in die experimentelle Entwickelungsgeschichte (Entwickelungsmechanik).* Wiesbaden: L. E. Bergmann, 1903.

Maienschein, Jane. "The Origins of *Entwicklungsmechanik.*" In *A Conceptual History of Modern Embryology,* edited by Scott F. Gilbert, pp. 43–61. Vol. 7 of *Developmental Biology: A Comprehensive Synthesis,* edited by Leon W. Browder. New York and London: Plenum Press, 1991.

———. *Transforming Traditions in American Biology, 1890–1915.* Baltimore: Johns Hopkins University Press, 1991.

Maienschein, Jane, Ronald Rainger, and Keith R. Benson, eds. "Special Section on American Morphology at the Turn of the Century," *Journal of the History of Biology* 14 (1981): 83–191.

Mannheim, Karl. "The Problem of Generations." In K. Mannheim, *Essays on the Sociology of Knowledge,* edited by Paul Kecskemeti, pp. 276–322. London: Routledge & Kegan Paul, 1952.

Marchand, Suzanne L. "Archaeology and Cultural Politics in Germany, 1800–1965: The Decline of Philhellenism." Ph.D. diss., University of Chicago, 1992.

Marshall, William. *Die Tiefsee und ihr Leben.* Leipzig: F. Hirt, 1888.

"Die mathematische und naturwissenschaftliche Lehrfächer im neuen Lehrplane für die Gymnasien und Realschulen des Königreichs Sachsen." *Zeitschrift für mathematischen und naturwissenschaftlichen Unterricht* 8 (1877): 460–74.

Maurer, Friedrich. "Georg Ruge." *Anatomischer Anzeiger* 54 (1921): 24–28.

"Die medicinischen Examina in Deutschland und die Frequenz der medicinischen Facultäten." *Deutsche Medizinische Wochenschrift* 3 (1877): 122–23.

Meyer, Georg Hermann von. *Lehrbuch der physiologischen Anatomie des Menschen.* Leipzig: Engelmann, 1856.

———. "Stellung und Aufgabe der Anatomie in der Gegenwart." *Biologisches Centralblatt* 3 (1883): 353–66.

Mills, Eric L. "Alexander Agassiz, Carl Chun and the Problem of the Intermediate Fauna." In *Oceanography: The Past,* edited by M. Sears and D. Merriman. New York: Springer-Verlag, 1980.

Mivart, St. George. "Notes on the Fins of Elasmobranchs, with Considerations on the Nature and Homologies of Vertebrate Limbs." *Transactions of the Zoological Society of London* 10 (1879): 439–84.

Mocek, Reinhard. *Wilhelm Roux–Hans Driesch: Zur Geschichte der Entwicklungsphysiologie der Tiere (Entwicklungsmechanik).* Jena, 1974.

Mollier, Siegfried. "Die paarigen Extremitäten der Wirbeltiere." *Anatomische Hefte* 3, pt. 1 (15 June 1893; 1894): 1–160.

Montgomery, William. "Germany." In *The Comparative Reception of Darwinism,* edited by Thomas F. Glick, pp. 81–116. Austin: University of Texas Press, 1974.

Mörike, Klaus D. *Geschichte der Tübinger Anatomie.* Contubernium: Beiträge zur Geschichte der Eberhard-Karls-Universität Tübingen, Bd. 35. Tübingen: Mohr, 1988.

Morrell, J. B. "The Chemist Breeders: The Research Schools of Liebig and Thomas Thomson." *Ambix* 19 (1972): 1–46.

Müller, Fritz. *Für Darwin.* Leipzig: W. Engelmann, 1864.

Müller, Irmgard, ed. *Nikolai Nikolajewitsch Mikloucho-Maclay: Briefwechsel mit Anton Dohrn.* Norderstedt: Verlag für Ethnologie, 1980.

Müller, Johannes. "Über die Metamorphose des Nervensystems in der Thierwelt." *Archiv für die Anatomie und Physiologie* (1828): 1–22.

———. *Handbuch der Physiologie des Menschen.* 4th ed. Coblenz: Hölscher, 1844.

Müller, Johannes, and Theodor Schwann. "Versuche über die künstliche Verdauung des geronnenen Eiweises." *Archiv für Anatomie, Physiologie, und wissenschaftliche Medizin* (1836): 69–90.

Nauck, Ernst Theodor. "Bemerkung zur Geschichte des physiologischen Institutes Freiburg i. Br." *Berichte der Naturforschenden Gesellschaft zu Freiburg im Breisgau* 40 (1950): 147–59.

———. *Zur Vorgeschichte der naturwissenschaftlich-mathematischen Fakultät der Albert-Ludwigs-Universität Freiburg im Breisgau: Die Vertretung der Naturwissenschaften durch Freiburger Medizinprofessoren.* Freiburg im Breisgau: Eberhard Albert, 1954.

Neuland, Werner. *Geschichte des anatomischen Instituts und des anatomischen Unterrichts an der Universität Freiburg im Breisgau.* Geschichte der Medizin in Freiburg i. Br., edited by Ludwig Aschoff, vol. 1. Freiburg im Breisgau: Hans Ferdinand Schulze, 1941.

Newman, H. H. "History of the Development of Zoology in the University of Chicago." *BIOS* 19 (1948): 215–39.

Nitsche, Hinrich. "Beiträge zur Kenntnisse der Bryozoen. V: Über die Knospung der Bryozoen." *Zeitschrift für wissenschaftliche Zoologie,* 25, supp. (1875): 343–402.

Nußbaum, Moritz. "Franz Leydig." *Anatomischer Anzeiger* 32 (1908): 503–6.

———. "Adolf Freiherr von La Valette St. George." *Anatomischer Anzeiger* 38 (1911): 29–30.

Nyhart, Lynn K. "The Disciplinary Breakdown of German Morphology, 1870–1900." *Isis* 78 (1987): 365–89.

———. "The Problem of the Organic Individual in Mid-Nineteenth-Century German Biology." Paper presented at the First International Summer School in the History and Philosophy of Science, Humboldt-Universität Berlin, June 1988.

———. "Writing Zoologically: The *Zeitschrift für wissenschaftliche Zoologie* and the Zoological Community in Late Nineteenth-Century Germany." In *Writing the Book of Nature: Textual Strategies, Literary Genres and Disciplinary Traditions in the History of Science,* edited by Peter Dear, pp. 43–71. Philadelphia: University of Pennsylvania Press, 1991.

———. "Beyond the Institutes: Alternative Settings for Zoological Research, 1880–1910." Paper presented at the annual meeting of the History of Science Society, Santa Fe, NM, 11–14 November 1993.

Oken, Lorenz. *Lehrbuch der Naturphilosophie.* 3 vols. Jena: Friedrich Frommann, 1809–11.

Olesko, Kathryn M. "Commentary: On Institutes, Investigations, and Scientific Training." In *The Investigative Enterprise: Experimental Physiology in Nineteenth-Century Medicine,* edited by William Coleman and Frederic L. Holmes, pp. 295–332. Berkeley and Los Angeles: University of California Press, 1988.

Oppenheimer, Jane. *Essays in the History of Embryology and Biology.* Cambridge: MIT Press, 1967.

———. "Curt Herbst's Contributions to the Concept of Embryonic Induction." In *A Conceptual History of Modern Embryology,* edited by Scott F. Gilbert, pp. 63–89. Vol. 7 of *Developmental Biology: A Comprehensive Synthesis,* edited by Leon W. Browder. New York and London: Plenum Press, 1991.

Ospovat, Dov. *The Development of Darwin's Theory: Natural History, Natural Theology, and Natural Selection, 1838–1859.* Cambridge and New York: Cambridge University Press, 1981.

Paulsen, Friedrich. *Die deutschen Universitäten und das Universitätsstudium.* Berlin: A. Asher, 1902.

Pauly, Philip J. "American Biologists in Wilhelmian Germany: Another Look at the Innocents Abroad." Paper delivered at the annual meeting of the History of Science Society, Chicago, 28–31 December, 1984.

Peter, Karl. "Franz Keibel." *Anatomischer Anzeiger* 68 (1929–30): 201–20.

Pfeffer, Wilhelm. "Carl Chun, Nekrolog." *Berichte über die Verhandlungen der königlich sächsischen Gesellschaft der Wissenschaften zu Leipzig: Mathematisch-physische Klasse* 66 (1914): 181–93.

Pistner, Robert Adam. "Würzburg als Handelsstadt." Inaugural diss., University of Würzburg, 1968.

"Aus dem preußischen Abgeordnetenhause 2." *Deutsche Medizinische Wochenschrift* 3 (1877): 117.

"Prospectus." *Archiv für Naturgeschchte* 1 (1835).

Purkýně, Jan Evangelista. "Über den Begriff der Physiologie, ihre Beziehung zu den übrigen Naturwissenschaften, und zu andern wissenschaftlichen und Kunst-Gebieten, die Methoden ihrer Lehre und Praxis, über die Bildung zum Physiologen, über Errichtung physiologischer Institute: Rede, gehalten bei der Eröffnung des physiologischen Institutes zu Prag am 6. October 1851." *Vierteljahresschrift für praktischen Heilkunde* (Prague) 33 (1852): 1–19.

―――. *Opera omnia.* 12 vols. to date. Prague, 1918–.

Querner, Hans. "Beobachtung oder Experiment? Die Methodenfrage in der Biologie um 1900." *Verhandlungen der Deutschen Zoologischen Gesellschaft,* pp. 4–12. Stuttgart: Gustav Fischer Verlag, 1975.

―――. "Darwins Deszendenz- und Selektionslehre auf den deutschen Naturforscher-Versammlungen." *Acta Historica Leopoldina* 9 (1975): 439–56.

―――. "Die Entwicklungsmechanik Wilhelm Roux und ihre Bedeutung in seiner Zeit." In *Medizin, Naturwissenschaft, Technik und das Zweite Kaiserreich,* edited by G. Mann and R. Winau, pp. 189–200. Göttingen: Vandenhoeck & Ruprecht, 1977.

―――. "Die Zoologie in der Ära von Eugen Korschelt." *Sudhoffs Archiv* 64 (1980): 313–29.

Rabl, Carl. "Carl Langer." *Anatomischer Anzeiger* 3 (1888): 79–80.

―――. "Theorie des Mesoderms." *Morphologisches Jahrbuch* 15 (1889): 113–252; 19 (1893): 65–144.

―――. "Über den Bau und die Entwicklung der Linse (3. Teil)." *Zeitschrift für wissenschaftliche Zoologie* 67 (1898): 1–138.

―――. "Gedanken und Studien über den Ursprung der Extremitäten." *Zeitschrift für wissenschaftliche Zoologie* 70 (1901): 476–557.

―――. "Über einige Probleme der Morphologie." *Anatomischer Anzeiger,* supp. 23 (1903): 154–90.

Radl, Emmanuel. *Geschichte der biologischen Theorien.* Vol. 2, *Geschichte der Entwicklungstheorien in der Biologie des XIX. Jahrhunderts.* Leipzig: Engelmann, 1909.

Raikov, Boris. *Karl Ernst von Baer, 1792–1876: Sein Leben und sein Werk.* Translated from the Russian and edited by Heinrich von Knorre. Leipzig 1968.

Rathke, Martin Heinrich. *Zur Morphologie: Reisebemerkungen aus Taurien.* Riga: Kymmel, 1837.

Raumer, Karl Georg von. *Contributions to the History and Improvement of the German Universities.* New York: F. C. Brownell, 1859.

Reibisch, J. "Karl Brandt zum Gedächtnis." *Wissenschaftliche Meeresuntersuchungen* 21, Abt. Kiel, n. s., 21 (1993): i–vi.

Reichert, Karl. "Bericht über die Fortschritte der mikroskopischen Anatomie im Jahre 1852." *Archiv für Anatomie, Physiologie, und wissenschaftliche Medizin,* supp. (1853).

―――. "Bericht über die Fortschritte in der mikroskopischen Anatomie im Jahre 1854." *Archiv für Anatomie, Physiologie, und wissenschaftliche Medizin,* supp. (1855).

―――. "Bericht über die Fortschritte der mikroskopischen Anatomie im Jahre 1855." *Archiv für Anatomie, Physiologie, und wissenschaftliche Medizin,* supp. (1856).

―――. "Über die neueren Reformen in der Zellenlehre." *Archiv für Anatomie, Physiologie, und wissenschaftliche Medizin,* pp. 86–151 (1863).

Richards, Robert J. *The Meaning of Evolution: The Morphological Construction and Ideological Reconstruction of Darwin's Theory.* Chicago: University of Chicago Press, 1992.

Richmond, Marsha. "Richard Goldschmidt and Sex Determination: The Growth of German Genetics, 1900–1935." Ph.D. diss., Indiana University, 1986.

———. "From Natural History Cabinet to Zoological Institute. The Institutionalization of Zoology in Munich University, 1850–1885." Paper delivered at the annual meeting of the History of Science Society, Madison, WI, 30 October–3 November, 1991.

Rinard, Ruth. "The Problem of the Organic Individual: Ernst Haeckel and the Development of the Biogenetic Law." *Journal of the History of Biology* 14 (fall 1981): 249–75.

Ringer, Fritz. *The Decline of the German Mandarins: The German Academic Community, 1890–1933.* Cambridge: Harvard University Press, 1969.

Risse, Günter B. "Wilbrand, Johann Bernhard." *Dictionary of Scientific Biography.*

Robinson, Gloria. "Carus, Julius Victor." *Dictionary of Scientific Biography.*

———. *A Prelude to Genetics: Theories of a Material Substance of Heredity, Darwin to Weismann.* Lawrence, KS: Coronado Press, 1979.

Rössler, Richard. "Beiträge zur Anatomie der Phalangiden," *Zeitschrift für wissenschaftliche Zoologie* 36 (1881–82): 671ff.

Rothschuh, Karl. *History of Physiology.* Translated by Günter Risse. Huntington, NY: Robert Krieger, 1973.

———. "Hyrtl contra Brücke: Ein Gelehrtenstreit im neunzehnten Jahrhundert und seine Hintergründe." In *Wien und die Weltmedizin,* edited by Erna Lesky, pp. 159–69. Vienna: Hermann Böhlaus, 1974.

Roux, Wilhelm. "Über die Verzweigungen der Blutgefässe des Menschen: Eine morphologische Studie." *Jenaische Zeitschrift für Naturwissenschaft* 12 (1878): 205–66.

———. "Über die Bedeutung der Ablenkung des Arterienstammes bei der Astabgabe." *Jenaische Zeitschrift für Naturwissenschaft* 13 (1879): 321–37.

———. *Die Kampf der Teile im Organismus: Ein Beitrag zur Vervollständigung der mechanistischen Zweckmässigkeitslehre.* Leipzig: Engelmann, 1881.

———. "Beiträge zur Entwickelungsmechanik des Embryo. Nr. 2: Über die Entwickelung der Froscheier bei Aufhebung der richtenden Wirkung der Schwere." *Breslauer ärztliche Zeitschrift* (22 March 1884).

———. "Beiträge zur Entwickelungsmechanik des Embryo. Nr. 1: Einleitung und Orientierung über einige Probleme der embryonalen Entwickelung." *Zeitschrift für Biologie* 21, n. s., 3 (1885): 411–524.

———. "Die Entwickelungsmechanik der Organismen, eine anatomische Wissenschaft der Zukunft." Festrede, 1889. Reprinted in Roux, *Gesammelte Abhandlungen über Entwickelungsmechanik der Organismen.* Vol. 2, pp. 24–54. Leipzig, Engelmann, 1895.

———. "Einleitung." *Archiv für Entwicklungsmechanik der Organismen,* 1 (1895): 1–42.

————. *Gesammelte Abdhandlungen über Entwickelungsmechanik der Organismen*. 2 vols. Leipzig, Engelmann, 1895.

————. "Ziele und Wege der Entwickelungsmechanik." *Ergebnisse der Anatomie und Entwickelungsgeschichte* 2 (1892): 414–45. Reprinted in Roux, *Gesammelte Abhandlungen über Entwickelungsmechanik der Organismen*. Vol. 2, pp. 54–94. Leipzig: Engelmann, 1895.

————. "Für unser Programm and seine Verwirklichung." *Archiv für Entwicklungsmechanik* 5 (1897): 1–80, 219–342.

————. *Gutachten über dringlich zu errichtende biologische Forschungsinstitute, insbesondere über die Errichtung eines Institutes für Entwicklungsmechanik für die Kaiser-Wilhelm-Gesellschaft zur Förderung der Wissenschaften*. Vorträge und Aufsätze über Entwickelungsmechanik der Organismen, Heft 15. Leipzig: W. Engelmann, 1912.

————. "Albert Oppel." *Archiv für Entwicklungsmechanik* 42 (1916–17): 261–66.

————. [Autobiography] In L. R. Grote, ed., Die Medizin der Gegenwart in Selbstdarstellungen. Leipzig: Felix Meiner, 1923.

Rudwick, Martin J. S. *The Meaning of Fossils: Episodes in the History of Paleontology*. 2d ed. Chicago: University of Chicago Press, 1985.

Runge, Uwe. *Johann Moritz David Herold (1790–1862): Leben und Werk*. Marburger Schriften zur Medizingeschichte, vol. 6. Frankfurt am Main: Peter Lang, 1983.

Russell, E. S. *Form and Function: A Contribution to the History of Animal Morphology*. London: John Murray, 1916. Reprint, Chicago: University of Chicago Press, 1982.

Sachs, J. J. "Flüchtige Reiseblicke." *Medicinischer Almanach für das Jahr 1837* 2 (1837): 57–140.

Saha, Margaret Somosi. "Carl Correns and an Alternative Approach to Genetics: The Study of Heredity in Germany between 1880 and 1930." Ph.D. diss., Michigan State University, 1984.

Sander, Klaus. "August Weismanns Untersuchungen zur Insektenentwicklung." In "August Weismann (1834–1914) und die theoretische Biologie des 19. Jahrhunderts," edited by K. Sander. *Freiburger Universitätsblätter* 24, no. 87–88 (July 1985): 43–52.

————. ed. "August Weismann (1834–1914) und die theoretische Biologie des 19. Jahrhunderts: Urkunden, Berichte und Analysen." *Freiburger Universitätsblätter* 24, no. 87–88 (July 1985).

Schäfer, Karl Theodor. *Verfasssungsgeschichte der Universität Bonn 1818 bis 1960*. Bonn: H. Bouvier/Ludwig Röhrscheid, 1968.

Scheele, Irmtraut. *Von Lüben bis Schmeil: Die Entwicklung von der Schulnaturgeschichte zum Biologieunterricht zwischen 1830 und 1933*. Berlin: Dietrich Reimer, 1981.

Schiebler, Theodor Heinrich. "Anatomie in Würzburg (1593–Gegenwart)." In *400 Jahre Universität Würzburg: Eine Festschrift*, edited by Peter Baumgart, pp. 985–1004. Quellen und Beiträge zur Geschichte der Universität Würzburg, vol. 6. Neustadt an der Aisch: Degener, 1982.

Schiefferdecker, Paul. "Der anatomische Unterricht: Ein ungehaltener Vortrag." *Deutsche Medicinische Wochenschrift* 8 (1882): 465–67.

Schiemenz, Paul. "Über das Herkommen des Futtersaftes und die Speicheldrüsen der Biene nebst einem Anhange über das Riechorgan." *Zeitschrift für wissenschaftliche Zoologie* 38 (1883): 71–135.

Schierhorn, Helmke. "Der Prosektor und seine Stellung in der Hierarchie anatomischer Institutionen, demonstriert vor allem an den Anatomien in Berlin, Halle, Leipzig, Rostock und Griefswald." *Anatomischer Anzeiger* 159 (1985): 311–46.

Schlemm, Friedrich. "Über die Verstärkungsbänder am Schultergelenk." *Archiv für Anatomie, Physiologie, und wissenschaftliche Medizin* (1853): 45–48.

Schmid, Günther. "Über die Herkunft der Ausdrücke Morphologie und Biologie: Geschichtliche Zusammenhänge." *Nova Acta Leopoldina*, n. s., 2 (1935): 597–620.

Schmid-Monnard, Carl. "Die Histogenese des Knochens der Teleostier." *Zeitschrift für wissenschaftliche Zoologie* 39 (1883): 97–136.

Schmidt, Heinrich, ed. *Was Wir Ernst Haeckel Verdanken.* 2 vols. Leipzig: Unesma, 1914.

Schmidt, Joseph Hermann. *Zwölf Bücher über Morphologie überhaupt, und vergleichende Noso-Morphologie insbesondere.* Berlin: T. C. F. Enslin, 1831.

Schmidt, Oscar. *Die Entwicklung der vergleichenden Anatomie: Ein Beitrag zur Geschichte der Wissenschaften.* Jena: Fromman, 1855.

Schmidt, W. J. "Einiges aus der Geschichte der Zoologie in Giessen." *Verhandlungen der Deutschen Zoologischen Gesellschaft.* Giessen, 1938. *Zoologischer Anzeiger*, Suppl. 11 (1938): 18–33.

Schneider, Anton. *Monographie der Nematoden.* Berlin: G. Reimer, 1866.

Schrader, William. *Geschichte der Friedrichs-Universität zu Halle. Zweiter Teil.* Berlin: Ferdinand Dümmler Verlagsbuchhandlung, 1894.

Schröer, Heinz. *Carl Ludwig: Begründer der messenden Experimentalphysiologie 1816–1895.* Stuttgart: Wissenschaftliche Verlagsgesellschaft, 1967.

Schuberg, August. "Das neue zoologisch-zootomisch Institut der königlichen Julius-Maximilians-Universität zu Würzburg." *Arbeiten aus dem zoologisch-zootomischen Institute zu Würzburg* 10 (1895): 1–12.

Schubring, Gert. "The Rise and Decline of the Bonn Natural Sciences Seminar." *Osiris* 5 (1989): 57–93.

Schultze, Max. "Zur Kenntniss der Leuchtorgane von *Lampyris splendidula*." *Archiv für mikroskopische Anatomie* 1 (1865): 124–37.

Schultze, Max, and M. Rudneff. "Weitere Mittheilungen über die Einwirkung der Überosmiumsäure auf thierische Gewebe." *Archiv für mikroskopische Anatomie* 1 (1865): 299–304.

Schulz, Hans. *Deutsches Fremdwörterbuch.* 7 vols. Strassburg: Trübner, 1913.

Schulze, Franz Eilhard. "Untersuchungen über den Bau und Entwicklung der Spongien." *Zeitschrift für wissenschaftliche Zoologie* 28 (1877): 1–48; 29 (1877): 87–122; 30 (1878): 379–420; 31 (1878): 262–95; 32 (1879): 117–57, 593–660; 33 (1879–80): 1–38; 34 (1880): 407–51; 35 (1881): 410–30.

Schumacher, Gert-Horst, and Heinz-Günther Wichhusen. *Anatomia Ros-*

*tochiensis: Die Geschichte der Anatomie an der 550 Jahre alten Universität Rostock.* Berlin: Akademie-Verlag, 1970.

Schuman, Howard, and Jacqueline Scott. "Generations and collective memories." *American Sociological Review* 54 (1989): 359–81.

Schwalbe, Gustav. "Über Wachstumsverschiebungen und ihren Einfluß auf die Gestaltung des Arteriensystems." *Jenaische Zeitschrift für Naturwissenschaft* 12 (1878): 267–301.

Schwann, Theodor. *Mikroskopische Untersuchungen über die Übereinstimmung in der Struktur und im Wachstum der Thiere und Pflanzen.* Berlin: Verlag der Sander'schen Buchhandlung (G. E. Reimer), 1839.

Selenka, Emil. "Zur Entwicklung der Holothurien (*Holothuria tubulosa* und *Cucumaria doliolum*): Ein Beitrag zur Keimblättertheorie". *Zeitschrift für wissenschaftliche Zoologie* 27 (1876): 155–78.

Semper, Carl. *Reisen im Archipel der Philippinen, Zweiter Theil. Wissenschaftliche Resultate.* Vol. 1, *Holothurien.* Leipzig: W. Engelmann, 1868.

———. "Kritische Gänge. III. Gang. Die Keimblätter-Theorie und die Genealogie der Thiere." *Arbeiten aus dem zoologisch-zootomischen Institut Würzburg* 1 (1874): 222–38.

———. *Die Verwandschaftsbeziehungen der gegliederten Tiere.* Würzburg: Stahel'sche Buch- & Kunsthandlung, 1875.

———. *Offener Brief an Herrn Prof. Haeckel in Jena.* Hamburg: W. Mauke's Söhne, 1877.

———. *Die natürlichen Existenzbedingungen der Tiere.* 2 vols. Leipzig: F. A. Brockhaus, 1880. Translated under the title *The Natural Conditions of Existence as They Affect Animal Life.* London: C. K. Paul, 1881.

———. "Zoologie und Anatomie: Eine Erwiderung auf Herrn v. Kölliker's Rede 'Die Aufgaben der anatomischen Institute.'" *Arbeiten aus dem Zoologischen Institut der Universität Würzburg,* 1883, pp. 29–40.

Servos, John W. "Research Schools and Their Histories." In *Research Schools: Historical Appraisals,* edited by Gerald Geison and Frederic L. Holmes. *Osiris* 8 (1993): 3–15.

Siebold, Carl Theodor Ernst. "Über den Generationswechsel der Cestoden nebst einer Revision der Gattung Tetrarhynchus." *Zeitschrift für wissenschaftliche Zoologie* 2 (1850): 198–253.

———. "Über die Acclimatisation der Salmoneer in Australien und Neu-Seeland." *Zeitschrift für wissenschaftliche Zoologie* 19 (1869): 349–75.

———. "Über das Anpassungsvermögen der mit Lungen athmenden Süßwasser-Mollusken." *Sitzungsberichte der math.-phys. Klasse der Akademie der Wissenschaften in München* 5 (1875): 39–54.

———. "Die haarige Familie von Ambros." *Archiv für Anthropologie* 10 (1877): 253–60.

Sloan, Phillip Reid. "Darwin's Invertebrate Program, 1826–1836: Preconditions for Transformism." In *The Darwinian Heritage,* edited by David Kohn, pp. 71–120. Princeton, NJ: Princeton University Press, 1985.

———, ed. Introduction to *The Hunterian Lectures in Comparative Anatomy, May and June 1837,* by Richard Owen. Chicago: University of Chicago Press, 1992.

Sobotta, Johannes. "Zum Andenken an Wilhelm von Waldeyer-Hartz." *Anatomischer Anzeiger* 56 (1922): 3.

Solger, B. "Hermann Welcker." *Anatomischer Anzeiger* 14 (1898): 102–12.

Spalding, Keith. *An Historical Dictionary of German Figurative Usage.* 52 fascicles, "A–Trommel." Oxford: Basil Blackwell, 1952–.

Spemann, Hans. "Zur Geschichte und Kritik des Begriffs der Homologie." In *Allgemeine Biologie,* edited by K. Chun and W. Johannsen, pp. 63–86. Vol. 3, pt. 4, sec. 1 of *Die Kultur der Gegenwart,* edited by Paul Hinneberg. Berlin; Leipzig: B. G. Teubner, 1915.

———. "Theodor Boveri." *Archiv für Entwicklungsmechanik* 42 (1917): 243–57.

———. *Forschung und Leben.* Edited by Friedrich Wilhelm Spemann. Stuttgart: J. Engelhorns Nachfolger Adolf Spemann, 1943.

Spencer, Herbert. "Progress: Its Law and Cause." In *Essays—Scientific, Political, and Speculative.* Library edition. Vol. 1, pp. 8–62. New York: Appleton, 1891.

Spitzer, Alan B. "The Historical Problem of Generations." *American Historical Review* 78 (1973): 1353–85.

Spix, Johannes. *Geschichte und Beurtheilung aller Systeme in der Zoologie nach ihrer Entwicklungsfolge von Aristoteles bis auf die gegenwärtige Zeit.* Naumberg: Schrag, 1811.

Stammer, Hanz-Jurgen. "Albert Fleischmann—Ein Nachruf." *Sitzungsberichte der physikalische-medizinische Sozietät Erlangen* 75 (1952): xx–xxxv.

———. "Ein kürzer Abriß der Geschichte der Erlanger Zoologie." *Verhandlungen der Deutschen Zoologischen Gesellschaft.* Erlangen, 1955. *Zoologischer Anzeiger,* supp. 19 (1956): 28–32.

Strasburger, Eduard. "Über die Bedeutung phylogenetischer Methoden für die Erforschung lebender Wesen." *Jenaische Zeitschrift für Naturwissenschaft* 8 (1874): 56–80.

Strasser, Hans, *Über die Luftsäcke der Vögel.* Leipzig: Engelmann, 1877.

———. *Zur Kenntnis der funktionellen Anpassung der quergestreiften Muskeln: Beiträge zu einer Lehre von dem kausalen Zusammenhang in den Entwicklungsvorgängen des Organismus.* Stuttgart: Ferdinand Enke, 1883.

Stürzbecher, Manfred. *Deutsche Ärztebriefe des 19. Jahrhunderts.* Göttingen: Musterschmidt, 1975.

Taschenberg, Otto. *Geschichte der Zoologie und der zoologischen Sammlungen an der Universität Halle, 1694–1894.* Halle: Max Niemeyer, 1894.

Tembrock, Günter. "Franz Eilhard Schultze und die Gesellschaft Naturforschender Freunde zu Berlin." *Sitzungsberichte der Gesellschaft Naturforschender Freunde zu Berlin* n. s. 6 (1966): 137–51.

Temkin, Owsei. "Concepts of Ontogeny and History around 1800." *Bulletin for the History of Medicine* 24 (1950): 227–46.

Thacher, James K. "Median and Paired Fins, a Contribution to the History of Vertebrate Limbs." *Transactions of the Connecticut Academy* 3 (1877): 281–310.

————. "Ventral Fins of Ganoids." *Transactions of the Connecticut Academy* 4 (1877): 233–42.

Todes, Daniel. "V. O. Kovalevskii: The Genesis, Content, and Reception of His Paleontological Work." *Studies in History of Biology* 2 (1978): 99–165.

Trienes, Rudie. "The Type Concept Revisited: A Survey of German Idealistic Morphology in the First Half of the Twentieth Century." *History and Philosophy of the Life Sciences* 11 (1989): 23–42.

Tuchman, Arleen. "Science, Medicine and the State: The Institutionalization of Scientific Medicine at the University of Heidelberg." Ph.D. diss., University of Wisconsin—Madison, 1985.

————. "From the Lecture to the Laboratory: The Institutionalization of Scientific Medicine at the University of Heidelberg." In *The Investigative Enterprise. Experimental Physiology in Nineteenth-Century Medicine*, edited by William Coleman and Frederic L. Holmes, pp. 65–99. Berkeley and Los Angeles: University of California Press, 1988.

Turner, R. Steven. "The Growth of Professorial Research in Prussia: Causes and Context." *Historical Studies in the Physical Sciences* 3 (1971): 137–82.

————. "The Prussian Universities and the Research Imperative, 1806–1848." Ph.D. diss., Princeton University, 1973.

————. "The Prussian Universities and the Concept of Research." *Internationales Archiv für die Sozialgeschichte der deutschen Literatur* 5 (1980): 68–92.

————. "Vision Studies in Germany: Helmholtz versus Hering." In *Research Schools: Historical Appraisals*, edited by Gerald Geison and Frederic L. Holmes. *Osiris* 8 (1993): 80–103.

Turner, R. Steven, Edward Kerwin, and David Woolwine. "Careers and Creativity in Nineteenth-century German Physiology: Zloczower *Redux*." *Isis* 75 (1984): 523–29.

Universität Berlin. *Index Lectionum*, 1833–54.

Universität Bonn. *Verzeichnis der Professoren*, 1826–42.

Universität Halle. *Catalogus Professorum*, 1822–45.

Universität Heidelberg. *Verzeichnis der Professoren und Privatlehrer*, 1818–45, 1880–90.

Universität München. *Vorlesungsverzeichnisse*, 1830–45, 1870–1910.

Universität Tübingen. *Vorlesungsverzeichnisse*, 1830–47.

Universität Würzburg. *Vorlesungsverzeichnisse*, 1820–44.

Uschmann, Georg. *Geschichte der Zoologie und der zoologischen Anstalten in Jena, 1779–1919*. Jena: VEB Gustav Fischer Verlag, 1959.

————. *Ernst Haeckel: Forscher, Künstler, Mensch*. Leipzig, Jena: Urania-Verlag, 1961.

————. "Haeckel, Ernst Heinrich Philipp August." *Dictionary of Scientific Biography*.

————. "Über die Beziehungen zwischen Albert von Kölliker und Ernst Haeckel." *Wissenschaftliche Zeitschrift der Friedrich-Schiller-Universität, mathematisch-naturwissenschaftliche Reihe* 25 (1976): 125–31.

Vierhaus, Rudolf, and Bernhard vom Brocke, eds. *Forschung im Spannungs-feld von Politik und Gesellschaft: Geschichte und Struktur der Kaiser-Wilhelm-/Max-Planck-Gesellschaft.* Stuttgart: Deutsche Verlags-Anstalt, 1900.

Vierordt, Carl. "Über die gegenwärtigen Standpunkt und die Aufgabe der Physiologie." *Archiv für physiologische Heilkunde* 8 (1849): 297–316.

Vogel, R. "Professor Dr. Friedrich Blochmann." *Jahreshefte des Vereins für vaterländische Naturkunde in Württemberg* 87 (1931): xxvii–xxxiii.

Volkmann, A. W. "Die Darwinische Theorie." *Bericht über die Sitzungen der Naturforschenden Gesellschaft zu Halle,* 1866. In *Abhandlungen der Naturforschenden Gesellschaft zu Halle* 10, pt. 2 (1868): 17–19.

———. "Über die Grenzen der organischen und unorganischen Natur." *Bericht über die Sitzungen der Naturforschenden Gesellschaft zu Halle,* 1867. In *Abhandlungen der Naturforschenden Gesellschaft zu Halle* 10, pt. 3 (1868): 3–5.

*Vorlesungsverzeichnisse der Universitäten Deutschlands, Oesterreichs und in der Schweiz.* Munich: *Hochschul-Nachrichten,* 1894–95.

Wagenitz, G. *Göttinger Biologen, 1737–1945.* Göttingen: Vandenhoek & Ruprecht, 1988.

Wagner, Rudolph. *Grundriß der Encyklopädie und Methodologie der medizinischen Wissenschaften.* Erlangen: Palm & Enke, 1838.

———. "Begründung meiner vom Prof. C. Ludwig in Zürich abgelehnten sogenannten 'Anmuthungen.'" *Zeitschrift für rationelle Medizin,* n. s., 5 (1854): 307–23.

———. "Neurologische Untersuchungen, Achte Fortsetzung: Über den Bau des Rückenmarks und die daraus resultirende Grundlage zu einer Theorie der Reflexbewegungen, Mitbewegungen und Mitempfindungen." *Nachrichten von der Georg-August-Universität und der königlichen Gesellschaft der Wissenschaften zu Göttingen* (1854): 89–104.

———. "Die Forschungen über Hirn- und Schädelbildung des Menschen in ihrer Anwendung auf einige Probleme der allgemeinen Natur- und Geschichtswissenschaft." *Abhandlungen der königlichen Gesellschaft der Wissenschaften zu Göttingen* 9 (1861): 153–204.

Waldeyer, Wilhelm. "Wie soll man Anatomie lehren und lernen: Rede gehalten zur Feier des Stiftungstages der militärärztlichen Bildungsanstalten am 2. August 1884." *Deutsche Medizinische Wochenschrift* 10 (1884): 593–96, 611–14.

———. "Albert von Brunn." *Anatomischer Anzeiger* 11 (1896): 481–83.

———. *Lebenserinnerungen.* Bonn: Friedrich Cohen, 1920.

Warner, John Harley. "Remembering Paris: Memory and the American Disciples of French Medicine in the Nineteenth Century." *Bulletin of the History of Medicine* 65 (1991): 301–25.

Weber, Eduard, and Wilhelm Weber. *Die Mechanik der menschlichen Gehwerkzeuge.* Göttingen, 1836.

Weber, Ernst Heinrich. *Über die Anwendung der Wellenlehre auf die Lehre vom Kreislauf des Blutes und insbesondere auf die Pulslehre.* Leipzig: W. Engelmann, 1889.

Wegelin, C. "Hans Strasser." *Anatomischer Anzeiger* 64 (1927): 193–99.
Wegner, Richard N. "Hermann Klaatsch." *Anatomischer Anzeiger* 48 (1915–16): 611–23.
Weindling, Paul Julian. *Darwinism and Social Darwinism in Imperial Germany: The Contribution of the Cell Biologist Oscar Hertwig (1849–1922)*. Forschungen zur neueren Medizin- und Biologiegeschichte, edited by Gunter Mann and Werner F. Kümmel, Bd. 3. Stuttgart and New York: Gustav Fischer, 1991.
Weismann, August. *Studien zur Descendenz-Theorie*, Leipzig: W. Engelmann, 1875.
———. "Über den Saison-Dimorphismus der Schmetterlinge." *Annali del Museo civico di Storia Naturale di Genova* 6 (1874): 209–307. Reprinted as the first part of *Studien zur Descendenz-Theorie*, by A. Weismann. Leipzig: W. Engelmann, 1875.
———. "Die Bedeutung der Zoologie für das Studium der Medicin." *Allgemeine Zeitung* (Augsburg), *Beilage* no. 95 (5 April 1979): 1401–3.
Weissenberg, Richard. *Oscar Hertwig, 1849–1922. Leben und Werk eines deutschen Biologen*. Leipzig: Barth, 1959.
Wiedersheim, Robert. *Das Gliedmassenskelet der Wirbelthiere*. Jena: Gustav Fischer, 1892.
———. *Lebenserinnerungen*. Tübingen: Mohr, 1919.
Witlaczil, E. "Entwicklungsgeschichte der Aphiden." *Zeitschrift für wissenschaftliche Zoologie* 40 (1884): 559–696.
Woodruff, A. E. "Weber, Wilhelm Eduard." *Dictionary of Scientific Biography*.
Wunderlich, Klaus. *Rudolf Leuckart, Weg und Werk*. Jena: Gustav Fischer, 1978.
Zaddach, Gustav. "Adolph Eduard Grube: Gedächtnissrede." *Schriften der physikalisch-ökonomische Gesellschaft zu Königsberg* 21 (1880): 113–30.
Ziernstein, Gottfried. "Friedrich Althoffs Wirken für die Biologie in der Zeit des Umbruchs der biologischen Disziplinen in Deutschland, der Erneuerung ihrer Forschung und Lehre an den Universitäten und des Rufes nach außeruniversitären Forschungsstätten, 1882 bis 1908." In *Wissenschaftsgeschichte und Wissenschaftspolitik im Industriezeitalter: Das 'System Althoff' in historischer Perspektive*, edited by Bernhard vom Brocke, pp. 355–73. Hildesheim: Lax, 1991.
Zloczower, A. *Career Opportunities and the Growth of Scientific Discovery in Nineteenth-Century Germany, with Special Reference to Physiology*. New York: Arno Press, 1981.

# INDEX

adaptation, 31, 340, 344: and Darwinism, 21, 141, 163–64, 176, 179–81, 190–93, 283, 310, 313–14; Haeckel on, 131–32, 189–90, 308; and inheritance, 131–32, 186, 189–90, 293, 308, 310; Kölliker on, 123, 128; law of, 113, 114; and scientific zoology, 169, 340, 344; and teleology, 21, 123, 128, 145, 181, 340. *See also* environment, effect on animals

Aeby, Christoph Theodor, 82, 224

Allen, Garland, 10, 333

Altenstein, Karl Freiherr vom Stein zum, 48n.25, 51

alternation of generations, 122, 125, 140, 148, 172

Althoff, Friedrich, 299, 301–5, 319n.25, 321–23, 354, 358; science policy of, 302, 303

Amsterdam, University of, 158, 161, 213

anatomy: careers in, 303, 351, 352; contrasted with zoology, 144, 207, 208, 306, 316, 333n.48; descriptive orientation, 220, 221, 226, 228; as a discipline, 11, 29, 144, 145, 207, 208; embryological orientation, 220; experimentalists in, 303; generations in, 27; institutional arrangements, 62, 229–31, 233, 235, 352; institutional relationship with physiology, 56, 59–61, 78, 79; intellectual relationship with physiology, 59, 77, 80, 86, 87, 89, 90; mission, 51, 52; morphology in, 18, 20, 30; orientations in, 219, 220, 228, 280; overcrowding, 352; professorships, 219, 238, 299, 336; research and teaching, relationship of, 49, 50, 62; teaching 51, 221–23, 225, 227, 228, 230, 231, 234, 235, 298, 299; as *Wissenschaft*, 47–50, 80, 89, 224, 225. *See also* comparative anatomy; mechanical anatomy; microscopic anatomy

animal behavior, 345

animals. *See* invertebrates; marine invertebrates; vertebrates

Appel, Toby, 7

*Archiv für Naturgeschichte,* 172

Arnold, Friedrich, 23, 64, 68, 78, 88, 89, 116, 208, 226, 227; defends morphology, 224, 225

Austria, 13

Baden, 13, 16, 19, 58n.48, 75n.28, 76

Balfour, Francis Maitland, 4n.2, 245, 252, 255–58

Bardeleben, Karl, 282n.5

Barfurth, Dietrich, 333n.48

Basel, University of, 13; zoology at, 273

Bavaria, 13, 16, 55, 58n.48, 76; university appointments, 326